THE GROW

The Growth of the Firm

The Legacy of Edith Penrose

Edited by

CHRISTOS PITELIS

OXFORD
UNIVERSITY PRESS

OXFORD

UNIVERSITY PRESS

Great Clarendon Street, Oxford OX2 6DP

Oxford University Press is a department of the University of Oxford.
It furthers the University's objective of excellence in research, scholarship,
and education by publishing worldwide in

Oxford New York

Auckland Bangkok Buenos Aires Cape Town Chennai
Dar es Salaam Delhi Hong Kong Istanbul Karachi Kolkata
Kuala Lumpur Madrid Melbourne Mexico City Mumbai Nairobi
São Paulo Shanghai Singapore Taipei Tokyo Toronto

with an associated company in Berlin

Oxford is a registered trade mark of Oxford University Press
in the UK and in certain other countries

Published in the United States
by Oxford University Press Inc., New York

British Library Cataloguing in Publication Data

Data available

Library of Congress Cataloging in Publication Data
The growth of the firm : the legacy of Edith Penrose / edited by Christos Pitelis.
p.cm.
"This volume has arisen from a special issue of Contributions to political economy (CPE)."
Includes bibliographical references and index.
1. Corporations—Growth. 2. Industrial organization (Economic theory) 3. International
business enterprises. 4. Technological innovations—Economic aspects. 5. Capitalism.
6. Human capital. 7. Economic development. 8. Penrose, Edith Tilton. I. Penrose, Edith
Tilton. II. Pitelis, Christos. III. Contributions to political economy.
HD2731 .G76 2002 338.7—dc21 2001052059
ISBN 0–19–924416–2 (Hbk)
ISBN 0–19–924852–4 (Pbk)

1 3 5 7 9 10 8 6 4 2

Typeset by Footnote Graphics, Warminster, Wilts
Printed in Great Britain
on acid-free paper by
T.J. International Ltd.,
Padstow, Cornwall

CONTENTS

1

ON THE GARDEN OF EDITH: SOME THEMES

CHRISTOS PITELIS*

The Judge Institute of Management Studies and Queens' College, University of Cambridge

I. PROLOGUE

In 1959 Edith Penrose wrote *The Theory of the Growth of the Firm* (hereafter *TGF*). The books reads like a story; it flows, it sounds eminently plausible and real life, it feels as if it describes just how things are perceived to be. Some readers might even be tempted to think that this is more or less what they too felt about the growth of firms, but could not or would not put it down in writing; in part because it is too realistic to be analytically useful. Yet, the more one reads carefully the more one realises that the book is written by one of the sharpest analytical minds ever to write in economics and management.[1] The book itself may well prove to be the single most influential in economics and management of the second part of twentieth century. Considering the competition,[2] this may sound like an overstatement. I do not believe it is. This introductory chapter, but more importantly the present volume as a whole, explains why.

This volume has arisen from a special issue of *Contributions to Political Economy* (*CPE*) guest edited by the present editor, which commemorated 40 years since the publication of *TGF*. The *CPE* volume included entries by eminent scholars on the topic, such as George Richardson, Robin Marris and Brian Loasby, and was preceded by an introductory chapter by Perran Penrose (Edith's son) and the editor; as well as a short preface by the latter. The chapters were in thematic order. They ranged from critical assessments of Penrose's contribution in the context of related contributions in economics (Richardson, Loasby, Marris) and strategic management (Foss), to

* I am grateful to numerous colleagues for comments and discussion pertaining to the issues covered in this introduction, notably Perran Penrose and Anastasia Pseiridis. My title is a variation on a title in a mimeo by the latter. Errors are mine.

[1] Arguably, much of what Penrose wrote would not qualify as theory in today's economics circles. She was fully aware of this! In a letter to Robin Marris in June 1984 she writes that 'theory nowadays means, at least in economics, mathematical models—and this is not my cup of tea. Much of the quasi-or-non-mathematical literature that some people call "theory" seems to me to be not much more than a kind of generalised description of "reality".' She describes her case as being 'interested in the analysis of relationships, but not in mathematical models'.

[2] By, for example, Arrow, Chandler, Debreu, North, Samuelson, Sraffa, Williamson and others.

extensions (and critical appraisal) of her work on the Hercules Powder Company case study (Kay), regional growth dynamics (Best) and (the interaction between) micro- and macroeconomic performance (Ghoshal, Hahn and Moran).

A number of important issues were only mentioned in passing. Examples are 'new ventures': the theory of multinational firm; the link between TGF and alternative per- spectives on the firm and industry organisation, notably that of transaction costs; the links and foundations of Penrose's thinking in the psychology literature; the links from her work to that of Schumpeter; the relevance and limitations of her thinking for global capitalism in the early twenty-first century. The present volume includes con- tributions by Elizabeth Garnsey on new ventures; Christos Pitelis on the multinational firm; Joel Thomas Ravix on Coase, Penrose and industry organisation; Margharita Turvani on foundations in psychology of Penrose's work; John Cantwell on Penrose, Schumpeter and the nature and effects of innovative activity in twenty-first century capitalism; and William Lazonick on the relevance and limitations of Penrose's con- tribution in the context of (today's) US corporate enterprise.

Overall, this is a more complete volume which, I believe, helps do some justice to Penrose's extraordinary contribution. I say some, because there is surely more to come. The book's ideas, and method are actually applicable to almost everything and one lesson we have learnt from it is that new applications are yet to come in areas such as economic institutions, the state, macroeconomics, and the many others mentioned in the various chapters of the present volume.[1]

I promised the publisher few introductory pages and I would be more than happy to now leave it at that and simply 'advise' the reader just to read the volume. To repeat in the introduction what the book said, and what almost every chapter in this volume does, would be both impossible and a waste of time. However, even in the short time since the appearance of the special issue of *CPE*, the literature on Penrose (and the resource-based theory of the firm whose motherhood is in part attributed to her) has expanded so much, that I feel it warrants some further comment.[2]

Given the current explosive growth of interest, these comments are likely to be dated as they are being written and ancient history as they are read! In part because of that, I will not try to be anywhere near exhaustive, just touch upon some themes from these recent debates, which I feel will be around for some time, for reasons that will become clearer later.

In addition to these comments on 'recent developments' there are essential issues pertaining to Penrose's contribution, which I feel need to be said, or at least (re)emphasised. They relate both to the content of her contribution and to the way it has been perceived and appreciated.[3]

---

[1] Foss (1996) for example, applies Penrosean insights to the development of scientific knowledge more generally.

[2] This is especially the case since, after some lengthy silence, major contributors in the economics and strategy area, notably Michael Porter (1999) and Oliver Williamson (1999), have taken (some) time to com- ment, critically, on resource-based contributions, including, up to a point, Penrose's.

[3] Also of interest is the exploration of the possibilities for generalising, extending, assessing and syn- thesising Penrose's with other contributions. This we do in the concluding chapter.

II. CONTENT

II.a. The argument

In short, *TGF* is about the *growth* of firms. Firms are 'flesh and blood' real-life *organisations*, not points on a cost curve. They consist of human and non-human resources, under administrative authoritative coordination and communication. Human, and especially managerial, resources are the most important. Resources provide multiple potential *services*. Firms use their resources to perform *activities* that result in products for sale in the *market* for a *profit*. Firms differ from markets, their *boundaries* defined by the reach of authoritative coordination and communication. For a multitude of reasons related to resource indivisibilities and the 'balance of processes', firms always have *'excess'* resources. The very performance of activities within firms creates new knowledge through specialisation, division of labour, teamwork and learning.[1]

The cohesive shell of the organisation is of essence in facilitating learning. As excess resources can provide services at zero marginal cost, they motivate entrepreneurs to apply them to new activities, engendering *endogenous innovation and growth*. The profitable marketing of new innovations requires entrepreneurial thinking, which is the identification and creation of markets. The external environment, markets and demand, are perceptions, 'images' in the entrepreneur's mind. Supply and demand are inextricably linked, as planned supply responds to perceived demand. There is a dynamic interaction between the perceived internal and external environments, which defines a firm's *productive opportunity*. The direction of expansion is motivated by the productive opportunity. There are limits to the growth of the firm, but not to its size. The conception and implementation of expansion requires managers whose firm-specific knowledge is a prerequisite for the successful planning and implementation of expansion, therefore, who are not available in the open market. This *limits the rate of growth* and hints to the pre-eminence Penrose attributes to the human resource, management.

Whilst we think these to be the heart of her main argument, the above ideas just touch upon a minor part of her overall contribution. She applied her insights to mergers, vertical integration, industry concentration, small firms and industry organisation more generally, as well as to business strategy and government competition policy. Her views on 'impregnable bases' that firms use to compete (to achieve sustainable competitive advantage in today's parlance), the existence of market 'interstices' (areas of no current interest to large firms, which small firms can occupy) and the idea that the process of expansion (growth) is by definition almost efficient, but the state (size) need not be (as firms may use size to restrict competition to obtain monopoly profits), and the related idea of competition as the 'god and the devil', because it is through it that firms succeed, but, through its restriction that in some cases they may maintain their (obtained through efficiency) (monopoly) power, are as innovative as they are topical and subjects of ongoing discussion.

---

[1] Learning by doing; to work with others; to recombine resources; learning what, how and why; learning to learn; hopefully learning to unlearn; learning oneself; and more generally as the ancient Greek philosopher and legislator Solon put it 'getting on always learning'.

However, in my view, the crux of the matter is in the original story. What is it in that story that has made the book so attractive and influential? Many hypotheses can be hazarded; to include Penrose's prose and the fact that the book reads like a story, with a beginning and an end, 265 pages of relentless pursuit of a single argument. However, the main reason is the argument itself. This is not an argument just about the theory of the growth of the firm; it is an argument about the theory of growth of everything. Extreme? Overstatement? I do not think so. Consider this. For anything and everything new to be even conceived, perceived, let alone implemented, one needs some prior knowledge, including the very capacity to obtain knowledge, i.e., to learn. There exist a variety of institutions that help to achieve this: families, schools, norms, customs and traditions, human interaction in society at large. Hayek's (1945) famous story was that this knowledge is dispersed and that there is at least in capitalist economies, an institution *par excellence*, the market, which facilitates its revelation and transmission and thus the co-ordination of individual plans and in society at large. This is a fundamental insight and in my view uncontroversial.[1]

Crucial in my view is that Hayek (1945) himself underplayed the importance of institutions—even his Juliette, the market, in *creating new knowledge*. In this sense, I believe, perhaps controversially, that Hayek did not serve the market cause. Markets can be fundamental in creating new knowledge.[2] Yet the important point is elsewhere. In Penrose it is *firms* that help create knowledge, indeed firms are better at doing this than markets are. If so, Penrose critiques Hayek (as firms involve planning) and complements him, as private firms and markets—thus the market system—together create new and transmit (dispersed) knowledge. This has crucial implications for the efficiency of the system as a whole. However, Penrose's contribution goes far deeper. If knowledge in general, or even a type of knowledge most suitable for production-related activities, is generated better within, than without, firms, and to the extent that this knowledge is for relevance or use to the society as a whole, everything and anything we conceive of or perceive and the lens through which we do so, is predicated upon the existence and functions of firms. Our very perception of life is a function of firms, not just in the widely accepted sense that firms provide employment, community, identity, etc.; but in the deeper sense of them providing us with a lens through which we perceive the wider environment and life as a whole. It is in this sense that Penrose's contribution is beyond the normally conceived (even by her) bounds, and where I believe further developments are likely to focus.

It is beyond the scope of this introduction to expand on this issue. The issues on which I focus below, however, while of significance in themselves, all relate to and are informed by, the aforementioned observation. The most significant of these issues are about the very nature and scope of economics, namely resource allocation versus

---

[1] That Hayek used this as a basis for a critique of (pure) central planning and whether that was right or wrong is of no concern to us here (but see Caldwell, 2000, for a discussion).

[2] Indeed even learning about others' plans may help modify someone else's plans; this is as much about transmitting and distributing existing knowledge as it is about creating new knowledge.

resource (wealth) creation and (relatedly) growth and (through) innovation and co-ordination. I start with the former.

II.b. Wealth creation versus resource allocation

It is known, and attested in most contributions in this volume (notably by Loasby) that one can effectively divide economics into two major camps; the one focusing on efficient allocation of resources, often assumed to be scarce, the other focusing on resource and wealth creation. Most classical economists, notably Adam Smith and Karl Marx, but also more recent contributors, such as Joseph Schumpeter (1942), have paid attention to the issue of resource and wealth creation. To varying degrees, these economists also dealt with the clearly related issue of resource allocation. Adam Smith, for example, arguably owes his place as the father-figure of modern (neo-classical) economics to his very analysis of the allocative and coordination role of the 'invisible hand', or the market. Yet, he believed in the labour theory of value, and attributed wealth creation to labour productivity engendered within firms. Smith is as much about markets as he is about the firms—his famous pin factory, from where his *An Inquiry into the Nature and Causes of the Wealth of Nations* starts. In the pin factory, labour productivity is achieved through specialisation, the division of labour and teamwork, which lead to, among other things, new inventions by those closer to the production process, i.e., labourers. The sources of labour productivity, inventions and the wealth of nations are thus to be found within firms. They are only limited by the size of the market.[1]

Penrose's place in this debate is prominent. As Loasby suggests in this volume, Penrose has been able to 're-invent' this classical tradition, which at the time of her writing was all but extinct. Already then, the battle on the nature and scope of economics has been won by the efficient allocation of (scarce) resources perspective of Jevons, Walras and their followers, what we now call the neoclassical perspective. According to this, the scope of economics as a science should appropriately be the analysis of efficient allocation of (scarce) resources. 'Re-inventing' the classical perspective was no small feat. Yet I believe Penrose went well beyond re-inventing and putting things together. To explain this, I need to first proceed with the argument.

II.c. Growth and co-ordination

The other major (and related) difference between the classical and neoclassical trad-itions concerns growth versus coordination. As Loasby and Richardson detail in their contributions to this volume, the neoclassical focus is on coordination, while the classical is on growth. Schumpeter's focus was also on growth. He and Penrose had to deal with the issue of the relevance of neoclassical theory to their concerns, and

[1] In Allyn Young's (1928) powerful insight, the size of the market itself is determined by specialisation and the division of labour; the latter both leads to further, more elaborate sub-division of labour and extends the size of the market. In a sense, it is specialisation and the division of labour that determines specialisation and the division of labour, thus productivity and the wealth of nations.

relatedly, the inevitable comparison between the two perspectives. They both chose to claim that both perspectives were useful but for different reasons, so their ideas did not necessarily and fundamentally question (the usefulness) of neoclassical thinking. In Penrose's famous 'garden' metaphor, neoclassical price theory does not deal with growth, which is Penrose's 'garden'. She felt that both neoclassicals and she could tend their respective gardens, and there was little use in trying to bring the two approaches together.

This is a big issue, dealt with in this volume by Richardson. He claims that in contrast to Penrose, one needs to deal with both coordination and growth, and that Penrose's contribution is arguably the very tool that can help this come about. I agree. Penrose's endogenous knowledge-based concept of 'productive opportunity' of entrepreneurs supplies us with the basis of perceived (and actively aimed for) demands, the fundamental dynamic *ex post* quasi-equivalence of supply and demand (in that supply is a response to entrepreneurial perceptions of demand), potentially powerful tools through which growth and coordination can be examined and achieved.[1]

While arguably less significant than the issue addressed by Richardson (of dynamic growth and coordination), the degree of compatibility and relative usefulness of efficient allocation versus dynamic growth is of interest and has occupied scholars in economics and (strategic) management.[2] The message is clear. Both allocative efficiency and growth are important. Whether the former is neoclassical price theory in its presently dominant form, of comparative static equilibria reached by rational utility maximising agents, is less clear to me.[3]

I think that a version of the neoclassical-type approach to firms and industry organisation can be of both relevance and use, in certain cases. Penrose's own analysis helps us to make it so; notably, her observation, mentioned above, that while the process of firm growth is efficient, the outcome (firm size) need not be, as it is competition that spurs growth and profits, but its very restriction is often required to maintain such profits. Indeed, reading the *TGF*'s last chapters on these issues, one may be excused for believing that they are both unrelated to the book's early ideas and rather early Bain-type market power-based industry organisation (see Rugman and Verbeke, 2001, for such an interpretation). This is not how I see things. A different interpretation is that once firms are large, they may be tempted to make use of their market power

[1] Where I am not clear is whether Penrose and Richardson are dealing with the same issue here. It is my hunch that Penrose herself would not have any quarrel with Richardson's arguments or my view presented above. Her focus was on the neoclassical model of the firm: i.e., of price–output decisions under assumptions about static profit maximisation, and different types of (rather unrealistic) industry structures. This, I would agree with Penrose, is harder to combine with a theory of dynamic firm growth of her type.

[2] In different terminology and contexts, for example, authors like Joseph Schumpeter and Michael Porter (1991, 1999) have suggested that both allocative (in Porter operational) efficiency and growth (in Porter strategy) are essential for a country's (firm's) success. Both claimed that even the countries or firms most efficient in resource allocation or operational efficiency will eventually lose out to successful innovators–strategists.

[3] When one considers the devastating critiques of this perspective by, among others, Hayek, Schumpeter, Allyn Young, Keynes, Sraffa and (in business strategy) Porter (1999), it is hard not to agree with Caldwell (2000) that the dominance of this perspective is a great mystery!

(even if gained through efficiency), to facilitate coordination and planning (which requires a degree of stability, especially after fast growth). This is to allow firms to 'take a breather', consolidate and 'think', to shield themselves from emerging competition from abroad or domestically in mature and declining industries, and also (why not?) allow them to receive some 'monopoly profit' they have worked so hard to earn.[1,2]

It is arguable that in mature industries one could consider a case of relative stability of market shares and oligopolistic structure, and look at these industries by taking a 'snapshot' at a particular point in time. Depending on the assumptions, the snapshot could show you a contestable market, a limit-pricing-based oligopoly, a profit-maximising collusive oligopoly, with or without strategic entry deterrence, or any of the existing variations in the IO type, on which see Pitelis (1991) or any recent IO textbook, e.g., Cabral (2000).[3]

In the context of this discussion, Penrose's endogenous knowledge perspective contributes fundamentally. It goes beyond Allyn Young (1928) by explaining why and how the size of the market is itself determined by specialisation and the division of labour (as well as vice versa of course). It provides the requisite perspective and knowledge to approach the synthesis between dynamic innovation = knowledge = productivity-growth perspective and dynamic coordination through the knowledge provided by firms' own operations and the perceived equivalence (in part achieved through firms' own conscious efforts) of supplies and demands. It helps to bring together resource creation and efficient allocation as explained above, but in a more fundamental way, given that it is the internally generated knowledge within firms that is required to supply firms with the tools to achieve (rather than assume they know how to achieve) efficient resource allocation.

It is in this sense that Penrose has contributed to everything. Knowledge is internally generated within firms, is an input to what they know, what they do, how they do it, every other issue we referred to and more. In addition, her approach is more than a re-invention, however important this re-invention is. Smith, Marx, Young, Hayek and Schumpeter have not dealt with endogenous innovation = knowledge-growth creation.[4]

---

[1] Note that in the Boston Consulting Group (BCG) matrix, this is exactly what consultants advise firms to do with their 'cash cows' (business units with high market shares in low growth industries), i.e., to 'milk' them!

[2] Some of these issues are raised by Cantwell and Lazonick (this volume) while Penrose herself has been explicit about the co-existence of profit through efficiency and profit through monopoly, claiming that both exist (see Penrose and Pitelis, this volume). She was also fully aware of what she called in one of her letters to a colleague 'my "Schumpeterian view" of capitalism, where competition does not rest on the importance of "competition at the margin" in an equilibrium framework'.

[3] It is no coincidence that even Marxists, such as Baran and Sweezy (1966) and post-Kaleckians, like Cowling (1982) have used similar theorising. Arguably, it is the scope that remains problematic, not some of the important ideas, e.g., entry deterrence, etc. Such ideas are most useful, and indeed the ability of the neoclassical framework to incorporate them within its methodological context, could in part explain its resilience and success (see Pitelis, 1991; Caldwell, 2000).

[4] This is also true of other eminent contributors who dealt with related issues, such as Ronald Coase, Alfred Chandler, Nicholas Kaldor, Michal Kalecki, Stephen Hymer, Gunnar Myrdall and Douglass North. This is crucial. In a world where 'there is nothing new under the sun', to come up with such a unique insight

footnote continued overleaf

II.d. The 'nature' of the firm

The other important issue on which Penrose can arguably make a significant contribution concerns the 'nature of the firm'. The very term of course is Coase's (1937) who
went on later (Coase, 1988) to distinguish between the 'nature of the firm' (why do
firms exist) and the essence of the firm ('running a business'). Clearly, Penrose has
dealt with the 'running' issue more effectively than any other economist. She is the
only one that combined the functions of the manager and entrepreneur in conceiving,
planning and implementing, i.e., in running a business (Pitelis and Wahl, 1998).
While this is rather uncontroversial, it can be questioned whether Penrose also dealt
with the issue of the nature.

Penrose did not deal directly with this issue. Like many others (see for example
Marris, this volume) she took it for granted that firms exist. Knowing her, this is
hardly surprising. Asking why firms exist, when it is clear as daylight that they do,
might have seemed to her like a waste of precious time. Clearly, Coase has shown that
this type of enquiry can prove most fruitful. He equated the 'nature' of the firm with
the 'employment relation' and attributed this to the efficiency benefits derived from
reductions in market transaction costs. Much ink has been spent already in repeating
what Coase said, so I will not do it here. Suffice it to note that, if we accept Coase's
view (not everybody does, of course), then Penrose's approach has immediate alternative, yet complementary implications. The 'employment relation' can be explained in
terms of efficiency gains, effected through productivity enhancements, through endogenous innovation = knowledge-growth (see Pitelis and Wahl, 1998).[1] Critically, the
very perception of when and how to reduce transaction costs can only be afforded
through intra-firm knowledge generation (see, among others Pseiridis, 1996; Fransman, 1994). To assume that firms know this beforehand is far too unrealistic.[2]

---

footnote 4 continued from previous page

is as good as it can get, especially when this is one that also helps explain the very process of perceiving
whether, why and how 'new' is new. An important aspect of this insight is the endogenous incentive to make
profitable use of 'excess resources'. This is far more subtle and powerful than the exogenous profit maximisation motive. While the two are clearly related—the latter in part motivating the former—making productive and profitable use of employed resources is a challenge, because it can help upgrade the resource,
because of the personal opportunity cost of inability to delegate, and for numerous socio-psychological
reasons not accounted for by an exogenously imposed profit motive (note, however, that Penrose had such a
broader explanation for the profit motive itself). Crucially the very process of thinking what to do about the
resources one employs, is itself innovation = knowledge.

[1] To the extent that (as I believe) dynamic (by definition given Penrose's framework) transaction costs can
also be of relevance here, it will be the overall dynamic transaction-cost reductions cum knowledge-induced
productivity benefits that could help explain the Coasean firm.

[2] In addition, it is totally in contrast to Oliver Williamson's (1975) insistence on bounded rationality. How
can one be boundedly rational and still get it always right in integrating, and only if in so doing one reduces
market transaction costs by more than one increases intra-firm transaction (managerial, administration)
costs has always defied me. Somehow I always felt that this is ultra-rationality of the highest order! Coase
(1993) himself considers bounded rationality (or any concept of rationality at all) as little more than nonsense. I submit that the Penrosean insight helps to address Williamson's 'inconsistency' and Coase's own
dislike of the rationality concept. In Penrose 'rationality', too, is endogenously generated through perennial
learning.

The Penrosean 'insight' leads us further afield. The nature of knowledge, tacit and hard to transmit across markets, is a serious blow to the very Coasean explanation of the 'nature'. To explain firms from a situation of no firms at all, one requires an entrepreneurial idea aimed to be put in practice. Selling this idea in the open market (or even sharing it with one's own employees) may be hard for at least two reasons. First, being tacit, it may be hard to transmit. Second, if in addition (which is possible) this idea also has public goods characteristics, explaining it to anyone can lead to it being expropriated. Among others, this can lead to a competitor at the best! (if the original conceiver even manages to go ahead with using it). So we have a two-pronged type of market failure, which, however, is not directly linked to transaction costs. Transaction costs enter the story if one suggests that in their absence one could conceive of contractual means of addressing the problem. This presupposes the counterfactual of a potential existing market, even when it does not exist. To claim, moreover, that transaction costs are the main reason, as opposed to the tacitness and public goods aspects of entrepreneurial ideas, is not evident at all. The control afforded to entrepreneurs on their ideas, in the cohesive shell of the firm, can be an adequate initial reason for not selling the idea in the open market. Note that this point is prior to, and complements, the idea that firms may have productivity benefits *vis-à-vis* markets, which can also be an adequate non-transaction-costs-related explanation of the 'nature' of firms. Such ideas may therefore critique the very *raison d'être* of the Coasean argument for the 'employment relationship'. (It may well be as dangerous to explain to employees-to-be what your idea is all about, as it is to tell it to anyone else.) If so, efficiency gains from transaction costs (or productivity or other sources of benefits) may not be an adequate explanation for employees voluntarily agreeing to work for employers. Other factors must be in place such as 'insuring the timid against risk', having a reputation for entrepreneurial flair (obtained, for example, in one's previous experience as an employer or a merchant), etc.[1]

These and other themes focusing on functions and/or characteristics of firms, such as co-ordination, culture, identity, continuity of association, learning (see for example Demsetz, 1988; Kogut and Zander, 1996) cannot by themselves explain the *differentia specifica* of the firm. All these functions and/or characteristics belong to varying degrees to other institutions too, including the market. For example, there are clearly all of the above-mentioned, and more, in a university; yet this is not a firm. Some arguments apply to different institutions and organisations, the family, the church, the market, too.[2]

---

[1] Some of the themes above would have surely been recognised by people in the area, in, for example, the work of Knight (1921), Marglin (1974, 1984) (on obtaining and protecting organisational knowledge), and also Liebeskind (1996) on knowledge protection.

[2] Arguably, firms can do the above better than other institutions. In certain cases, this may be true. If so, it is Penrose's ideas that explain why and how, as suggested above. However, doing some things better does not explain what things, what for and what, if anything, explains the 'division of labour' between different institutions and organisations. Penrose contributed to all these. Firms make products for sale in the market for a profit. Partly anticipating George Richardson's (1972) seminal contribution, Penrose explicitly suggested that firms are more suitable to markets when there co-exist similar and complementary activities (see Ravix, this volume).

II.e. Impact and some existing critiques

In one respect, the book's timing could not have been better.[1] In the late 1950s there were important contributions in the Bain (1956) tradition of Industry Organisation (e.g., Modigliani, 1958). Alfred Chandler was working on his *Strategy and Structure* (1962), Stephen Hymer was completing his 1960 dissertation of the multinational firm, and there was the great discussion of the 'managerial revolution' and the managerial theories of the firm (Baumol, 1959, 1962; Williamson, 1964; Marris, 1964; see also Slater, 1980). Penrose was dealing with issues very close to Chandler's and Hymer's (see Pitelis, this volume) and Bain's influence in the book is clear.[2]

The Bain–Modigliani work was then the only game in town in neoclassical IO circles, and I would be surprised if the TGF would have drawn the early attention it did in the absence of the 'managerial theories' and Marris's in particular.[3] For me the essence of the story is this. Managerial theories depended on the assumption of separation of ownership from management and control, with managers managing but also acquiring control of the firm. Assuming that managers' objectives diverged from owners' and that the former had a preference for growth and size-related variables, not short run profit *per se*, different theories of the firm could follow (for example Baumol's sales revenue maximisation and Williamson's discretionary expenditures) with different implications of the short-run profit maximisation neoclassical model.

Now, clearly these two are neoclassical-type static comparative equilibrium models of rational utility maximising agents.[4] The utility function and the ensuing equilibrium are different. All these have little to do with Penrose. However, the focus is on the same things, i.e., the importance of management and growth. This might have helped Penrose's book earn some early notoriety in the context of that discussion. This contributed to the recognition of the 'Penrose effect', later to be incorporated into wider neoclassical models of growth, on which see Slater (1980), Foss (this volume), and Penrose and Pitelis (this volume).

While the timing could not have been better for gaining early notoriety, to what extent that was good for the book more generally is an open question. To some extent it seems that that notoriety gave the book the status of a 'classic', defined by some as a book that everybody cites but few read (see Lazonick, this volume).[5] I certainly plead guilty to this.[6]

---

[1] So often with important issues and ideas, there seems to be a clustering around specific times. The late 1930s were such a period, with Coase, Hayek, Keynes, Schumpeter, to mention the best known.

[2] Bain is explicitly mentioned in the book. The affinity of her work to Chandler's was recognised later in her preface to the 1995 edition.

[3] Marris (this volume) gives his personal account of the relationship between Edith Penrose's work and his own seminal contribution and book (1964).

[4] For Penrose's influence in neoclassical thinking more generally, see Slater (1980), Penrose and Pitelis (this volume). In advising Penrose on how to deal with such developments in the second edition of her book, Fritz Machlup writes in a letter to her, in 1978, that this 'task is a little more demanding. Particularly in your case where the mathematicians picked up the ball that you had thrown, and ran away with it in all possible directions.'

[5] For Slater (1980) instead, *TGF* was already then, in the late 1970s, a classic actually being read!

[6] My *Market and Non-Market Hierarchies* (1991) dealt with the genesis, growth and failure of capitalist

footnote continued overleaf

I believe many people did just that, thus, in effect, leaving the book unappreciated. Today the situation is different, but not in all cases. Following the rise and rise of the resource/knowledge-based perspective to the theory of the firm, more and more contributors build directly on the book. Kor and Mahoney (2000) discuss this point in more detail.[1]

Penrose's contribution has been criticised extensively. Many of these criticisms are, so to speak, from within the family. Penrose has conceded some such criticisms, notably by Robin Marris concerning the role of capital markets and finance, and revisited some issues, e.g., her treatment of growth and profits, in response to such criticisms (see Penrose and Pitelis, this volume). Most contributors in this volume point to limitations of her theory and views in general, in the context of specific issues or recent developments (see especially Marris, Kay, Cantwell and Lazonick). All I wish to note here is that many such limitations and critiques are relevant and I believe could well have been made by Penrose herself. Clearly one cannot deal with everything and predict everything. What is important is to have a conceptual lens through which to perceive, explain and predict; and I think Penrose's lens is the best we have.

There are two criticisms of the resource-based theory, and up to a point, of Penrose, that I wish to discuss here, no less because they come from two figures whose views

footnote 6 continued from previous page

institutions: markets, firms, states and international organisations. It attributed these to attempts by 'principals' to further their benefits derived from specialisation, division of labour and teamwork, by removing constraints to its realisation, including dynamic transaction costs. In this context, the existence of such institutions was attributed to their differential abilities, while their co-existence to differential advantages (e.g., markets in exchange, firms in production, states in legitimacy, ideology and the provision and enforcement of 'law', international organisations in the provision of international 'public goods' and their distance from intra-country sectoral rivalries). In this context, I explicitly discussed limits to firm growth, in terms of product markets, labour markets, managerial markets, financial and technological constraints. Clearly all these are very Penrosean themes. However, Penrose's appearance in the book was in terms of conventional interpretation of the 'Penrose effect'. What little I read from it was interpreted in the lens of received wisdom (another Penrosean theme). I had understood precious little. As a result, endogenous innovation-knowledge and growth was absent there too with all that that entails.

[1] A current dispute, instead, focuses on the extent to which Penrose's book was influential in the early resource-based contributions; some like Barney (2000, quoted in Rugman and Verbeke, 2001) believe it was not, even that it should not be. However, as discussed in Foss (this volume) and Rugman and Verbeke (2001), among others, at least a variant of modern resource-based theory is about rents in equilibrium, which is far from Penrose's main theme. As I have already mentioned, the last part of the book, which deals with the issue of firms size, artificial barriers to entry, etc., as well as her references to 'impregnable bases' can be interpreted in 'rents in equilibrium' terms. Clearly, however, there is little question that that was not her main theme and interest. It is arguable (indeed clear) that this variant has not been influenced by Penrose. While this helps to establish a claim to originality, I believe these views could have been better, had they been influenced by Penrose. For illustration purposes, consider the recent critique of Barney (1991), by Priem and Butler (2001A, 2001B), that, among others, the 'Barney-type resource-based view' has exogenous value and no theory of the firm. Barney (2001) admits the exogenous value point, and cites himself (Barney 1991) in evidence that he had been aware of this. Now Penrose's theory is if anything at all, exactly about endogenous value creation! In addition, her perspective lends itself to a theory of the firm, as discussed. If so, I see little reason for the quarrels. I also see the potential tension between growth (and value creation) and rents in equilibrium, as a variant of the issue of growth and allocation-coordination I discussed earlier in a wider context. What I said there is of input to this current debate. Both value creation and 'rents in equilibrium' can have their uses, but I believe the latter's use can be better appreciated if one's starting point is value creation. In this sense, I take the opposite view to Rugman's and Verbeke's in that I feel Penrose's contribution could and should be of input to all contributors in this area.

currently inform the two major 'competing' perspectives on firm (strategy) to Penrose's, Michael Porter and Oliver Williamson. Both have dealt with resources and competences more recently in Porter (1999) and Williamson (1999). Porter is largely dismissive, the main reason being what he sees as vague concepts. Williamson has wider-ranging critiques, a major one being the apparently tautological nature of the perspective and the lack of operationalisability and supporting evidence.[1]

Penrose's contribution has not been criticised as tautological; this would have been absurd. The predictive and prescriptive aspects of her theory are both operationalisable and testable, and they have been operationalised and tested. This answers Williamson's critique on this point. For Kor and Mahoney (2000) the persistence of critiques on operationalisation and evidence is just misinformed. They cite an extensive list of empirical studies that test and mostly support the Penrosean views.

Williamson's reference to operationalisability and evidence is familiar; he has made similar comments on the 'power' perspective (Williamson, 1996). The reason I wish to pursue this issue just a fraction more, is because I believe that Porter's own work refutes both Porter and Williamson, but also, and most importantly, because I believe that most of the available evidence is still seriously limited in value, for which I agree with Williamson that more and better operationalisation and empirical work is needed on all fronts.

Concerning first the critique on evidence, it should be noted that, considering the length of time for which resources/competence-based ideas have been around, they have not done at all badly on this front (see Kor and Mahoney, 2000). It has arguably taken longer for transaction costs to start becoming operationalised and tested, and this is not really surprising. Resources, especially tangible ones, are operationalisable directly and to some extent, intangible assets too (e.g., through data on patents). The case with transaction costs is more difficult, as extensive empirical work had to await the introduction of a separate concept, leading to transaction costs, by Williamson, i.e., asset specificity. This is a more roundabout method of testing the implications of transaction costs, and it is worrying that no less than Coase (1991) himself, but also Demsetz (1995) have expressed doubts on the ubiquity and importance of Williamsonian asset specificity. Concerning my point on Porter, in 1987 he had, I think, written a classic article on conglomerate diversification, which in many respects is the best available test of the Penrosean theory of diversification. The prescriptive implication of Penrose's theory on this is to diversify where there is perceived evidence of the applicability of existing resources to the new venture. By transferring knowledge (thus skills and competences) there will be value added. Porter's study provides empirical support for the claim that by transferring skills and by sharing activities, diversification

---

[1] Porter does not refer to Penrose and it is unlikely he has read the book, Williamson's critique seems to be reinforcing Porter's and both need to be addressed. The issue of tautology is also explicitly dealt with in the discussion between Barney and Priem and Butler. I believe Barney is effective in countering this criticism. He refers to other apparent tautologies, including Coase's, and claims that the issue is whether one can provide operational context and testing of hypotheses to the apparent tautologies. I agree with this. Allyn Young's famous 'tautology' (that the division of labour is determined by the division of labour) is clearly not a tautology at all, if you have the theory to explain it.

can add value and succeed. The finding itself, and even the terminology (skills, activities) are almost as Penrosean as one can get.[1]

Where I side with Williamson is the nature of the evidence. What we crucially need now is to operationalise power, Porter-type, transaction costs and resource/knowledge-based ideas and test them simultaneously in the same general estimated econometric equation. There has been precious little progress on that. One reason is that most authors tend to test for their preferred theory. This, however, is inadequate, especially when (as is the case more often than not) different theories can lead to the same prediction. Which theory has the highest explanatory power? Unless one sees both coefficients, one's support for one's own theory's implications means next to nothing, just that one has no reason to reject one's hypothesis. What if other hypotheses have similar implications but even higher explanatory power? We will simply not know, until we proceed to the type of empirical work I mentioned above.[2] The lack of econometric work (including time series analyses) among management scholars, who are now dealing with these concepts, is a major drawback.

III. POSTSCRIPT

The approach and contribution of Edith Penrose to economics and management scholarship has already earned her a place in the pantheon of the very best. The scope of her contribution, however, is only just being realised. The chapters that follow show some of the possible directions that could take. They are testament to our belief that, enormous as it already is, the appreciation of Penrose's contribution is only in its beginnings. As we have already suggested, this is largely because Penrose's view on (the growth of) the firm, is only one of the many possible applications by which her underlying epistemology, her theory of knowledge, can be put to fruitful use.

REFERENCES

BAIN, J. S. (1956). *Barriers to New Competition*, Cambridge, MA, Harvard University Press.
BARAN, P. and SWEEZY, P. (1966). *Monopoly Capital*, Harmondsworth, Penguin.

[1] Clearly there are resource-based aspects in Porter's other contributions, for example, in the *Competitive Advantage of Nations* (Porter, 1990, see also Foss, 1996), but also his recent work on clusters. On clusters, too, Penrose's theory is of essence, and Porter's evidence arguably in support of her views. Richardson's (1972) classic conceptual foundation of clustering explicitly built on Penrose. Indeed seen in this context of knowledge generation, informing decisions about cooperation and clustering, the Penrosean relevance on clusters is even more important (see Pseiridis, 2001). I feel Porter should read Penrose. Given his extraordinary qualities for prescription, and notwithstanding critiques, such as Grant's (1991), I feel this could serve both.

[2] This is needed even for variants of the same theory. Take the case of the literature on the multinational firm. As explained in Pitelis (this volume), there are at least three transaction-cost-related explanations of foreign direct investment (FDI) versus market-type modes of entry. To test which is better you need FDI on the left-hand side of an equation and (proxies for) the (three or more) competing determinants (specific assets, public goods-type intangible assets, intangible assets with tacit knowledge dimensions) on the right hand side. We have nothing of the sort yet, and we need it desperately. This is probably the main reason why all theories persist and each is able to claim star status.

BARNEY, J. B. (1991). Firm resources and sustained competitive advantage, *Journal of Management*, Vol. 17, No. 1, 99–120.

BARNEY, J. B. (2001). Is the resource-based 'view' a useful perspective for strategic management research? Yes, *Academy of Management Review*, Vol. 26, No. 1, 41–56.

BAUMOL, W. J. (1959). *Business Behaviour, Value and Growth*, New York, Macmillan.

BAUMOL, W. J. (1962). On the theory of the expansion of the firm, *American Economic Review*, Vol. LII, 1078–87.

CABRAL, L. M. B. (2000). *Introduction to Industrial Organisation*. Cambridge, MA, MIT Press.

CALDWELL, B. (2000). 'Hayek: right for the wrong reasons?', presidential address, History of Economics Society, July 2.

CHANDLER, A. D. (1962). *Strategy and Structure*, Cambridge, MA, MIT Press.

COASE, R. H. (1937). The nature of the firm, *Economica*, Vol. IV, 386–405.

COASE, R. H. (1988). *The Firm, The Market, and The Law*, Chicago, IL, University of Chicago Press.

COASE, R. H. (1991). The nature of the firm: meaning, and the nature of the firm: influence, in *The Nature of the Firm: Origins, Evolution and Development*, edited by O. E. WILLIAMSON and S. G. WINTER, Oxford, Oxford University Press.

COASE, R. H. (1993). Coase on Posner on Coase, *Journal of Institutional and Theoretical Economics*, Vol. 149, No. 1, 90–8.

COWLING, K. (1982). *Monopoly Capitalism*, London, Macmillan.

DEMSETZ, H. (1988). The theory of the firm revisited, in *Ownership, Control and the Firm: the Organization of Economic Activity*, Vol. I, Oxford, Basil Blackwell.

DEMSETZ, H. (1995). *The Economics of the Business Firm: Seven Critical Commentaries*, Cambridge, Cambridge University Press.

FOSS, (1996). Research in strategy, economics, and Michael Porter, *Journal of Management Studies*, Vol. 33, No. 1, 1–24.

FRANSMAN, M. (1994). Information, knowledge, vision and theories of the firm, *Industrial and Corporate Change*, Vol. 3, No. 3, 713–57.

GRANT, R. M. (1991). Porter's 'Competitive Advantage of Nations': an assessment, *Strategic Management Journal*, Vol. 12, 535–48.

HAYEK, F. A. (1945). The use of knowledge in society, *American Economic Review*, Vol. XXXV, September, 519–30.

KNIGHT, F. (1921). *Risk, Uncertainty and Profit*, New York, Houghton Mills.

KOGUT, B. and ZANDER, U. (1996). What firms do? Coordination, identity, and learning, *Organisation Science*, Vol. 7, No. 5, September–October, 502–18.

KOR, Y. Y. and MAHONEY, J. T. (2000). Penrose's resource-based approach: the process and product of research creativity, *Journal of Management Studies*, Vol. 37, No. 1, 99–139.

LIEBESKIND, J. P. (1996). Knowledge, strategy, and the theory of the firm, *Strategic Management Journal*, Vol. 17, Winter Special Issue, 93–107.

MARGLIN, S. (1974). What do bosses do? The origins and functions of hierarchy in capitalist production, *Review of Radical Political Economics*, Vol. 6, Winter, 60–112.

MARGLIN, S. (1984). Knowledge and power, in *Firms, Organizations and Labour: Approaches to the Economics of Work Organization*, edited by F. H. STEPHEN, London, Macmillan.

MARRIS, R. (1964). *The Economic Theory of 'Managerial' Capitalism*, London, Macmillan.

MODIGLIANI, F. (1958). New developments on the oligopoly front, *Journal of Political Economy*, Vol. 66, 215–32.

PENROSE, E. T. (1959). *The Theory of the Growth of the Firm*, Oxford, Basil Blackwell.

PENROSE, E. T. (1995). *The Theory of the Growth of the Firm*, 3rd edn, Oxford, Oxford University Press.

PITELIS, C. N. (1991). *Market and Non-Market Hierarchies*, Oxford, Basil Blackwell.

PITELIS, C., ed. (1999). *Contributions to Political Economy*, No. 18, Special Issue on Edith Penrose.

PITELIS, C. N. and WAHL, M. (1998). Edith Penrose: pioneer of stakeholder theory, *Long Range Planning*, Vol. 31, No. 2, 252–61.

PORTER, M. E. (1987). From competitive advantage to corporate strategy, *Harvard Business Review*, May/June.

PORTER, M. E. (1990). *The Competitive Advantage of Nations*, Basingstoke, Macmillan.

PORTER, M. E. (1991). Towards a dynamic theory of strategy, *Strategic Management Journal*, Vol. 12, 95–117.

PORTER, M. E. (1999). Michael Porter on competition, *Antitrust Bulletin*, Winter, 841–80.

PRIEM, L. R. and BUTLER, J. E. (2001A). Is the resource-based 'view' a useful perspective for strategic management reseach?, *Academy of Management Review*, Vol. 26, No. 1, 22–40.

PRIEM, L. R. and BUTLER, J. E. (2001B). Tautology in the resource-based view and the implications of externally determined resource value: Further comments, *Academy of Management Review*, Vol. 26, No. 1, 57–66.

PSEIRIDIS, A. N. (1996). 'Edith Penrose's *Theory of the Growth of the Firm* and Transaction Costs: A Comparison and an Attempted Synthesis', mimeo, University of Cambridge.

PSEIRIDIS, A. N. (2001). 'Competence Clusters as an Industrial Strategy', doctoral thesis, University of Cambridge, forthcoming.

RICHARDSON, G. (1972). The organisation of industry, *Economic Journal*, Vol. 82, 883–96.

RUGMAN, A. M. and VERBEKE, A. (2001). 'A note on Edith Penrose's Contribution to the Resource-based View of Strategic Management', mimeo.

SCHUMPETER, J. (1942). *Capitalism, Socialism and Democracy*, 5th edn, London, Unwin Hyman (1987).

SLATER, M. (1980). Foreword, in E. PENROSE, *The Theory of the Growth of the Firm*, New York, Sharpe.

WILLIAMSON, O. E. (1964). *Economics of Discretionary Behaviour: Managerial Objectives in the Theory of the Firm*, Englewood Cliffs, NJ, Prentice Hall.

WILLIAMSON, O. E. (1975). *Markets and Hierarchies: Analysis and Antitrust Implications*, New York, Free Press.

WILLIAMSON, O. E. (1996). Efficiency, power, authority and economic organization, in *Transaction Costs: Economics and Beyond*, edited by J. GROENEWEGEN, Dordrecht, Kluwer.

WILLIAMSON, O. E. (1999). Strategy research: governance and competence perspectives, *Strategic Management Journal*, Vol. 20, 1087–108.

YOUNG, A. (1928). Increasing returns and economic progress, *Economic Journal*, Vol. XXVIII, No. 152, 527–42.

2

EDITH ELURA TILTON PENROSE: LIFE, CONTRIBUTION AND INFLUENCE

PERRAN PENROSE[1] and CHRISTOS PITELIS[2]*

[1]Centre of International Business and Management Studies, Judge Institute of Management Studies, University of Cambridge and [2]Judge Institute of Management Studies and Queens' College, University of Cambridge

Forty years since the publication of *The Theory of the Growth of the Firm*, Edith Penrose's contribution to economics and business studies is currently receiving increasing acknowledgement and appreciation. Her ideas arguably dominate recent thinking, at least on the issue of the theory of the firm and strategic management. This introductory paper presents Penrose's life and assesses her contribution and influence so far.

I. EDITH PENROSE—HER LIFE

Edith Elura Tilton was born on 15 November 1914 at Sunset Boulevard in Los Angeles. Her father, George Tilton, was a road engineer for the Department of Public Works, and was descended from early immigrants to the United States. Her mother, Hazel Sparling, was a descendant of a passenger on the *Mayflower*. Edith had two brothers, Harvey and Jack. It was a close and supportive family, living in a small town community. For much of the time, the family followed George as he surveyed the new Californian road network (he headed the party that surveyed Highway 1 along the Pacific coast). As a small child Edith spent much of her time in road camps, packing along narrow trails, receiving a basic education in small shacks that served as classrooms. For a child the life was exciting and not without danger: she was about to engage in a close dialogue with a rattlesnake when her mother shot the snake through the head.

The family settled eventually in San Luis Obispo, where she went to school. She went on to graduate at the top of her class in San Luis Obispo High, and entered the University of California at Berkeley. She gained a scholarship after her first semester, and more or less by accident decided to study economics. She met David Denhardt, who was slightly older than she was, while a student, and married him. She completed

* We are grateful to numerous people for discussions pertaining to the contents of this paper, not least to Edith herself. We are also grateful to Ronald Coase, Roger Sugden and the contributors to this volume. Whilst the result of a joint effort section I of this paper is mainly accredited to Penrose and sections II and III to Pitelis.

her studies in Berkeley in 1936 with a BA in economics, and the couple moved to Northern California where Edith took a job as a social worker with a Depression Relief agency. The first tragedy in a life that was to see further tragedy came when David, who was a lawyer and an aspirant District Attorney, was killed in a hunting accident when Edith was four months' pregnant. Circumstances surrounding the incident suggest the possibility that it was not an accident, but the mystery was never resolved.

One of her teachers at Berkeley was the economist E. F. Penrose ('Pen'), who was born of Cornish parents in Plymouth, UK, in 1896. Like Edith, Pen came from a relatively humble background, had served in the trenches in the First World War, managed to finance himself through Cambridge after the war, and had come to Berkeley via Japan and the Stanford Food Research Institute. Edith acted as a *de facto* assistant to Pen, who was 20 years her senior, and attended his summer extension classes in economics.

In 1939 Pen moved to the International Labour Office (ILO) in Geneva at the suggestion of John Winant, who was at the ILO and who was later to become US Ambassador to Britain. Edith took a job as a researcher in the Economics and Statistics Section of the ILO, and travelled to Geneva. Life there was tense during this period, and Pen and Edith were involved in assisting Jews to escape from Germany through Switzerland amid real uncertainty as to whether Germany would invade Switzerland. The ILO then moved to Montreal, and Pen and Edith made the first part of the journey by bus across France in front of the German invasion. During this period she wrote *Food Control in Great Britain*, published in 1940, which analysed the problems of production, distribution and consumption of food in wartime Britain.[1]

When Winant was appointed to London, Pen went with him as Economic Adviser, and Edith served on his staff as a researcher. Pen was deeply involved in the negotiations between the UK and the US, including those between John Maynard Keynes and Harry Dexter White, over the post-war economic order (described in *Economic Planning for the Peace*, Princeton, 1953, a fascinating account of missed opportunities). Edith's association with Pen, who was in a web of economists working on post-war planning, had brought her into contact with many of the most prominent economists of the day. She had been greatly influenced by Schumpeter, whom she met once (Pen knew him), and as a young woman came into contact with Keynes, Meade, D. H. Robertson, Austin Robinson, H. D. Henderson, Robbins, Jewkes, all before, as she used to say later, seriously taking up economics!

In 1945 another tragedy occurred: Edith's brother Jack, a pilot in the US Airforce in Italy, was shot down and killed. Her other brother, Harvey, was also an airforce pilot, and in 1952 his plane went down over Alaska and he perished before help could arrive. Pen and Edith married in 1945.

Immediately after the war Edith and Pen joined the US delegation to the United Nations, again following Winant, who died in 1947. Pen spent a year at the Institute

---

[1] There would be only a tenuous connection between this work and her subsequent work. However, the analysis includes the organisation of the food industry and its tendency towards amalgamation and cartelisation.

of Advanced Studies in Princeton, and in late 1947 moved to Baltimore where he took a chair in Human Geography. In 1946 Edith bore a son, Trevan, who died at the age of 18 months, and is buried in Baltimore. Edith bore two further sons.

The move to Johns Hopkins was an important one in Edith's development as an economist. She began her master's and doctoral studies, which she completed in 1951. (At her PhD oral examination she was asked by Professor Clarence Long [who later became a US Congressman] to give a brief discourse on the economics of the shmoo, which was a creature in the comic strip L'il Abner and which reproduced instantly, gave milk and eggs, and on request would lie down and die and become a cooked ham. The shmoo was a free good. Edith had not read the comics and had no idea what a shmoo was. She was very annoyed with Long for asking such a question and let him know it. She passed unanimously.) She then became Lecturer and Research Associate at Johns Hopkins where she and Pen stayed until 1959, with extensive stays out of the country. At Hopkins Edith's supervisor was Fritz Machlup, who co-directed with G. H. Evans Jr a research project on the growth of firms. Her description of how she became involved in the project is given in an illuminating interview with David King in the British edition of Parkin and King's (1992) economic textbook:

I had no special interest in firms, but a Professor there [Johns Hopkins] had a large grant to do studies of the growth of firms, and he asked a group of us to participate. I didn't mind what I specialised in, but I had to earn some money and the growth of firms seemed interesting. So I elected to work on the theory of the growth of the firm and it took me nine months of reading and especially thinking before I realised that the traditional theory of the firm, in which I, like other economists had been trained, was not relevant to the problem of the growth of firms.

For her fieldwork Edith was attached to the Hercules Powder Company as a Fellow under the College–Business Exchange Program, and this was the beginning of research which eventually led to *The Theory of the Growth of the Firm*.[1] Machlup remained a lifelong friend and mentor. Indeed, Edith's intellectual development and the rigour of her thought were very much a product of her association with first, Pen, and second, Fritz Machlup, both of whom were demanding in their requirement of clear and uncluttered analysis. But Pen was also very wise in the way institutions worked, and mistrusted in the increasingly specialised trends in economics. Although her conclusions were sharply at odds with textbook theories, she never sought confrontation: 'Would anyone . . . try to reconcile a football game with a cricket match just because they are both ball games?'.

Pen became increasingly disillusioned with the US during the 1950s. The Committee for UnAmerican Activities chaired by Senator McCarthy had targeted academic institutions, and Johns Hopkins was caught up in the net as Owen Lattimore, the eminent sinologist and Mongolia specialist, was accused 'of losing China' by McCarthy. Edith and Pen played central roles in his defence. Although the university was generally supportive, the experience prompted Pen to take sabbatical leave until retirement, first at the Australian National University in Canberra (1955), where

---

[1] A fascinating paradox is how Machlup, a doyen of neo-classical economics, should have been partially responsible for a work so far removed from the mainstream.

Edith continued work on *The Theory*, and then at the University College of Arts and Sciences in Baghdad (now the University of Baghdad) (1957–59), from which sprang a lifelong interest in the Arab world. After Pen's retirement in 1960, they made many return trips to Iraq until they were expelled from the country in the mid 1960s. Later they jointly wrote *Iraq: International Relations and National Development* (1978), and were an important influence on a generation of Iraqi economists, many of whom are in exile. Pen and Edith travelled extensively in the Middle East, teaching at the American University of Beirut, and at the universities in Cairo and Khartoum.

It was a natural development of Edith's work on the growth of the firm that she became interested in the international firm and the oil industry. Although it is often believed that her work on multinational firms is a parallel interest rather than a direct offshoot of her work on the growth of firms, she regarded her study of the international oil companies as an extension of the theory of the growth of the firm internationally (see next section). Indeed, Edith's economic preoccupations were frequently a response to the situations in which she found herself. Although in one sense this characteristic meant that she did not follow a given path over time, in another sense it contributed to the way she approached theory, from observing the real world and trying to make sense of it.

When I went to Baghdad in the late 1950s . . . I found that no economist had published any analysis of the international oil companies in spite of the fact that the oil industry was a very large and vitally important industry, accounting for a large proportion of international trade and run by some of the largest companies in the world. I happily took advantage of this splendid opportunity for empirical research.

Her interest in developing countries dates from this period, in particular the role that international firms have as key investors in developing countries, and, by extension, problems of development in a wider sense. Again, her theoretical concerns arose from observation of the real world:

essentially this is all common sense, which is especially important when you deal with theory, for it is not always easy for beginning students to understand the relation of the 'theory' they study to what they see as the 'real world'.

The path to Johns Hopkins was a fortuitous one: so was the path to London. In 1959 Edith and Pen drove across the Syrian desert, through Turkey and on to England in an old Hillman estate car so that she could attend an interview at Cambridge. *The Theory* had not yet been published but was circulating in manuscript. She failed to be appointed, but successfully applied for a joint readership in economics with reference to the Middle East at the London School of Economics and the School of Oriental and African Studies (SOAS).

In 1964 she took the Chair of Economics with special reference to Asia at the SOAS which she held until 1978. During this period her preoccupation was very much with oil and the multinationals. But it should also be noted that she was a remarkable teacher. She made visits to many parts of the world, notably the Middle East (Baghdad, American University of Beirut, Cairo, Khartoum, Amman) but also Delhi,

Tanzania (where she provided advice on the government's treatment of foreign companies, which was not well received),[1] Indonesia, and many other countries. Her economic interests were very much centred in the real world, perhaps influenced by her formative years at the heart of wartime and post-war economic policy making, and she had close contacts with the energy industry.

In 1978 she retired and took up her position as Professor of Political Economy at INSEAD, where she was also Associate Dean for Research and Development: she had been one of the very few female full professors in London, and INSEAD was also largely a male preserve. Here her interests moved back to the firm, and management, and she was in an environment which recognised her achievement. Pen died in 1984. Edith and Pen had had a long and fulfilling marriage, and had been entirely self-sufficient in each others' company. His death, though not unexpected, was a serious blow, as each had provided for the other the geographical centre of their existence: location did not matter. On his death she retired from INSEAD moving to Waterbeach, near Cambridge, to live near her sons.

In her retirement she led a very active life. She was on several governing bodies, including the council of the Overseas Development Institute, and the board of the Commonwealth Development Corporation. She sat on the Pharmaceuticals Committee for many years, having in the early 1970s provided advice to the labour government on pharmaceutical industrial policy. She also undertook consultancy work, including advisory work for the Iranian government on compensation claims resulting from the nationalisation of the oil industry. She argued that there was little in the economic literature on the subject apart from a straight discounted cash flow approach regardless of circumstances. There was, therefore, a lack of economic principles on which to base resolution of conflicts arising between private and public interests where the law only recognised private interests, in the particular case of nationalisation in the public interest where exhaustible national resources are involved.

It was in the period after her retirement that the pace of recognition of her early work gained momentum. She received honours from American, British, and other European universities, and a steady stream of visitors arrived in Waterbeach. She had always been modest about her work, and gained much pleasure from the belated recognition accorded to it. The stimulus of the renewed interest in her work set her thinking again about theories of the firm, both in terms of business and management, and in terms of the poverty of the neo-classical model. She became interested in how firms were changing, having toyed with the idea of a theory of the death of the firm, the idea metamorphosing into the metamorphosis of the firm. Edith used to quote advice given to her by John Winant: 'the important thing is to lose your illusions and not become disillusioned'. She did not become disillusioned. She was one of the most original economic thinkers of the century, but it seemed as though she was unaware of

---

[1] She took the view that economic independence of a country is not necessarily threatened by extensive foreign investment, although it could be under certain circumstances, and that the main question was whether state intervention could increase the net contribution that multinational firms can make to the economy.

her achievement, and she never felt that her work had been ignored: she merely passed on to different work. She believed in the goodness of people, as Pen had done, and retained a vigorous interest in other people's business, as all those who sought her advice (as well as many who did not seek it) will remember.

She had a slight stroke in 1994, after which her thought processes slowed, but she lost none of her acuity or interest in life. She continued to enjoy robust conversations with her visitors, even better if they came in the evening for a few drinks. She was irked by medicines which were incompatible with alcohol. The night before she died she called the doctor, complaining of not feeling well, and he gave her something to take. As he left, she ran out after him into the road in her nightclothes to make sure that the prescription would not conflict with her evening whisky. She died from heart failure in her sleep the next day, in October 1996, shortly before her 82nd birthday.

II. CONTRIBUTION

As the short account of her life shows, Penrose's contribution to economics was wide ranging: from 'food control' through the patent system, to the theory of the (growth of the) firm, the multinational enterprise, the theory of industry organisation, the international oil industry, the economics of Arab countries, international economic relations, and more. Within this context, she has proposed theories of competition, innovation, mergers and acquisitions, small firms and networking and competition policy. In her early work, for example on the economics of the International Patent System (1951),[1] Penrose questioned the alleged positive benefits of the international extension of the patent system to social welfare, in particular of the developing countries. That work was described as 'novel and controversial . . . several dogmas which legal experts have held in great respect are exposed to the bright searchlight of a skilled economic analyst and are shown to be untenable' (Machlup, 1951, p. viii). Despite such praise, Penrose's early work has not been particularly noticed. However, it helps to expose some of her early interests in 'monopoly' and 'innovation' and social welfare, which were later to assume dominant position in her thinking.

Penrose's major claim to be remembered as an economist is undoubtedly *The Theory of the Growth of the Firm*, first published in 1959. The result of her work with the Hercules Powder Company, was completed in 1956, and published in 1960. It 'was originally intended for inclusion in *The Theory of the Growth of the Firm*, but omitted to keep down the size of the book' (1960, p. 1). Given this, it is safe to consider the two pieces as part of an integrated whole and consider its main arguments.

On various occasions, Penrose (1959, 1985, 1995) describes her experiences in seeking guidance from existing economic theory to address the issue of her concern, the growth of a real life firm. She found very little. In brief, economic theory of the firm had no firms in it. For mainstream economic theorists 'the "firm" was primarily a set of supply and demand functions' (Penrose, 1985, p. 6):

---

[1] See also Machlup and Penrose (1950) and Penrose (1973).

a part of the wider theory of value, indeed one of its supporting pillars, and its vitality is directed almost exclusively from its connection with this . . . general system for the economic analysis of the problem of price determination and resource allocation . . . the 'equilibrium of the firm' is, in essence, the 'equilibrium output' for a given product (or given group of products) from the viewpoint of the firm. It does not pretend to be an 'equilibrium' of the firm if the firm is represented in any other way, or if any other considerations affect it than those permitted in the theory of price and output. Hence if we become interested in other aspects of the firm we ask questions that the 'theory of the firm' is not designed to answer. In that theory the 'growth' of the firm is nothing more than an increase in the output of given products, and the 'optimum size' of the firm is the lowest point of the average cost curve for its given product (1959, p. 11).[1]

She went on to suggest that the adequacy of traditional explanations of limits to growth (limitations to management, the market and/or uncertainty) do not stand up to scrutiny and concluded that such a theory of the firm cannot be easily adapted to 'the analysis of the expansion of the innovating, multi-product, "flesh and blood" organisation that businessmen call firms' (p. 13).

II.a. The Penrosean firm and the market

The firm in Penrose is a collection of productive resources (human and non-human) under administrative coordination for the production of goods and services for sale in the market for a profit (Penrose, 1959, 1985, 1995). Administrative coordination and 'authoritative communication' define the boundaries of the firm. Very much like Coase (1937), albeit without having at the time been influenced by his classic 1937 article, Penrose maintained the distinction between the firm and the market.

The essential difference between economic activity inside the firm and economic activity in the 'market' is that the former is carried on within an administrative organization, while the latter is not (1995, p. 15).[2]

The boundary of the firm is what distinguishes it from the market and therefore it must 'exist' whether or not it is 'real' . . . (1995, p. xvi).[3]

From the resources within the firm, human resources, and in particular managerial resources, are most important. A reason for this is that any expansion requires 'planning', which can be done by the firm's own management, which itself is firm-specific and not available in the open market.

There are two major categories of 'causes' of growth; those external to the firm and

---

[1] Despite her critique of the traditional theory, Penrose chose not to quarrel with the theory of the 'firm' as part of the theory of price and production, 'so long as it cultivates its own garden and we cultivate ours' (1959, p. 10).

[2] Penrose goes on to point out the 'central managerial direction' within firms which she calls a 'court of last resort', much in line with latter-day arguments to that effect by Masten (1996) and Williamson (1996).

[3] Despite glowing references to G. B. Richardson's 1972 'splendid and pioneering article' (1995, p. ix) in which 'he challenges the whole notion of a firm/market dichotomy, pointing out that there are three means of coordination: direction, cooperation and market coordination and that firm networking blurs the boundaries of firms; that the firm in reality is not the island in a sea of market transactions, but itself part of a network . . .' (1995, p. xviii), Penrose maintained that 'firms and markets are both, in their different ways, networks of activity, but the difference between them is crucial to an understanding of the nature of the economy as a whole' (1996, p. 1717).

those internal. Penrose suggests that external causes, for example raising capital, demand condition, etc., albeit of interest 'cannot be fully understood without an examination of the nature of the firm itself' (1995, p. 532). The problem as she saw it was 'the internal incentives to and limits on growth—a theory of the growth of the firm that does not relate to fortuitous externals events' (1955, p. 532).

There are two basic reasons why there are endogenous to the firm incentives for growth, which moreover are self-reinforcing, leading to opportunities for further expansion. First is the claim that the execution of any plan requires resources which are in excess of those strictly necessary for the execution of the plan. Second, upon completion of a plan, managerial resources will be released. Crucially, moreover, 'the services that the firm's management is capable of rendering will tend to increase between the time when the plan is made and the time when the execution is completed' (1955, p. 533).

Penrose attributes the ubiquitous presence of unused resources to arguments by Babbage, Austin Robinson and Sargent Florence such as the 'balance of processes' or 'the principle of multiples', which suggest that

if a collection of invisible productive resources is to be fully used, the minimum level of output at which the firm must produce must correspond to the least common multiple of the various outputs obtainable from the smallest units in which each type of resource can be acquired. . . . This output will tend to be greater the larger the variety of resources and the more diverse the units in which they come. A firm would have to produce on a vast scale if it were to use fully the services of all the resources required for much smaller levels of output (1955, p. 533).

In addition,

most productive services . . . are capable of being used in many different ways and for many different purposes. Hence a firm in acquiring resources for particular purposes—to render particular services, also acquires a range of potential productive services, most of which will remain unused (1955, p. 534).

Managerial services are of particular importance in this context, also because they are available to the firm only in limited amounts:

executives with experience within any given firm can only be found within that firm . . . The production of these services requires time and this limits the scope of a firm's expansion plans at any given time, but permits continuous extension of these plans through time (1955, pp. 534–5).

However, the completion of expansion plans creates and releases resources. It creates resources because 'all personnel in the firm will gain additional experience as time passes' (1955, p. 538). It releases resources because 'not only is there likely to be a generalized improvement in skill and efficiency but also the development of new and specialized services' (1955, p. 538).[1]

This increase in knowledge not only causes the productive opportunity of the firm

---

[1] Importantly, 'the unused services thus created, do not ordinarily exist in the visible form of idle man-hours but in the concealed form of unused abilities' (1995, p. 538).

to change in ways unrelated to changes in the environment but also contributes to the 'uniqueness' of the opportunity of such individual firms (1959, p. 53). This is particularly true given that not all knowledge is 'objective' (transmittable); some takes the form of 'experience', which is hard to transmit.

Unused productive services are, for the enterprising firm, at the same time a challenge to innovate, an incentive to expand, and a source of competitive advantage. They facilitate the introduction of a combination of resources—innovation—within the firm (1959, p. 85). In addition, they are 'a selective force in determining the direction of expansion' (1959, p. 85). Once it is recognised that firms are not to be defined in terms of products, but instead of resources, and given the potential versatility of the latter, demand conditions cannot limit a firm's expansion. In this sense, and in the absence of traditional managerial diseconomies, which Penrose questions,[1] there is no limit to the size of the firm, but rather to its rate of growth.

Whilst focusing on the internal environment, Penrose does not ignore the external one. In her introduction to The Hercules Powder Company case, she claims that

growth is governed by a creative and dynamic interaction between a firm's productive resources and its market opportunities. Available resources limit expansion; unused resources (including technological and entrepreneurial) stimulate and largely determine the direction of expansion. While product demand may exert a predominant short-term influence, over the long term any distinction between 'supply' and 'demand' determinants of growth becomes arbitrary (1960, p. 1).

It is not an easy task to summarise all of Penrose's major ideas. The following points, however, may help to recapitulate and complement the above account.

- Firms are bundles of resources, under internal direction, for use of goods and services, sold in markets for a profit. Their boundaries are defined by the area of coordination and 'authoritative communication'.
- Firms differ from markets; transactions in the latter do not take place within 'administrative coordination'.
- Resources render (multiple) services. The heterogeneity of services from resources gives each firm its unique character. Effective use of resources takes place when resources are combined with other resources.
- Human, and in particular managerial resources are of essence, because expansion requires planning and managerial resources that enable the firm to plan are firm-specific, they cannot be acquired in the market.
- The cohesive shell of the firm helps to create knowledge. This can be 'objective' (transmittable) or experience (hard to transmit). Experience renders managerial services firm-specific.
- Unused resources always exist; they are released after the completion of an expansion and they are created through experience and new knowledge. They are an internal stimulus to growth and innovation, and determine in part the direction of expansion.

[1] See also Demsetz (1995).

- Firms are not defined in terms of products, but resources and (so) 'diversification' is the normal state of affairs in firm expansion.
- There are economies of growth, quite apart from any economies of size.
- There are limits to growth, but not to size, and they are determined by the rate at which experienced managerial staff can plan and implement plans. The services of 'inherited' managerial resources control the amount of new managerial resources that can be absorbed, and thus limit the rate of growth of firms.
- The external environment is an 'image' in the mind of the entrepreneur. Firms' activities are governed by their 'productive opportunity', i.e. all the productive possibilities that its entrepreneurs can see and take advantage of.
- Entrepreneurs are in search of profits; firms desire to increase total long-term profits 'for the sake of the firm itself and in order to make more profit through expansion' (1959, p. 29). In the long run, growth and profits are equivalent as criteria for selecting investment programmes.
- There exists a dynamic interaction between the external and internal environments, which creates opportunities for diversification.
- Intra-firm specialisation leads to higher common multiples, and thus to greater specialisation.

In the long run, Penrose felt that the profitability, growth and survival of firms depend on them establishing relatively *impregnable* bases from which to adapt and extend their operations in an uncertain, changing and competitive world. A new (productive or technological) base requires the firm to achieve a 'competence' in some significantly different area of technology.

II.b. From firm growth, to industry organisation and competition policy

The above account is, by and large, well known by now and much of it can be found in other contributions in this volume. Less is known of Penrose's use of these ideas in explaining vertical integration, mergers and acquisitions, industrial concentration and the scope for small firms, and competition policy. Penrose's ideas on all these are complex and hard to present in a short space, but some elements are presented here.

First, vertical integration. For Penrose, firms integrate vertically in part because they may be able to produce more cheaply for their own requirements (Penrose, 1956, 1959). However, they have to set this against the diversion of resources from potentially more profitable activities. Mergers and acquisitions can be motivated in part by the need to acquire productive services. Targets are likely to complement or supplement the acquiring firm's existing activities. Concentration in a growing economy emerges when the larger firms as a group grow faster than the smaller firms and (therefore) the economy as whole (Penrose, 1956, 1959). Larger and older firms have a 'competitive advantage' over smaller firms in terms both of non-monopolistic advantages (size, experience, access to funds, etc.) but also due to sheer 'monopolistic power' (1956, p. 64). In a growing economy, however, and given limits to firm growth, it is unlikely that large firms can take advantage of all opportunities open to them, allowing potentially profitable opportunities for smaller firms. These relatively unprofitable activities for larger firms are the 'interstices' of the economy. Limits to the rate of growth of large firms, and big business competition will tend to lead to a decline in concentration, albeit not the absolute size of large firms.

On competition, Penrose observed that a strong case can be made for the big firm and for 'big business competition' especially 'with respect to the rate of development of new technology and new and improved products . . .' (1959, p. 160). The 'basic dilemma' is that competition induces innovation but 'competition is at once the god and the devil' (p. 265) in that the growth of firms may be efficient but the consequent size may lead to industry structures which impede growth. Penrose's complex perception of reality and her overall perspective on firm growth and industry organisation shows vividly when she considers 'monopoly and competition in the petroleum industry' (1964). Echoing (but critically) Schumpeter (1942), Chandler (1962) and Demsetz (1973), she observes that the firms'

. . . efficiency in production and distribution, in inventions and technological advance, could not account for the dominant position they achieved. Their record in finding, producing and distributing oil and its products is indeed impressive, but efficiency in this respect would not have been enough to secure their dominance. Hence the story of the rise of the great companies deals as much with financial power, commercial and political negotiations and intrigue, with cartel agreements, marketing alliances, price maintenance arrangements, price wars and armistices, mergers and combination, actions to avoid taxes, and the national and international political interests of governments, as it does with the economics of production and distribution. This statement does not necessarily imply any condemnation of the companies (1964, p. 155).[1]

The above points provide a broad picture of what Penrose covered in her early work (on the growth of the firm). This part of her work has received growing interest and is here, too, the main focus of attention. However, Penrose's subsequent work has also been important, particularly her work on the theory of the multinational firm, the international oil industry, and the political economy of international relations and the economics of the Arab world.

II.c. The multinational corporation and the political economy of industrial relations

Penrose's flirtation with the multinational firm and foreign direct investment has been long and enduring. Many of her major publications are on this topic. This includes an *Economic Journal* article in 1956, and interestingly the homonymous entry in the *New Palgrave* (Penrose, 1987). In all her latter-day writings, notably her Uppsala, 1994, lecture on the 25 years since *The Rate of Growth* book, in her 1995 introduction to the book's third edition and in her entry to the *International Encyclopedia of Management* (1996), Penrose includes a section on multinationals.

Throughout her writings on this topic, Penrose considers the multinational as the natural outcome of the very pressures for growth:

[1] For those viewing Penrose's theory as yet another efficiency-based explanation of firms and markets, this is a strong blow.

The processes of growth, the role of learning, the theory of expansion based on internal human and other resources, the role of administration, the diversification of production, the role of mergers and acquisitions are all relevant. . . . It is only necessary to make some subsidiary 'empirical' assumptions to analyze the kind of opportunities for the profitable operations of foreign firms that are not available to firms confining their activities to one country, as well as some of the special obstacles (1995, p. xv).

The same is emphasised in her 1987 entry to the *New Palgrave*.

There are differences between national and international firms, but the differences are not such as to require a theoretical distinction between the two types of organizations, only a recognition that national boundaries make an empirical difference to their opportunities and costs (1987, p. 563).

The emphasis changes in her 1996 paper.

International borders make enough difference to justify separate treatment of international firms. The differences arise from the additional obstacles (or advantages) relating to culture, language and similar considerations (1996, p. 1720).[1]

Penrose's reluctance to acknowledge any fundamental difference in the nature of multinationals, as opposed to the relative importance of national boundaries, could partly explain the fact that her work on this topic has been unnoticed. There are other considerations too, which, however, are beyond the scope of this paper; see Pitelis and Sugden (2000) and Kay (1999) for more on this.

Her analysis of the political economy of international relations, notably the relation between multinationals and developing countries, similarly, does not take any position other than the one she was able to come to believe after a careful, objective, dissecting analysis. Her analyses of transfer pricing, dumping and protectionism are an anathema to mainstream views, yet extremely modern in the context of 'new international trade' and related theories (see for example, Krugman, 1986, 1990). To mention just a few points, in 1962 she suggests that 'restriction on the repatriation of profits under some circumstances may be a useful means of ensuring, for a whole, continued foreign investment . . .' (1962, p. 138). In 1973, she suggests that 'infant firms', not just infant industries, arguments be 'accepted as an exception to the doctrine of "free investment"' (1973, p. 8), while in 1990 she suggests that 'dumping is endemic in the system, an integral part of the competition among large, diversified, research-based, integrated companies' (1990, p. 185).

II.d. *Epistemology*

Penrose did not deal explicitly with the issue of methodology. Yet her 'endogenous growth' approach to organisations and institutions, her insistence that in order to appreciate the external, one has to start from the internal, the 'nature', of the object under investigation, the dynamic interaction between the 'nature' and the external, her own experience with research, both at the Hercules Powder Company, and later in the Arab world, her views of mainstream theory, and, lastly, her approach to the link

---

[1] In the entry to the *New Palgrave*, Penrose also comments on Coase (1937) and Hymer (1976) to the effect that the advantages of transaction costs do not distinguish the multinational from the domestic firms.

between 'theory' and evidence, can all provide scope for discussion of methodology in (social) science. She appears to have based her work on the belief of a dynamic inter-action between induction and deduction, however, in the context of history-based, path-dependent evolutionary change, shaped by conscious actions by economic actors. She critiqued some biology-based theories of the firm for failing to account for human motivation.

To treat the growth of the firm as the unfolding of its genetic nature is downright obscurantism. To treat innovation as chance mutations not only obscures their significance, but leaves them essentially unexplained, while to treat them directly as purposive attempts of men to do something makes them far more understandable (1952, p. 818).

In her view,

there would seem to be a genuine complementarity between theory and history (1989, p. 11).

Without theoretical analysis of cause and consequence one has no standard against which to appraise the significance of any given set of observations, for this significance is a question of what difference the observations make to what might otherwise have been the historical interpretation . . . Some of the 'theory' may be little more than dressed-up common sense deduction from common observations and therefore not even recognized as such, but much of it has a deeper significance (1989, p. 10).

Last but not least, 'theory is needed precisely *because* reality is so complicated' (1989, p. 11).

II.e Edith's last years

Edith's main interest in her time at Waterbeach following her retirement was the issue of collaboration and 'networking'. She was impressed by the work of G. B. Richardson, his emphasis on the 'dense network of cooperation and affiliation', his attempt to distinguish market from direction and from competition in terms of complementarity and similarity of activities. Building on her earlier work she viewed corporate alliances or cooperative arrangements, as driven 'not necessarily by monopolistic intent but as a means of gaining mutual access to resources such as technology, regional markets and information services' (1996, p. 1722). She struggled a lot with the idea whether such networks, clusters, webs, etc., require a different (to hers) theory of the firm, pointing in particular (in private discussions) to the single person, single computer terminal 'firm' in IT industries. Edith's conclusion and (perhaps only compromise) was that

The individual companies do not lose their 'independent' identity; but the administrative boundaries in each of the linked firms may become increasingly amorphous and the effective extent to which any individual firm exercises control is often not at all clear (1996, p. 1722).

II.f Assessment

There can be no doubt that Penrose's contribution to economics has been immense. By focusing within the firm, emphasising the production side, and by pointing to the knowledge-creation process through specialisation and division of labour, Penrose has, at the very least, managed to single-handedly reinvent, help revive and at the same time complement the 'wealth creation' approach to economics. This was reinventing and building on the work of, among others, Adam Smith, Allyn Young, Alfred Marshal, Friedrich Hayek and Joseph Schumpeter. In so doing, she has not only created a knowledge-based theory of the firm, but also provided the foundations of a theory of the knowledge-based economy (Loasby, 1999). While less noticed, Edith's ideas on a host of other issues discussed in this section were no less significant.

III. INFLUENCE

Penrose's ideas on the theory of the firm have been noticed, discussed and praised. In this sense, they have undoubtedly been influential. However, it is arguable that this influence has fallen short of a recognition of her overall contribution as described above, let alone of the realisation and application of the full potential of her ideas for mainstream economic theory. There can be various reasons for this and some specu-lative thoughts will be presented below.

In the review of her the *Theory of the Growth of the Firm* for the *Economic Journal*, Marris (1961) predicted that the book would prove to be one of the most influential books of the decade. In his entry to the *New Palgrave*, he added that 'this proved an understatement' (Marris, 1987, p. 831). It is interesting to note that Marris based this last view on the then available literature. This was primarily limited to 'managerial theories of the firm' (see Marris, 1996) and in particular to the so called 'Penrose effect' (that administrative, especially managerial, limits to the rate of growth exist), as well as to the potential impact of this effect on macroeconomic growth (Uzawa, 1969).

III.a. Penrose and the neoclassical theory

An important focus of managerial theories was on the extent to which managerially run firms could pursue objectives different to short-term profit maximisation such as, for example the maximisation of sales revenue (Baumol, 1959, 1962), discretionary expenditures (Williamson, 1964) of growth maximisation (Marris 1964), and what are the implications of such behaviour for 'managerial capitalism'.

Penrose's own role in this context was seen in terms of her providing justification for the motivation for growth and the 'Penrose effect'. Concerning the former, and following a critique by Marris (1961) that her treatment of profits and growth was 'woolly', as well as Slater's (1980) work, Penrose admitted that profits and growth could not be treated as 'equivalent criteria for the selection of investment pro-grammes' (Penrose, 1985, p. 8). Nevertheless, she maintained that she found

the assumption that managers of firms try in general to make as much money as seems practical to be not only the most useful, but in fact the only general assumption from which reasonably general conclusions can be drawn (1985, p. 12).[1]

---

[1] It is also interesting to read the follow-up to this statement: 'That I adopt a "weak" form of the profit-maximizing hypothesis is only a recognition of the fact that even in the strong form, maximum profits cannot be uniquely determined *ex-ante* in the face of uncertainty; that no *ex-post* outcome can be unequivocally identified as the maximum that would have been obtained; and that managerial and entrepreneurial attitudes towards uncertainty differ greatly among firms. The problem of "testability" and "refutability" I leave to others, who have long been puzzling over it' (1985, p. 12).

In this paper, Penrose also conceded the limitations of her theory on the issue of capital markets, and the relation between diversification of consumers' demand and that of new products. She praised Marris' theoretical contributions on this front. She was also able, however, to point out that the remarkably similar and independent work of Chandler (1962) provided support for her theory of growth, while Williamson's (1971) analysis of the M-form organisation provided powerful support for her view that growing firms 'expand their ability to manage growth efficiently, with minimum interference with on-going operation . . .' (1985, p. 11).

Penrose quotes, not disapprovingly, various applications of the Penrose effect, which 'has been applied in a number of contexts, and even to my surprise to agricultural products' (1985, p. 9).

As already shown, in the context and way it has developed, the 'Penrose effect' is of utmost importance. It simultaneously describes and determines firms' (limits to) endogenous growth and the (receding) boundaries of the firm. Out of context, the 'Penrose effect' could be seen as just another reason why there can be constraints to 'optimal growth'. Arguably, Penrose's name in mainstream economics textbooks is synonymous to the concept of managerial limits to growth. Stripped from its context, that was also the one contribution that was not hard to formalise in models of firm growth and optimal investment. Besides Marris' (1964) seminal contribution, notable examples are Slater (1980) and Gander (1991). A problem with this literature, however, as noted by Foss in this volume, was that it allowed the mainstream to consider that

the Penrose effect was just a minor detail in the neoclassical analysis of optimal investment and that 'Penrose's critique of equilibrium economics should not really be taken seriously, as her ideas were fully compatible with extended notions of equilibrium'.

A conclusion from the above is that while Penrose's work has been noted, used and influenced mainstream thinking, it was done in a way which misrepresented, misunderstood, even ripped apart her major insights. As noted, these were the endogenous, production side growth advantages associated with the knowledge-creation process through specialisation and division of labour in an evolving, cohesive shell called a firm. This is not efficient allocation of scarce resources under conditions of perfect knowledge, it is not static, it is not equilibrium; in a word it is not neoclassical. Accordingly, it is hardly surprising that while

the total effect of Edith Penrose's work was that of destruction of the neoclassical model of the firm, followed by reconstruction . . . In the following years, and despite the wide recognition the work received, classroom microeconomic theory, and also classroom industrial organisation, often seemed to continue as if nothing had happened (Marris, 1987, p. 831).

Twelve years on not much has changed! Various reasons why this is the case can be put forward; some are discussed by Loasby and Richardson in this volume. Here we provide a number of speculative ideas. An easy observation is that, in part, it was due to Edith herself. She had tended not to ever look back; she had often chosen not to link her early work with latter-day developments; and, importantly, she had never really tried to promote her ideas.[1]

Another reason in this category is that her totally uncompromising attitude rendered her impossible to pigeon-hole. Her views on efficiency and monopoly, on international trade, on big business competition, while being the result of, and fully consistent with her analysis, could often ring simultaneously Schumpeterian, Hayekian and marxian. All these are linked together in their focus on dynamics and resource creation (as opposed to efficient resource allocation). Yet it is true that they are so ideologically diverse that it can render anyone who holds such views simultaneously, too much to handle.

A second, related, reason is that Penrose chose to 'cultivate' her own 'garden', insisting that the neoclassical theory had its uses, but it was simply unsuitable for her concerns, and that it was not useful to try to integrate the two. Her choice not to 'quarrel' with the neoclassicals would help explain why, in their turn, they chose not to quarrel with her. If, as Richardson observes in this volume, the focus of the neoclassicals was coordination, while that of Penrose was growth, it could be argued, in contrast to her ideas, that bringing the two together is both essential and a major challenge. Richardson claims exactly that and offers his views as to the way forward.

While we do not doubt that co-ordination and growth could (should) be usefully integrated, it is arguable that any attempt to integrate Penrose with the neoclassical theory could prove problematic. Importantly, this is due to the issue of knowledge. The neoclassical perspective focuses on the efficient allocation of scarce resources given the assumption of perfect (even if asymmetric) information. In Penrose's theory knowledge, in the form of experience, is not just non-transmittable but it can simply not be known in advance both because of uncertainty and also because knowledge is being created in the context of an evolutionary process. In addition, knowledge is not really scarce in the conventional sense. Its use by someone need not necessarily and always exclude somebody else from using it, and the exchange of knowledge can actually help to enhance it. A theory that starts by assuming full, pre-existing knowledge is clearly unsuitable to deal with these issues. But these, as Hayek (1945) has shown, *are* the issues.

In addition to the above, as already noted, and in contrast to the neoclassical theory, Penrose does not have conventional, rational optimising agents, does not focus on efficient allocation of scarce resources alone, and does not look for an equilibrium. In all—and while we do not doubt the significance of, and need to bring together, co-ordination and growth—we claim that neoclassical theory could simply not fathom Edith's ideas and remain neoclassical. Trying to incorporate her ideas could amount

---

[1] In reviewing a paper, which was building explicitly on her work (thus helping to revive it), Penrose recommended outright rejection, among other reasons, because 'he adds nothing to what I and others have said for years and takes eight pages to do so'. Thankfully the paper was later published in another journal, and went on to become influential. To these, one could add, that while clearly connected in her mind (as pointed out earlier), her arguably diverse interests were certainly not compatible with short-term influence maximisation. But then again she was not a short-term person.

to committing mass suicide, not likely in a profession populated by self-interested optimising agents.[1]

III.b. Penrose and the resource/knowledge-based perspective

While, however, that is the state of affairs concerning neoclassical economics, the past 15 years have experienced a major resurrection of Penrose's work in other fields, notably organisation economics and (strategic) management. In these fields the resource-based, competence-based, or knowledge-based theory of the firm, with or without explicit acknowledgement of her work, has revisited all of her main points. Already in 1985 Edith became aware of these emerging developments.

Some of the rapidly growing literature and research on strategy and strategic management . . . tend, by their very nature, to merge what I have called the 'resources approach' with the 'environment approach' and are likely to produce empirical results useful in testing a theory of the growth of firm (p. 15).

Indeed it was all happening! What Penrose seems to have in mind here is not just the 'industry attractiveness', Porter-type approach to strategy (Porter, 1980), but also the then emerging resource-based, competence-based, (dynamic) capabilities-based and/or knowledge-based approach to strategy. All these were building consciously or unconsciously on the work of Penrose, as well as Richardson, Chandler (1962), Demsetz (1973) and many others, including the very founding fathers Adam Smith, Alfred Marshall and others. Being in a business school, Penrose is likely to have come across some of this literature already. We know, for example, that she had read David Teece's (1982) piece on the multiproduct firm, where he builds on Penrosean and transaction cost ideas.[2]

Since then, the literature has expanded by leaps and bounds. This resource and knowledge-based perspective is now arguably the dominant one in strategic management and organisational studies. The major journals in the area are editing special issues on the topic (*Strategic Management Journal*, 1996; *Organization Science*, 1996) and it is now hard to find a new issue in any major journal in this field without reference to these issues and Penrose. Every aspect of the firm, including the multinational, is currently influenced by this work, see Pitelis and Sugden (2000). Another area where Penrosean ideas emerge is the new theory of macroeconomic endogenous growth, see Romer (1989). Such theories recognise the role of human resources and 'endogenous innovation' in explaining macroeconomic growth. However, it is arguable that these theories are not endogenous in the Penrosean sense of a path-dependent, evolutionary endogenous growth process as, in effect, they treat the aforementioned factors as essentially given, thus exogenous. It is beyond the scope of this paper to review this huge amount of literature; recent critical surveys and readings are by Foss (1997, this volume) and Pitelis and Pseiridis (1999), to mention just two.

---

[1] The outcome, however, could certainly have been an equilibrium one!

[2] Penrose had provided a detailed commentary on Teece's paper in December 1980 (Edith Penrose's personal files).

Edith was fortunate to see and enjoy this new wave of recognition. While in Waterbeach, she was approached by contributors in this field asking for her comments and contribution. Sure enough she went back to the issue. Her 1996 paper on 'growth of the firm and networking' revisits the issues she raised and comments on the current state of play. She recognised the importance of transactions costs issues, which she considered as one of the 'two major types of explanation for the growth of firms in a market economy' (1996, p. 1717). The second is the resource-based view she originated. Interestingly, she felt that 'the two approaches are not mutually exclusive' (1996, p. 1717). Her view is shared by Coase. In a letter to the guest editor, Coase comments on Penrose's views as follows:

I do not regard her views as an alternative view to mine in *The Nature of the Firm*, but as a necessary addition to it. As I indicated in my Yale lectures ' . . . there has been insufficient attention to the role of the firm in "running a business"'.

Coase's recognition of the need to go back to production, to 'running a business', represents an important vindication of Penrose's ideas. All the same, there is an important sense in which the transaction costs perspective of Coase, as developed by Williamson (1975, 1981) and applied by North (1981), is far easier to assimilate in neoclassical thinking. It assumes full knowledge and optimising agents and focuses on static equilibria. While there is scope for rendering transaction costs theorising dynamic, and for integrating transaction costs with Penrosean ideas, it is arguable that this cannot be done without either stripping Penrose from her fundamental insights or stripping neoclassical theory from its very neoclassicism. A battle of paradigms may well be involved, and no less than the outright domination of one may be required. In the absence of a full return to the wealth-creation tradition, it could well be that Penrose's ideas will find it difficult to be fully appreciated, acknowledged, and assimilated by mainstream economic theory.

IV. POSTSCRIPT

Penrose's contribution to economics and management has been seminal. Her ideas are currently going from strength to strength, dominating recent thinking on knowledge-based perspectives on the firm and the economy. The influence of her ideas so far has been impressive and yet nowhere near their full scope and potential. In part, this is because Penrose's 'garden' represents an altogether different paradigm, fundamenally at odds with neoclassical theorising. Unfortunately, this is likely to delay further a full recognition of her fundamental contributions to economics, management, and (social) science at large.

REFERENCES

BAUMOL, W. J. (1959). *Business Behaviour, Value and Growth*, New York, Macmillan.
BAUMOL, W. J. (1962). On the theory of the expansion of the firm, *American Economic Review*, Vol. 52, 1078–87.

CHANDLER, A. D. (1962). *Strategy and Structure*, Cambridge, MA, MIT Press.

COASE R. H. (1937). The nature of the firm, *Economica*, Vol. 4, 386–405.

DEMSETZ, H. (1973). Industry structure, market rivalry and public policy, *Journal of Law and Economics*, Vol. 16, 1–9.

DEMSETZ, H. (1995). *The Economics of the Business Firm: Seven Critical Commentaries*, Cambridge, Cambridge University Press.

FOSS, N. J. (1997) (ed.). *Resources, Firms and Strategies*, Oxford, Oxford University Press.

GANDER, J. P. (1991). Managerial intensity, firm size and growth, *Managerial and Decision Economics*, Vol. 12, 261–6.

HAYEK, F. A. (1945). The use of knowledge in society, *American Economic Review*, Vol. 35, 519–30.

HYMER, S. H. (1976). *The International Operations of National Firms: A Study of Foreign Direct Investment*, Cambridge, MA, MIT Press.

KAY, N. M. (1999). The resource-based approach to multinational enterprise, in *The Nature of the Transnational Firm*, edited by C. PITELIS and R. SUGDEN, 2nd edn, London, Routledge.

KRUGMAN, P. R. (1986). Introduction: new thinking about trade policy, in *Strategic Trade Policy and the New International Economies*, edited by P. R. KRUGMAN, Cambridge, MA, MIT Press.

KRUGMAN, P. R. (1990). *Rethinking International Trade*, Cambridge, MA, MIT Press.

MACHLUP, F. (1951). Foreword, *The Economics of the International Patent System*, Baltimore, MD, Johns Hopkins Press, pp. vii–ix.

MACHLUP, F. and PENROSE, E. T. (1950). The patent controversy in the nineteenth century, *Journal of Economic History*, Vol. 10, No. 1, pp. 1–29.

MARRIS, R. (1961). Book review of *The Theory of the Growth of the Firm. Economic Journal*, Vol. 71, No. 1, 110–13.

MARRIS, R. (1964). *The Economic Theory of 'Managerial' Capitalism*, London, Macmillan.

MARRIS, R. (1987). Penrose, Edith Tilton, in the *New Palgrave, A Dictionary of Economics*, London, Macmillan, 831.

MARRIS, R. (1996). Managerial theories of the firm, *International Encyclopaedia of Business and Management*, London, Routledge, London, 3117–25.

MASTEN, S. (1996). Empirical research in transaction cost economics, in *Transaction Costs Economics and Beyond*, edited by J. GROENEWEGEN, Dordrecht, Kluwer.

NORTH, D. C. (1981). *Structure and Change in Economic History*, London, W. W. Norton.

PARKIN, M. and KING, D. (1992). *Economics*, London, Addison Wesley.

PENROSE, E. F. and PENROSE, E. T. (1978). *Iraq: International Relations and National Development*, Westview, Benn.

PENROSE, E. T. (1940). *Food Control in Great Britain*, Geneva, IO.

PENROSE, E. T. (1951). *The Economics of the International Patent System*, Baltimore, MD, Johns Hopkins Press.

PENROSE, E. T. (1955). Research on the business firms: limits to growth and size of firms, *American Economic Review*, Vol. XLV, No. 2, 531–43.

PENROSE, E. T. (1956). Foreign investment and the growth of the firm, *Economic Journal*, Vol. LXVI, 220–35.

PENROSE, E. T. (1959). *The Theory of the Growth and the Firm*, Oxford, Oxford University Press [1995, 3rd edn].

PENROSE, E. T. (1960). The growth of the firm—a case study: the Hercules Powder Company, *Business History Review*, Vol. XXXIV, No. 1, 1–23.

PENROSE, E. T. (1962). Some problems of policy towards direct private foreign investment in developing countries, *Middle East Economic Papers*, Lebanon, American Research Bureau, American University of Beirut, 121–39.

PENROSE, E. T. (1964). Monopoly and competition in the international petroleum industry, in *The Yearbook of World Affairs*, Vol. 18, London, Stevens.

PENROSE, E. T. (1973). International patenting and the less developed countries, *Economic Journal*, Vol. 83, 768–86.

PENROSE, E. T. (1985). The theory of the growth of the firm: twenty five years after, *Acta Universitatis Upsaliensis: Studia Oeconomicae Negotiorum* (Uppsala Lectures in Business), vol. 20, 1–16.

PENROSE, E. T. (1989). History, the social sciences and economic theory with special reference to multinational enterprise, in *Multinational Enterprise in Historical Perspective*, edited by A. TEICHOVA, M. LEVY-LEBOYER and H. NUSSBAUM, Cambridge, Cambridge University Press.

PENROSE, E. T. (1990). Dumping, 'unfair' competition and multinational corporations, *Japan and the Word Economy*, Vol. 1, 181–7.

PENROSE, E. T. (1996). Growth of the firm and networking, in *International Encyclopaedia of Business and Management*, London, Routledge.

PITELIS, C. N. and PSEIRIDIS, A. N. (1999). Transaction costs versus resource value?, *Journal of Economic Studies*, Vol. 16, 121–40.

PITELIS, C. N. and SUGDEN, R. (2000). The (theory of the) transnational firm: the 1990s and beyond, in *The Nature of the Transnational Firm*, edited by C. PITELIS and R. SUGDEN, 2nd edn, London, Routledge.

PORTER, M. E. (1980). *Competitive Strategy*, New York, NY, Free Press.

RICHARDSON, G. (1972). The organisation of industry, *Economic Journal*, Vol. 82, 883–96.

ROMER, P. (1989). Capital accumulation and the theory of long-run growth, in *Modern Business Cycle Theory*, edited by R. BARRO, Cambridge, MA, Harvard University Press.

SCHUMPETER, J. (1942). *Capitalism, Socialism and Democracy*, London, Unwin Hyman [5th edn, 1987].

SLATER, M. (1980). The managerial limitation to a firm's rate of growth, *Economic Journal*, Vol. 90, 520–8.

TEECE, D. J. (1982). Towards an economic theory of the multiproduct firm, *Journal of Economic Behavior and Organization*, Vol. 3, 39–63.

UZAWA, H. (1969). Time preference and the Penrose effect in a two-class model of economic growth, *Journal of Political Economy*, Vol. 77, 628–52.

WILLIAMSON, O. E. (1964). *The Economics of Discretionary Behaviour: Managerial Objectives in a Theory of the Firm*, Englewood Cliffs, NJ, Prentice Hall.

WILLIAMSON, O. E. (1971). Managerial discretion, organization form and the multidivision hypothesis, in *The Corporate Economy*, edited by R. MARRIS and A. WOOD, London, Macmillan, 343–86.

WILLIAMSON, O. E. (1975). *Markets and Hierarchies*, New York, NY, Free Press.

WILLIAMSON, O. E. (1981). The modern corporation: origins, evolution, attributes, *Journal of Economic Literature*, Vol. 19, No. 4, 1537–68.

WILLIAMSON, O. E. (1996). Efficiency, power, authority and economic organization, in *Transaction Costs Economics and Beyond*, edited by J. GROENEWEGEN, Dordrecht, Kluwer.

3

MRS PENROSE AND NEOCLASSICAL THEORY

G. B. RICHARDSON
St John's College, Oxford

Mrs Penrose's theory of the growth of the firm is widely known and accepted. It has not, however, been fully integrated into standard microeconomics. This was no simple oversight on Mrs Penrose's part; in the foreword to the third edition of her book, she states quite explicitly, in relation to neoclassical theory, that it is not, in her opinion 'useful to attempt to "integrate" the two approaches'. My own view is that some integration is not only useful, but necessary, for I believe that only by incorporating Penrosean ideas can some 'traditional theory' be put right. I shall consider, in this paper, two parts of our subject of which this seems to me true: first, that which deals with the adjustment of supply to demand, and, second, that which seeks to account for the necessity of intra-firm planning within market economies.

I should like to begin, as old men too often do, autobiographically, with an account of how Mrs Penrose entered into my own intellectual life. I met her only twice. She dined with me in Oxford some 40 years ago, about the time that *The Theory of the Growth of the Firm* was first published. When the book was republished, in 1995, she kindly sent a copy to me and subsequently invited me to stay in her house in Waterbeach. Although she could not recall our earlier meeting, rapport was rapidly established between us and, after whisky, a good dinner and a bottle of wine, we talked at length—talked of life in general, but of economics not at all. Nevertheless, despite this absence of direct professional exchange between us, I was influenced by Mrs Penrose's work. *The Theory of the Growth of the Firm* was published in 1959 and my own book, *Information and Investment*, was published in 1960. At that time, her work did not influence mine, nor mine hers. Had I first read her book, mine would have been different. In considering the matters to which an entrepreneur has to attend, I distinguished between *market conditions* and *technical conditions*. The former related to the projected activities of others in the system—such as customers, competitors and suppliers—while the latter had to do with the production possibilities created by the existing state of technology. A principal argument of the book was that the availability of necessary information about *market conditions* depended on the particular market organisation in operation, and therefore on the number of competitors, differentiation of the product, ease of entry, the existence of inter-firm agreements and so on. To put the matter very briefly, I maintained that firms were able to obtain needed market information either by relying, in varying degrees, on agreements or understandings with customers or suppliers or whatever degree of stability in the business environment was afforded by the natural 'imperfections' of the system, including the fact that firms, having different perceptions and capabilities, could not all respond to market opportunities equally well and equally quickly. The availability of *technical information*, I maintained, did not depend on the organisation, it being 'perfectly legitimate, quite independently of whatever economic system or market structure we presume is in force, to postulate that entrepreneurs are acquainted with a certain "state of the arts", or given "production functions" '.

When writing *Information and Investment*, therefore, I was eager to show how the prevailing economic organisation would affect the producer's ability to make informed estimates about the demand for his output, and about the supply of needed inputs, but was content to set aside the relationship between organisation and the process of production itself. By talking about 'the entrepreneur' and 'production functions', I effectively reduced this process to the selection, by one all-seeing mind, of a least-cost mixture of ingredients to be miraculously 'combined' according to the best recipes currently available. I must not, of course, be too hard on myself; one cannot discuss everything at the same time and, being anxious to concentrate on one topic, I bundled the other away by adopting the prevailing theoretical treatment of it. By implication, moreover, I admitted the role of the organisation in production, by stressing the differential capability of firms to respond to opportunities. I later sought to redeem myself by publishing, in 1972, an article, *The Organisation of Industry*, in which the relationship of the process of production to both intra-firm and inter-firm organisation is explicitly considered.

I observed, in this article that 'production has to be undertaken (as Mrs Penrose has so very well explained) by human organisations embodying specifically appropriate experience and skill'. In doing so I wished to distance ourselves from the assumption underlying much economic theory to the effect that everyone could know, or do, everything. I accepted, as Hayek had maintained, that the working of the economic system could be explained only on the basis that knowledge, far from being 'perfect', was uncertain and distributed. The ability of firms to perceive and respond to a profit opportunity depended, as Penrose was abundantly to make clear, on the specific capabilities they embodied. These capabilities determined the areas of activity in which they could expect to enjoy a comparative advantage and thus limited both the direction and the magnitude of their response.

The Theory of the Growth of the Firm appealed to me, no doubt, because the ideas within it were already implicit in, and essential to, my own thinking, although they did not receive from me the full development and support that Mrs Penrose provided. More interesting is the way in which Mrs Penrose's concerns differed from my own. In the foreword to the third edition of *The Theory of the Growth of the Firm*, Mrs Penrose refers to the 'traditional theory of the firm' as used in 'predicting the response of an economy to exogenous disturbances' and in proving 'the superior welfare efficiency of competitive markets'. 'Few economists' she observed, 'thought it necessary to enquire what happened inside the firm—and indeed their "firm" had no "insides" so to speak.' Theoretical economists, she went on to say, 'saw reality differently from other people and asked different questions about it'. This is perhaps why she then discouragingly concluded, as I mentioned earlier, that it was not in her opinion 'useful to attempt to "integrate" the different approaches'.

I. CONDITIONS FOR INFORMED INVESTMENT DECISIONS

This is where our paths diverged. It seemed to me that the accepted theory of the working of markets was deficient and could only be put right through the incorporation of a more realistic representation of firms. In *Information and Investment*, I had argued that the accepted theory of market adjustments was erroneous. I denied that there could ever be a tendency to equilibrium under the market conditions which were prescribed as those most favourable to it, i.e. perfect competition. I maintained that the required adjustments would be possible only if there existed, neither 'perfect knowledge' nor 'perfect mobility' in the sense that everyone knew and could do everything, but precisely those differences in perception and capability that characterised Penrosean firms. A profit opportunity that was open to all was in effect open to no one; only if firms differed in their perceptions and capabilities, so that some were able to respond better and faster than others, could a general profit opportunity create particular profit opportunities for particular firms and thus permit the orderly adjustment of supply to demand. I had in effect convinced myself (but few others) that we could provide a satisfactory account of the process of adjustment, of the path to equilibrium, only by incorporating Mrs Penrose's theory of the firm. Her aim was to provide a realistic account of the internal working of that centrally important institution, in which she examined the factors which determined the direction and magnitude of its response to perceived market opportunities. My concern, in *Information and Investment*, was to identify the conditions which would, by permitting informed investment decisions, promote the process of resource allocation in a market economy. And I was persuaded that unless we were to replace assumptions about 'the entrepreneur', 'given production functions', and 'perfect knowledge' by assumptions about the need for organisation, differential capabilities, and limitations to growth, then this could not be done. Perhaps I should, after all, have discussed economics when I stayed at Waterbeach, and have taken the opportunity to tell Mrs Penrose about the argument contained within *Information and Investment* and the underpinning which her work (unknown to me when I was writing) had unintentionally provided for it. I hope that she would have agreed with me that markets would not work best (or indeed at all) given perfect competition and perfect knowledge, but depended for their proper functioning precisely on the presence of those properties of real world enterprises, i.e. their specific capabilities and limited potential response, and from these 'imperfections', standard theory had chosen to abstract.[1]

II. PRICES, PLANNING AND THE RATIONALE OF THE FIRM

I wish now to turn to another theory, about which Mrs Penrose had little to say directly, but for which she provided explanatory tools. A large literature exists about the different roles played, in the process of resource allocation, by (as Ronald Coase puts it) 'administration' within firms, on the one hand, and 'the price mechanism' on the other. Why, the question was put, cannot the allocation of resources not be left entirely to the price mechanism? Why is there a need for administration, and why, therefore, firms?

The question, put this way, seems to me confused. It postulates, one may first observe, a simple dichotomy—between 'firm' and 'market'—that does not exist. As I pointed out in the 1972 article mentioned above, firms are related in the real world, by a 'dense network of cooperation and affiliation', which, until recently, economists tended to ignore. Some tasks, which require the undertaking of complementary activities, can be accomplished, it is obvious enough, only by cooperation, in the sense that one person will do one thing only if assured that some other person will do another. Cooperation is as necessary as competition for the working of market economies, a fact the obviousness of which may explain why it is often ignored. The forms which cooperation can take are various; it takes place both between firms and within them. A firm, indeed, is a system in which cooperation is organised in a particular way.

There is another false dichotomy implicit in the question. Prices influence the planned allocation of resources, whether through 'administration' within firms or through cooperation between them. 'The price mechanism' does not work without human agency. It works because prices influence plans, and in such a way as that these plans result in a general economic order unintended by those who make the plans. Most obviously prices influence decisions to produce goods 'for the market', but they similarly influence, in a regime of economic freedom, the decisions of those who enter

---

[1] By temperament and by training, I should never have found it easy to forswear the 'attempt to integrate the different approaches'. In writing this contribution, I had occasion to revisit an article entitled *The Limits to a Firm's Rate of Growth* which I wrote in 1964 and had since forgotten. It illustrates the difference in Mrs Penrose's approach and my own. Although it refers to what 'Mrs Penrose has effectively brought to our attention', its starting point was the results of an enquiry by the Oxford Economists' Research Group, of which I was a member, into the circumstances that controlled a firm's rate of growth. Most of the business men we interviewed, but not all, said that lack of managerial capacity was the crucial limitation. This finding clearly accorded with, and can be best explained by, Mrs Penrose's analysis. My aim, in the article, apart from that of presenting our findings, was to provide a 'theoretical formulation' in terms of a traditional graphical representation of profitability, and of the cost of finance, as both functions of planned expansion, my wish being to take account of the notion of a managerial constraint in our consideration of general economic issues then under discussion, such as the sensitivity of investment to interest rate changes. I do not know if Mrs Penrose ever read this article, which she might have thought as introducing a degree of formalisation greater than the subject matter could easily bear. Whether successful or not, however, it further exemplifies my desire to promote the 'integration' which she did not recommend.

into cooperative arrangements, whatever their form, and sell their services for limited periods and on terms which are competitively determined. 'The price mechanism' is not superseded, as Coase would seem to suggest, by the processes of resource allocation that take place within firms, or through cooperation between them, but works through mediating the decisions associated with it.

It seems to me, therefore, that Coase's famous question 'why do we need firms?', needs re-statement. We may ask 'why do we need cooperation?', or 'why does cooperation sometimes take the organised form we associate with the firm?'. We need cooperation when *closely complementary* activities have to be matched, when each of them, specified both in terms of character and timing, will be effective only when the others in the set are undertaken also. Those working together in factory production, for example, must continuously match their activities in order that they are jointly effective. A firm buying special purpose inputs, such as a software programme for a particular reference work on CD ROM, must cooperate with those who can produce them—by entering into contracts—but need not do so for general purpose inputs, such as electric light bulbs, which it can rely on buying on the spot market.

These considerations account for the fact that cooperation is necessary, but do not help us to explain the particular forms that, in different circumstances, it will most conveniently take. In particular, they do not enable us to specify the circumstances in which cooperation is best organised within a firm. It is to help answer this question that the notions of 'transactions cost', 'moral hazard' and 'relational contracts' have been introduced. They are certainly relevant to the question of how cooperation, in different circumstances, should be organised, but I believe that Mrs Penrose identified the most fundamental reasons for the peculiar effectiveness of the form of organisation, the firm, which she chose to study.

At one level of abstraction, the process of production has been represented as the 'combination' by an 'entrepreneur' of available resources, or inputs, in proportions appropriate to known technical possibilities and the prices at which they can be acquired. But, as Mrs Penrose pointed out, 'it is never resources themselves that are the "inputs" in the production process, but only the services that resources can render. The services yielded by resources are a function of the way in which they are used . . .'. Specifically, the services provided by members of a cooperating group will depend on how it is organised. The capabilities of any particular cooperating group— the scope and effectiveness of the activities it can undertake—will depend both on the skills of its members and on their inter-relationship. Irrespective of the contractual arrangements associated with this inter-relationship, is the need for it to be stable enough for members of the group to learn to work with each other; this is a crucial factor in Mrs Penrose's theory of the growth of the firm, the rate at which new managers can be integrated into the team being the principal constraint. A firm is more than a group of persons linked contractually at a particular time for a particular purpose; indeed, it can have a continuing existence while those who work for it, and the activities it undertakes, both change. It embodies particular capabilities which it can sustain, and can be modified, over time. We need firms for the same reason that we need many other institutions which preserve, and enable to be adapted, the knowledge, attitudes and practices upon which society depends.

If one is to think of any goods on the market, a shirt for example, it is clear that an almost unlimited number of processes are directly or indirectly associated with its production, the undertaking of these being distributed among firms according to their special capabilities and the prevailing economies of scale. And if one is to explain the number of processes which a firm will itself choose to undertake, the so-called 'make or buy' decisions, these are the two considerations which must be given primacy. If one were to assume that all firms were equally able to undertake any productive process, because they have access to the same technology and are confronted by the same factor prices, and are equally able to expand sufficiently to exploit the available economies of scale, then transaction costs might decide the issue. But, as Mrs Penrose showed, firms at any one time have different capabilities and only a limited capacity to expand efficiently. Inevitably, therefore, there will be an advantage to a firm in buying in whatever, given its peculiar capabilities and capacities, it is not suited itself to produce. If what it seeks to buy is commonly used and reliably available 'on the market', then transaction costs are unlikely to be significant. If it has to be made specially, then the firm will have to enter a contract with its supplier, a contract which—for reasons that are well known—may be difficult to devise to the satisfaction of both parties. It is in these cases that the costs and risks of transacting with an independent enterprise may oblige a firm to make, rather than buy, even although it lacks the fully appropriate capability.

III. ORIGINALITY AND SIGNIFICANCE

All of us, when writing about our subject, ask ourselves (or certainly should ask ourselves) whether what we have to say is in any degree new or important. How then would we judge, from this point of view, both Mrs Penrose's theory of the firm and the work I have done which is associated with it? It was at one time commonplace in Cambridge, I am told, to say that 'it is all in Marshall' and indeed, in respect of what I have been saying in this paper, a good case can be made for this assertion. Marshall would not have needed Mrs Penrose to tell him that—and now I quote him— 'Knowledge is our most powerful engine of production' and 'Organisation aids knowledge'. Indeed, he goes on to say that 'it seems best sometimes to reckon Organisation apart as a distinct agent of production' (Marshall, 1920, pp. 138–9). What Mrs Penrose so successfully did was to remind a different generation of economists of these facts, of which she gave such a rich and convincing analysis, and bring back into recognition an important feature of reality which our conceptual apparatus had put out of focus. Perhaps Dr Johnson was right when he said that 'men more frequently need to be reminded than informed'.

I doubt also if Marshall would have dissented from my own argument that there could be no tendency towards equilibrium under perfect competition. He did not himself use the term but chose to refer to 'much free competition' or terms equivalent to this. His instinct was to distrust Walrasian theory, which his pupil Keynes thought nonsense.[1] Nevertheless, he did not analyse, or even indicate, the part played by 'imperfections', of knowledge or of competition, in providing firms with sufficiently sure prospects for investment to make possible the adjustment of supply to demand. A modern neoclassical theorist, but not Marshall, might defend the perfect competition model as merely a parable that serves to illustrate how, under free enterprise, prices would tend to the level of production costs as determined by current technology. The danger here is that the conditions of perfect competition—many producers, homogeneous products, perhaps also perfect foresight (whatever that might mean)— then come to be represented as optimal for the successful process of adjustment, which they are very clearly not. And this error, as it seems to me, has led in turn to treating many interventions—such as the attempt to regulate commodity markets—as evidence of wickedness or confusion, the implication being that the processes of adjustment would work perfectly well if we just left them alone.

For these reasons, I should like to think that what I had to say in my book *Information and Investment*, associated as it is with what Mrs Penrose said in her *The Theory of the Growth of the Firm*, was in some degree, both new and significant in terms of identifying conditions needed for the effective working of a market economy. I should also credit Mrs Penrose specifically with leading us to understand firms, more fully than before, as embodying and transmitting through time attitudes, habits, experience, knowledge and skills, both general and specific, without which no economy can work successfully. It might be said that she helped to establish the 'knowledge based' theory of the firm. This now widely used term, however, seems to me too narrowly epistemological. The capability of a firm depends not only on knowledge, whether 'knowledge how' or 'knowledge that', but as much on qualities attributable to its members and its organisation that are of a different kind. Doing requires more than knowing; more conventional virtues, such as courage and patience, as required. And an organisation cannot be accounted for in terms of embodying knowledge; to be effective, for example, it has, while constraining its members from actions which do not further its purposes, give them at the same time the freedom and encouragement needed to respond creatively to opportunities and to change. These considerations lead us very far afield, but are of some practical concern; perhaps the difficulties experienced by Russia in moving from communism to capitalism have reminded us that institutional inheritance, as well as technology and resources, affect the process.

REFERENCES

COASE, R. H. (1937). *The Nature of the Firm*, Economica 1937.
HAYEK, F. A. (1937). *Economics and Knowledge*, Economica 1937.
MARSHALL, A. (1920). *Principles of Economics*, London, Macmillan.

---

[1] In a letter to (of all people!) John Hicks, in 1934. 'Walras's theory and all others along those lines are little better than nonsense.' Quoted in Skidelsky, 1992.

PENROSE, E. (1959). *The Theory of the Growth of the Firm*, Oxford, Oxford University Press.
RICHARDSON, G. B. (1960). *Information and Investment*, Oxford, Oxford University Press.
RICHARDSON, G. B. (1964). The limits to a firm's rate of growth, *Oxford Economics Papers*.
RICHARDSON, G. B. (1972). The organisation of industry, *Economic Journal*, Vol. 82, 883–96.
SKIDELSKY, R. (1992). *John Maynard Keynes*, London, Macmillan.

4

THE SIGNIFICANCE OF PENROSE'S THEORY FOR THE DEVELOPMENT OF ECONOMICS

BRIAN J. LOASBY

Department of Economics, University of Stirling

To appreciate the significance of Edith Penrose's (1959) *Theory of the Growth of the Firm*, it is best (as it often is) to begin with Adam Smith. In this instance what matters is Smith's agenda for economics and its treatment by his successors. When Smith's name is invoked by mainstream economists, it is typically as the formulator of the proposition that an economy of self-interested individuals can generate a coherent, and even, in a carefully specified sense, an optimal pattern of activity. This attribution is not entirely wrong, but it is dangerously misleading as a guide to Smith's concerns. It is misleading even when considered in isolation because it suggests that modern conceptions of rational choice equilibrium adequately represent Smith's proposition; but it is fundamentally misleading because that proposition is removed from its context, which is described with exemplary clarity at the outset of Smith's (1976B) *Inquiry into the Nature and Causes of the Wealth of Nations*.

I. THE DIVISION OF LABOUR AND THE PROBLEM OF COORDINATION

Smith begins by identifying the principal source of growth in the wealth of any nation as an increase in the productivity of labour which, though made possible by the accumulation of capital, is a consequence of the increasing division of labour. Among the effects of specialisation are increases in dexterity and the saving of time in moving between different tasks, which may be represented by comparisons of sequential equilibria; but what distinguishes Smith's treatment from preceding accounts is his assertion that 'the very different genius which appears to distinguish men of different professions, when grown up to maturity, is not upon many occasions so much the cause, as the effects of the division of labour' (Smith, 1976B, p. 28). The division of labour is thereby transformed from a principle for the efficient allocation of a given collection of skills, as in a production set, into a principle for organising the development of skills, and also of different kinds of knowledge, thereby generating the possibility of novel production sets. As the attention of individuals becomes more closely focused, some of them become inspired to devise new machinery; and as increasing specialisation leads to an ever-increasing variety of focus, so it tends to encourage the development of an ever-increasing variety of machinery.

Smith's presentation, which leads from the familiar account of the pin factory, chosen, as he explains, because the whole process of the division of labour in pin-making can be 'placed at once under the view of the spectator' (p. 14), to the complex patterns of international specialisation that provide the necessary support to the standard of living in a civilised country (pp. 22–4), is intended to demonstrate the range and power of this single explanatory principle. Smith is here following his own advice on the most effective method of giving 'an account of some system' (Smith, 1983, pp. 145–6). His emphasis on a single causal explanation to encompass specialisation within a single business, the separation of activities between businesses, the emergence of distinguishable industries and the growth of international trade may now be judged to have diverted attention from the differences between these categories of organisation; but it serves a valuable purpose if it warns us not to forget what is common to them all—in particular that they are all systems for the generation of new knowledge, new skills and new technology. This is a theory of 'development from within', in Schumpeter's (1934, p. 60) sense of a process arising from the character of the economic system.

It is this process of developing knowledge that creates the need for some means of coordinating these increasingly diverse activities, while still encouraging the generation of novelty which inevitably is in some degree disruptive of existing patterns of coordination. We might indeed reflect, with Schumpeter (1934, p. 80), that in the absence of novelty activities would eventually settle into a stable pattern of routines, and thus that coordination would not be a problem worth analysing. But in the presence of multifarious novelty it is not helpful to pose the coordination problem as one of efficient allocation, to be viewed from the perspective of central planners—a practice that has outlasted the fashion for central planning. It may be worth reminding ourselves that, in his famous observation of the relationship between each individual and the butcher and baker, Smith (1976B, pp. 26–7) posed the problem as one for the participants to solve. The division of labour generates such a differentiation of knowledge that no person or group can gain access to all the knowledge that would be required to construct an efficient comprehensive plan for society. The 'man of system' (Smith, 1976A, pp. 233–4) fails to comprehend the complexities that he presumes to resolve into order.

In Smith's setting, coordination is a problem of differentiated, dispersed and localised knowledge, as Hayek (1937, 1945) later insisted; and the division of labour tends to increase the differentiation and dispersion. There is a counteracting force, which Smith draws to our attention: this is the particular form of specialisation practised by those 'whose trade it is, not to do anything, but to observe everything; and who upon that account are often capable of combining together the powers of the most distant and dissimilar objects' (Smith, 1976B, p. 21), thus creating locally coherent structures. The importance of those who make such novel connections is presented in language that recalls Smith's theory of scientific progress expounded much earlier (though not published in his lifetime) in *The History of Astronomy* (Smith, 1980); and this is appropriate, because that theory is a psychological explanation of the creation of new structures of knowledge by the invention of new 'connecting principles' with which to classify and link phenomena.

Smith accepted Hume's (1978) conclusion that there was no way of proving the truth of any claim to empirical knowledge, and followed Hume in changing the question to that of the process by which human beings come to accept the truth of such claims. The result is a theory which is proto-Popperian in relying on a series of conjectures that are imposed on events, and which are subject to falsification—sometimes after a lengthy interval—by new observations, though it does not rest on a belief, such as Popper held, that conjectures might, though it cannot be proved, correspond to the truth. The absence of final truth implied the absence of any final destination for the process of generating new knowledge through the division of labour; and that in turn, had clear implications, not only for the insufficiency of any concept of co-ordination as an equilibrium allocation within a closed system, but also, more fundamentally, for the kind of theorising that was appropriate. Smith's economics is not an axiomatic but a causal system.

Smith had a vestigial theory of the firm, in which anyone who 'employs a great number of labourers, necessarily endeavours, for his own advantage, to make such a proper division and distribution of employment, that they may be enabled to produce the greatest quantity of work possible' (Smith, 1976B, p. 104). In addition, 'the plans and projects of the employers of stock' (and therefore of labour) lead to 'acuteness of understanding' (Smith, 1976B, p. 266), which tends to increase further the productivity of labour. Say (1964) was responsible for developing the role of employers as both co-ordinators of work and originators of improvement, and Babbage (1963) drew attention to the reductions in cost that could be achieved by a careful balancing of processes that had different rates of output.

Though Menger (1976 [1871]) disputed the priority that was given to the division of labour in Smith's theory of economic development, he shared both Smith's belief in the central importance of the growth of knowledge (which in *The History of Astronomy* had given rise to the division of scientific labour) and his preference for causal explanation, linking the two in his definition of a good as something which men had learnt to place in a causal relation with one or more of their needs. The origins of these needs are not analysed; the emphasis is on the discovery of ways to satisfy them. Goods are thus a product of human knowledge; and Roscher, to whom Menger dedicated his *Principles*, had identified the creation of new goods as 'a characteristic task of the modern manufacturer' (Marshall, 1920, p. 280). Human action is driven by human purpose, and human purpose results in new goods and new skills.

It is not therefore surprising that Menger also shared Smith's view that co-ordination was a problem to be addressed by individual human action. He clearly recognised the prevalence of what are now called transaction costs, and the consequent general divergence—not yet absorbed into price theory—between buying price and selling price; and he explained how differences between products in the costs of arranging transactions, or, in Menger's more evocative language, differences in marketability, influence their suitability for commodity production. That goods which were selected for commodity production, and thus for efforts to expand their markets, would therefore be subject to knowledge-enhancing division of labour and the innovative ideas of Roscher's manufacturers is an inference that he failed to draw. Nor did he develop his analysis of marketability into an exposition of the process of market-making, focusing solely on the use of the most marketable commodities as intermediaries in exchange; thus the supersession of barter by monetary exchange becomes both a consequence of the division of labour and, by increasing the extent of the market through its reduction of transaction costs, a cause of further division of labour.

Because Menger regarded the growth of knowledge as the fundamental cause of human progress, it was impossible for him to treat knowledge as anything but highly imperfect. Attempts to meet human needs must therefore include provision for future needs which cannot be precisely identified; and Menger accordingly gave attention to the provision of various kinds of reserves, among which money was particularly important. Unfortunately, he did not recognise the importance of market-making as a means of creating reserves in the form of possibilities for future exchange.

II. THE DIVORCE BETWEEN GROWTH AND COORDINATION

The sequence thus sketched is a very incomplete account of the development of economics during the century following the publication of the *Wealth of Nations*. Over this period there had also emerged an increasingly distinctive treatment of coordination, which led to Jevons' (1871, p. 255) clear statement of efficient allocation, on the assumption of complete relevant knowledge, as 'the great problem of economy', and then to Walras's (1954) attempt to produce an ultimate solution to the issue of economy-wide coordination. We should remember that Walras intended his proof of the existence of a competitive general equilibrium to be the first major objective in a three-stage research strategy. He then hoped to demonstrate how that equilibrium configuration could in practice be attained, and finally to develop a theory of economic growth, in the form of continuous adjustment to changing data. (He did not, however, apparently aspire to explain the causes of growth.) Both the second and the third stages relied on tatonnement, which Walras conceived as a real-time and real-world process of trial and error—indeed the only available real-world process.

He began the second stage by modelling the simplest kind of equilibration, represented by the exchange of well-defined commodities, the total stock of which is fixed, in highly organised markets, as exemplified by the Paris Bourse. However, since he failed to see the relevance of Menger's concern for what we now call transaction costs, he did not ask how and why highly organised markets come into existence. It is therefore not surprising that he gave no thought to unorganised markets, where there is no recognised location in which all potential trading partners can be found and no agreed product specification—in which, indeed, there is no well-defined search space. That was left for Chamberlin (1933), who was unfortunately thought to be doing something else, and through the continued rejection of his repeated protests became one of the tragic figures in the history of economics.

Instead, Walras turned directly to the equilibration process in production, and he began auspiciously. He recognised that most goods are produced in advance of any specific order, and offered an explicit theory of entrepreneurship, which in one respect is better than Kirzner's (1973); for his entrepreneurs did not perceive the truth, but made conjectures about the prospective demand for their products. Production decisions were based on fallible expectations, and producers quoted a price at which they hoped to dispose of their output. If they could not sell it all, then they would reduce their price; if, on the other hand, they could not supply all their customers, the price would rise. Any divergence between the price realised and that which producers had expected when making their plans then influenced the next set of production decisions, and if demand and supply curves were both stable and well behaved, the economy, Walras claimed, moved towards an equilibrium.

However, because this process of equilibration depended on trading at disequilibrium prices, which necessarily redistributed income, there was no general reason—even ignoring the absorption of resources by the costs of repeated transactions—why the equilibrium to which the economy moved should be the equilibrium that could be calculated from the original data. But if this was so, then what precisely was the point of starting with a calculation, or even a proof of existence, of a state that was rather unlikely to come about? Walras avoided this problem until 1899; then he resolved it by removing the threat of path dependency, which he knew he could not handle, from his model, at the cost of removing his model from the world. In place of a real-world process of trial and error, in which producers' incomes depended on the success of their conjectures, Walras introduced an exchange of pledges between producers and prospective purchasers, which could be renegotiated without cost, and production would not begin until the quantities that the producers promised to make was precisely matched to the demand at the prices they promised to charge (Walker, 1997). Time-dependent production was thus assimilated to highly organised anticipatory exchange. The expedient of settling everything by costless contracting before the world is allowed to start turning has become a core convention of general equilibrium theory, and it has served for many years to preserve the apparent internal coherence of the theoretical structure, by incorporating dates, locations and states of the world into the specification of each commodity; but Walras made no further attempt to explain change (since that required a response to real-time disequilibrium), thereby effectively abandoning his ambition of creating a real-world economics.

This theoretical choice by Walras might be seen as the single defining act of twentieth century economics. That is not to claim that, had Walras made the opposite choice, economics would have developed along very different lines; on the contrary, I believe that the reasons that led to his decision would in any event have caused most theorists to reveal their preference for formal consistency over more representational theory, and for axiomatic reasoning over causal sequences. But the opportunity cost of this choice was the suppression of endogenous growth theory in the tradition of Adam

Smith. Henceforth, allocative efficiency was divorced from the generation of knowledge and wealth.

Marshall refused to follow Walras, for that would have entailed the rejection of his hopes, not just for economics, but for the improvement of human conditions, both through the development of improved preferences (for wants were endogenous too) and through improvements in productivity. Though he produced a simple model of general equilibrium, he recognised that a fully specified equilibrium required data about future possibilities that were not yet imagined (1920, p. 379), and preferred a partial equilibrium apparatus that would allow him to investigate the consequences of particular kinds of change.

Moreover, Marshall, like Smith, was sceptical of the Cartesian method that Walras used. He had been thoroughly trained in Euclidean geometry at a time when this was widely believed to exemplify the possibility of demonstrating empirical truth by axiomatic reasoning, only to be faced with the equally coherent axiomatic system of non-Euclidean geometry; and when he turned to economics he discovered that Cournot's analytical system 'led inevitably to things which do not exist and have no near relation to reality' (Marshall, 1961, p. 521). His response to the impossibility of justified knowledge was to follow Hume and Smith (though apparently unaware of these precedents) in turning to the processes by which people acquire knowledge.

In doing so he took imaginative advantage of Darwin's theory of evolution to complement Smith's emphasis on the power of the division of labour to generate knowledge, and his two distinctive contributions to growth theory were based on combinations of ideas from Smith and Darwin. First, he added to Smith's account of the development of capabilities through specialisation the importance of variety within each specialism—the opposite, it will be noted, of the homogeneity which is required to deliver the allocative efficiency that is associated with perfect competition. Secondly, he explicitly linked coordination to specialisation, and biology to social organisation, by propounding the bold new connecting principle that 'the development of the organism, whether social or physical, involves an increasing subdivision of functions between its separate parts on the one hand and on the other a more intimate connection between them' (Marshall, 1920, p. 241). Like Smith and Menger, Marshall looked for causal explanations that could be empirically corroborated.

As a consequence of this close association between the processes of differentiation and integration, the coordination problem first appears in Marshall's *Principles* in the guise of aids to knowledge through appropriate forms of organisation (Marshall, 1920, pp. 138–9). These include single firms, groups of firms within a trade (which provide variety) and firms in complementary trades (which provide information and ideas). In Smith's tradition of coordination from within, each kind of organisation is developed, formally and informally, by the participants, and each provides the knowledge base not only for the development of new products and processes, but also for a convergence towards equililibrium prices and quantitites. Changes in organisation are actually included in Marshall's (1920, p. 318) definition of increasing return, which for him is the key to economic development, as it was for Smith.

This conception of economic progress and its implications for equilibrium theory were summarised by Allyn Young (1928). 'New products are appearing, firms are assuming new tasks, and new industries are coming into being. In short, change . . . is qualitative as well as quantitative. No analysis of the forces making for equilibrium . . . will serve to illumine this field, for movements away from equilibrium, departures from previous trends, are characteristic of it. . . . Every important advance in the organisation of production . . . alters the conditions of industrial activity and initiates responses elsewhere in the industrial structure which in turn have a further unsettling effect' (Young, 1928, pp. 528, 533). Sraffa had already identified the conflict between increasing returns and equilibrium analysis, and had opted (without explanation) for equilibrium, first in the form of perfectly competitive equilibrium, which entailed the denial of increasing returns (Sraffa, 1925) and then in the form of monopolistic equilibrium, which admitted the notion of cost as a decreasing function of output (Sraffa, 1926), but still excluded the continual reorganisation and the generation of novelty that was inherent in the theory which had been developed from Smith's principles.

Sraffa's advice was followed, and led to the exploration of the relationships between various forms of market structure and allocative efficiency, on the assumption that resources, preferences and technology were known (or, at best, changed only in ways that could be anticipated). Growth of output was then merely a matter of comparing equilibrium positions, and was not—indeed, could not be—analysed as a process. The only significant exception to this theoretical emphasis was provided by Schumpeter's (1934) theory of economic development through the creation of new combinations; but it should be remembered that this was a second edition, translated into English, of a book published in German in 1912, and that its influence on the practice of Anglo-Saxon economists was scarcely perceptible.

III. PENROSE'S ACHIEVEMENT

Edith Penrose received an orthodox training in economics during this period of eclipse for Smith's conception of the subject—though what contemporaries noticed was not this eclipse but the flow of new theories in microeconomics, macroeconomics and the economics of welfare which appeared to have direct and substantial implications for policy (Shackle, 1967). Like many of her contemporaries, and many present-day economists, she was not aware of what had been achieved by her predecessors. It is consequently a remarkable tribute to the quality of her thinking that almost all the ideas within Smith's tradition that have been noted in this chapter are incorporated into her *The Theory of the Growth of the Firm* (1959). The central notion is a combination of Smith and Marshall: the division of labour, both within and between firms, leads to the development of skills and the perception of possibilities, while firms within a similar line of business will develop somewhat different skills and perceptions. Enterprise grows out of management, as Say and Marshall had argued; and it is driven by human purpose, seeking to discover and exploit causal relationships by producing new goods for new markets, in accordance with Menger's theory. Preferences are at least partly endogenous, as Marshall believed. Firms are continually looking for ways to achieve a more effective balance of processes, as Babbage recommended, and changing their organisation, in conformity with Marshall's law of increasing return, to align their increasing knowledge with their productive opportunity. The firm is a form of organisation which aids knowledge, as Marshall had perceived, and its structure and resources provide an ordered cluster of reserves against future threats and opportunities, such as Menger thought a natural consequence of purposeful human action. The combination of all these elements into a single coherent and comprehensive account of the firm as a major institution for the generation and coordination of knowledge makes her book, even now, the obvious and unchallenged foundation for any comprehensive treatment of the generation of variety which is essential for evolutionary economics. (Foss's chapter provides a more detailed exposition.)

All that seems to be missing from what had previously been written is a significant treatment of external organisation and of the continuous restructuring of industry which was emphasised by Allyn Young, and these omissions may be attributed to a deliberate focus on the individual firm, and in particular on the relationship between its internal organisation and the process of internal learning, which may be regarded as one of the two most significant novel elements in her work. The evidence for the significance of this focus is the increasing interest in studying the influence of organisation on learning. As guidance to such studies Penrose (1959, pp. 53–4) reminds us that 'learning' is not simply (as in the Bayesian systems favoured by orthodox economists) the acquisition of 'information' but a process of interpretation, which is reminiscent of Smith's (1980) account of the growth of knowledge through the invention and application of fallible connecting principles. Thus learning may lead to foresight or oversight, as abundantly illustrated in the case studies reported by Garud et al. (1997). A theory of economic development that respects both human abilities and the historical record must rest on conjecture and exposure to refutation rather than rational expectations.

The second of the most significant novel elements in Penrose's theory is her distinction between productive resources and productive services. This entails a rejection of the standard concept of a production function in which inputs are assimilated to factors of production, in favour of an analytical scheme in which resources become a distinct subject of analysis and their application is problematic, not only because the opportunities for their use have to be perceived (as Kirzner would say) or imagined (as Shackle would prefer) but because the suitability of a resource to a particular application can never be guaranteed in advance. Here too the principle of purposeful action is fallible conjecture.

Penrose's terminology of resources and services seems to have been generally superseded by that of 'capabilities' and 'activities', which was introduced by Richardson (1972) in an article that was directly inspired by her work and which must be a close contender with Young's (1928) exposition of increasing returns for the title of the best article ever written on industrial organisation. By classifying capabilities along the dimensions of similarity and complementarity Richardson produced an analytical framework by which to explain, not only which activities are likely to be collected within a single organisation and which coordinated by market transactions (though rarely of the anonymity required for perfect competition), but also which activities are likely to be coordinated by collaborative arrangements, of varying degrees of intensity, between businesses that are formally independent.

In the course of that article, Richardson refers to Ryle's (1949) distinction between 'knowing that' and 'knowing how', between knowledge of facts and propositions and the possession of performance skills. In standard economics, knowledge is always of the first kind, even when the knowledge is of production sets. Ryle's distinction not only provides a link, as yet little developed, with Hayek's (1952) theory of evolutionary psychology (in which it is cited) but also offers a means of expanding on Penrose's own treatment of the knowledge that is embedded in the firm (see Loasby, 1998). It is clear that, in Winter's (1991, p. 185) phrase, firms are organisations that know how to do things, and that this knowledge is not easy to extract from its organisational context. 'Knowledge how' is often difficult to represent in language, symbols or diagrams, and therefore difficult to codify or transmit; and the coordination of various kinds of 'knowledge how' which have been developed in different contexts may be extremely difficult to manage by intellectual procedures. We frequently observe that the coordination of knowledge within two merged companies is achieved by parting with those whose knowledge has been acquired within the context of one of the partners.

IV. PENROSE'S RESEARCH STRATEGY

The quality of Penrose's achievement is not in dispute. How did it come into existence? We have shown how it might, in principle, have been assembled from ideas already in print, as she was later very ready to acknowledge; but we have also shown that these ideas had been substantially discarded and were no longer presented to students of economics. By what mental processes she came to reinvent them no one can now say, and it is doubtful whether she could herself have given a satisfactory account. But what we can do, with the aid of her own recollections (Penrose, 1985, 1995), is to outline the circumstances in which they came to be developed. In the 1950s, Fritz Machlup and G. H. Evans at Johns Hopkins University set up a research group to study the growth of firms. This was primarily an empirical enquiry, but it was thought appropriate for someone to consider the relevant theory, and Penrose undertook this task. She soon discovered that there was virtually no relevant theory—for reasons which have already been briefly explained. The question then was what to do about it.

It is instructive to compare Penrose's situation with that of Philip Andrews, who nearly 20 years earlier had also participated in empirical research into business behaviour which produced results that had no counterpart in contemporary theory. The crucial difference was that these enquiries, by the Oxford Economists' Research

Group, were motivated directly by that theory, and in particular by the predictions about investment, pricing and output decisions that were embodied in macro-economic theory and the theory of imperfect competition. One might see this as a revisitation of Marshall's encounter with the false consequences of Cournot's (1927) theory; indeed on pricing and output the parallel is almost exact.

When theory and evidence are in conflict, it is always possible to reject the evidence, or to supplement the theory in a way that preserves its essential features. (Neither is necessarily wrong.) Both options were tried: businessmen and economists were not talking the same language, and there were many ways of achieving and describing optimisation; or businessmen were incompetent, and should be replaced by those who could apply the prescriptions of sound theory. It is fair to say that these options were not favoured by the economists who had participated in the research; but it was Andrews who followed the lead of Hall and Hitch (1939) and persisted in seeking to develop a theory within which the observed behaviour could be explained as purpose-ful and intelligent (Andrews, 1949). He was fortunate in having as a fellow member of the group D. H. MacGregor, who understood Marshall's industrial theory and the cautious use of equilibrium logic in the service of causal explanation that it required. Andrews's work appeals specifically to this tradition, though it is much more con-cerned with the coordination of economic activity than with growth (which is perhaps not surprising in view of its origins in the late 1930s). He was, however, not fortunate in confronting a theoretical environment in which Marshallian imprecision and hesita-tion had been superseded by analytical clarity and confidence; his message was not wanted, and neither was his presence in Oxford (Lee, 1993).

Penrose makes sympathetic references to Andrews; but her situation was different. On the one hand, she had no one like MacGregor to direct her towards the relevant elements in Marshall's work, and though Machlup could supply Austrian insights, business organisation was an empty chapter in any guide to Austrian theory. However, since modern theory made no attempt to deal with the growth of firms it was at least conceivable that a theory that was designed to deal explicitly and exclusively with that growth could avoid any direct confrontation with standard doctrine. The division of labour between theories of coordination and theories of growth thus provided an intellectually productive opportunity. This was the rhetorical basis of Penrose's book. She begins by making clear the absence of any intention to criticise 'the theory of the firm' (not even observing, as Denis O'Brien (1984) did later, that in this theory there is no firm) and throughout the book avoids any formal analysis of decisions about investment, price and output, which Andrews could not escape. She argues that a theory that is consciously restricted to explaining the price and output of a single product at a given time (an interpretation, we may note, that is more congenial to Machlup than to general equilibrium theorists) requires a different kind of abstraction from that which is consciously restricted to explaining how a single firm, which is capable of producing a range of products, may grow over time (Penrose, 1959, pp. 14–15).

Penrose thus follows the policy previously adopted by Schumpeter of erecting an impenetrable barrier between the theory of coordination and the theory of development, which incidentally allowed Schumpeter to combine his acclaim for Walras as the greatest of analytical economists with his own hopes for high distinction. Williamson (1985) subsequently took similar precautions in claiming that his theory of transaction costs, which explained the principles on which the coordination of economic activities was divided between markets and hierarchies, left untouched the standard analysis of what activities will be undertaken. Was Marshall, then, wrong in trying to maintain contact between the theory of development and the theory of coordination?

V. THE DIFFERENTIATION AND INTEGRATION OF ECONOMIC THEORY

It is difficult to deny that the rapid development of formal theory to its present state would not have happened if economists had insisted on maintaining this link. Yet, as with any choice, the abandonment of the link had its costs, which were substantial. The freedom to define the problem of coordination in an analytically tractable way led to the apparent conclusion that central planning was potentially—and, within the context of the theory, practically—superior, with unfortunate consequences not only for policy but in encouraging a widespread sympathy for the supposed principles of economic organisation in Communist systems which almost everyone now admits was totally unjustified. Paradoxically, recent theories, which in effect confer on individuals degrees of knowledge that were for many years assigned to governments, have led to a faith in market systems that also goes well beyond justification, as well as employing a conception of 'market theory' in which there are no markets in the real-world sense of institutional arrangements and continuing relationships of very varying degrees of continuity and intensity.

Coase (1988, p. 3) complained that in economic theory we have 'exchange without markets', as well as 'consumers without humanity' and 'firms without organisation', and attributed these deficiencies, with their serious consequences for both understanding and policy, to the neglect of transaction costs. This attribution is not entirely wrong, but we should note, as Langlois (1992) has pointed out (and as Coase would doubtless agree), that in a predictable environment transaction costs fall to a very low level, since transactions are patterned into routines that require no thought and minimal exertion. It is change that gives rise to transaction costs—as to all economic problems (Knight, 1921, p. 313; Hayek, 1945, p. 523); it is the division of function, and the distinctive capabilities and ways of organising and presenting knowledge that it engenders, that both create the need for integration and makes it problematic. To understand the problem of coordination—which includes, let us not forget, the problem of avoiding means of coordination which suppress desirable variety—we need to approach it in the context of economic growth, which, as Smith, Menger and Marshall have told us, is the context of the growth of knowledge. We are very fortunate to have Penrose's work to guide our approach.

Though it is not put quite in this way, that is the message of Richardson's contribution to this collection. He took seriously Hayek's (1937) challenge to economists to explain how people within an economy could acquire the knowledge that was necessary to support the equilibria which the economists had deduced, and came to the conclusion that the market structures within which these equilibria were presented were quite incapable of permitting access to such knowledge. His (1960) analysis concentrated on market knowledge, explicitly assuming away the problem of production skills, which he later addressed from a complementary perspective (Richardson, 1972). He now demonstrates how Penrose's account of the development of each firm's capabilities may be used to expand his criticism of standard theories (including their recent extensions) and also to expand his earlier account of the ways in which co-ordination may be imperfectly achieved; in many situations, only a few businesses will have the skills to compete. There is, of course, always the possibility of competition from an unsuspected quarter, but coordination must always be fallible if novelty is to be permitted.

The division of labour between the analysis of coordination and the theory of growth has given economists the freedom to develop each within its own frame of reference. Growth-free microeconomics has provided continuing opportunities for expertise and ingenuity, which have encompassed transitions to imperfect competition, back to general equilibrium, to new classical macroeconomics, to game theory and even to what is called endogenous growth theory, finding new productive services for the key analytical resource of rational choice equilibrium; by ignoring this conception of coordination as closed-system logic Penrose was able to found (or refound) the knowledge-based theory of the firm, and, by simple but crucial extension, the knowledge-based theory of the economy. Is this, then, to use Richardson's (1972) analytical framework for explaining the organisation of industry, a situation in which dissimilar capabilities are best kept apart? The objection to an absolute separation is simply, but fundamentally, that, as Hayek insisted, the central difficulties of real-world coordination arise precisely from the knowledge basis of the economy. Scarcely less fundamental is the empirical observation that, as Hayek did not perceive, the means of coordination are derived from the same basis. Firms, and networks of firms, are essential, as Richardson reminds us in his chapter, for 'embodying and transmitting through time attitudes, habits, experience, knowledge, and skills, both general and specific, without which no economy can work successfully'. He is also right to remind us that much more than 'knowledge', narrowly conceived, is required—as Smith and Marshall knew very well but modern economists are professionally debarred from acknowledging.

The coordination problem in standard economics is 'a Gordian knot, carefully tied up in advance, however, by the very man who is going to cut it'—with the sword of a timeless system (Shackle, 1972, pp. 10, 127); its relevance to practical problems is supposedly demonstrated by the double proposition that agents are rational and markets always clear. The fact that this double proposition is not open to discussion cannot itself be discussed. Richardson's (1960) criticisms, reinforced by Penrose's emphasis on the distinctiveness and, often, the tacitness of capabilities, have not been met, and cannot be met, for two reasons. The first is that the relevant factors cannot be incorporated within the analytical system. Plans require more than prices, and the organisation of production requires capabilities which are not capable of specification in any standard model. The second is the analytical method, which is a grand system of tautologies. Working through tautologies is not a useless occupation; proofs of existence or of non-existence can guide search, and the consequences of premises are often not apparent, and sometimes surprising. But tautologies are not appropriate for explaining the transmission and modification through time of knowledge, skills and moral sentiments. They are useful resources, but among the productive services which they cannot perform is the provision of causal explanations.

The generation of knowledge is a non-logical (but not illogical) process: imagination and novelty entail the generation of new premises. If there is to be an adequate theory of this process it must be causal rather than strictly deductive. Elements of standard treatments of coordination may nevertheless be adapted for new purposes, which, as we have seen, are often old purposes. For example, I have argued previously, with particular reference to Penrose's work (Loasby, 1991) that a concept of equilibrium, similar to that suggested by Hahn (1973), in terms of theories and policies may be useful in analysing the imprecise and impermanent boundaries (which we may call institutions) that coordinate the creation and application of knowledge. Simon's (1957) 'decision premises' may be construed in a similar way, and applied to the decomposition of complex processes both within and between organisations; the particular configuration of decision premises within a Penrosian firm at a particular time coordinates the development of its capabilities and the (fallible) identification of productive opportunities. The prime significance of Penrose's work is that it both reminds us of the central importance of the growth of knowledge in economics and simultaneously provides the basis for an appropriate way of investigating and understanding this growth, the problems of coordination to which it gives rise and the processes of coordination which help to shape it.

REFERENCES

ANDREWS, P. W. S. (1949). *Manufacturing Business*, London, Macmillan.

BABBAGE, C. (1963). *On the Economy of Machinery and Manufactures*, 4th edn, London, Frank Cass.

CHAMBERLIN, E. H. (1933). *The Theory of Monopolistic Competition*, Cambridge, MA, Harvard University Press.

COASE, R. H. (1988). *The Firm, the Market and the Law*, Chicago, University of Chicago Press.

COURNOT, A. A. (1927). *Researches into the Mathematical Principles of the Theory of Wealth*, translated by N. T. BACON, New York, Macmillan. [First published in French 1838.]

GARUD, R., NAYYAR, P. R. and SHAPIRA, Z. B., eds (1997). *Technological Innovation: Oversights and Foresights*, Cambridge, Cambridge University Press.

HAHN, F. H. (1973). *On the Notion of Equilibrium in Economics*, Cambridge, Cambridge University Press. [Reprinted in HAHN, F. H. (1984). *Equilibrium and Macroeconomics*, Oxford, Basil Blackwell.]

58 B. J. LOASBY

HALL, R. and HITCH, C. J. (1939). Price theory and business behaviour, *Oxford Economic Papers*, 2, 12–45. [Reprinted in T. WILSON and P. W. S. ANDREWS, (eds) (1951). *Oxford Studies in the Price Mechanism*, Oxford, Clarendon Press.]

HAYEK, F. A. (1937). Economics and knowledge, *Economica*, New Series, Vol. 4, 33–54.

HAYEK, F. A. (1945). The use of knowledge in society, *American Economic Review*, Vol. XXXV, 519–30.

HAYEK, F. A. (1952). *The Sensory Order*, Chicago, University of Chicago Press.

HUME, D. (1978). *A Treatise of Human Nature*, edited by L. SELBY-BIGGE, 2nd edn, Oxford, Clarendon Press.

JEVONS, F. R. (1871). *The Theory of Political Economy*, London and New York, Macmillan.

KIRZNER, I. M. (1973). *Competition and Entrepreneurship*, Chicago, University of Chicago Press.

KNIGHT, F. H. (1921). *Risk, Uncertainty and Profit*, Boston, Houghton Mifflin.

LANGLOIS, R. N. (1992). Transaction-cost economics in real time, *Industrial and Corporate Change*, Vol. 1, 99–127.

LEE, F. S. (1993). Introduction: Philip Walter Sawford Andrews 1914–1971, in *The Economics of Competitive Enterprise: Selected Essays of P. W. S. Andrews*, edited by F. S. LEE and P. E. EARL, Aldershot, Edward Elgar, 1–34.

LOASBY, B. J. (1991). *Equilibrium and Evolution*, Manchester, Manchester University Press.

LOASBY, B. J. (1998). The organisation of capabilities, *Journal of Economic Behavior and Organization*, Vol. 35, 139–60.

MARSHALL, A. (1920). *Principles of Economics*, 8th edn, London, Macmillan.

MARSHALL, A. (1961). *Principles of Economics*, 9th (variorum) edn, London, Macmillan.

MENGER, C. (1976 [1871]). *Principles of Economics*, translated by J. DINGWALL and B. F. HOSELITZ, New York, New York University Press.

O'BRIEN, D. P. (1984). The evolution of the theory of the firm, in *Firms, Organization and Labour: Approaches to the Economics of Work Organization*, edited by F. H. STEPHEN, London, Macmillan.

PENROSE, E. T. (1959). *The Theory of the Growth of the Firm*, Oxford, Basil Blackwell.

PENROSE, E. T. (1985). The theory of the growth of the firm twenty-five years after, *Acta Universitatis Upsaliensis, Studia Oeconomiae Negotiorum* 20.

PENROSE, E. T. (1995). *The Theory of the Growth of the Firm*, 3rd edn, Oxford, Oxford University Press.

RICHARDSON, G. B. (1960). *Information and Investment*, Oxford, Oxford University Press [2nd edn, 1990].

RICHARDSON, G. B. (1972). The organisation of industry, *Economic Journal*, Vol. 82, 883–96. [Reprinted in RICHARDSON, G. B. (1990), *Information and Investment*, 2nd edn, Oxford, Oxford University Press.]

RYLE, G. (1949). *The Concept of Mind*, London, Hutchinson.

SAY, J. B. (1964). *A Treatise on Political Economy*, 4th edn, New York, Augustus Kelley [1st edn, 1803].

SCHUMPETER, J. A. (1934). *The Theory of Economic Development*, Cambridge, MA, Harvard University Press.

SHACKLE, G. L. S. (1967). *The Years of High Theory*, Cambridge, Cambridge University Press.

SHACKLE, G. L. S. (1972). *Epistemics and Economics*, Cambridge, Cambridge University Press.

SIMON, H. A. (1957). *Administrative Behavior*, New York, Free Press.

SMITH, A. (1976A). *The Theory of Moral Sentiments*, edited by D. D. RAPHAEL and A. L. MACFIE, Oxford, Oxford University Press.

SMITH, A. (1976B). *An Inquiry into the Nature and Causes of the Wealth of Nations*, edited by R. H. CAMPBELL, A. S. SKINNER and W. B. TODD, Oxford, Oxford University Press.

SMITH, A. (1980). The principles which lead and direct philosophical enquiries: illustrated by the history of astronomy, in *Essays on Philosophical Subjects*, edited by W. P. D. WIGHTMAN, Oxford, Oxford University Press.

SMITH, A. (1983). *Lectures on Rhetoric and Belles Lettres*, edited by J. C. BRYCE, Oxford, Oxford University Press.

SRAFFA, P. (1926). The laws of returns under competitive conditions, *Economic Journal*, Vol. 36, 535–50.

SRAFFA, P. (1998 [1925]). On the relations between cost and quantity produced, translated by J. EATWELL and A. RONCAGLIA, *Italian Economic Papers*, Vol. 3, 323–63. [First published in Italian in *Annali di Economia*, Vol. 2, 277–328.]

WALKER, D. A. (1997). *Walras's Market Models*, Cambridge, Cambridge University Press.

WALRAS, L. (1954). *Elements of Pure Economics*, translated by W. JAFFÉ, London, Allen and Unwin. [First edition published in French, 1874.]

WILLIAMSON, O. E. (1985). *The Economic Institutions of Capitalism: Firms, Markets, Relational Contracting*, New York, Free Press.

WINTER, S. G. (1991). On Coase, competence, and the corporation, in *The Nature of the Firm: Origins, Evolution and Development*, edited by O. E. WILLIAMSON and S. G. WINTER, Oxford, Oxford University Press.

YOUNG, A. (1928). Increasing returns and economic progress, *Economic Journal*, Vol. 38, 527–42.

5

EDITH PENROSE AND ECONOMICS

ROBIN MARRIS

Emeritus Professor of Economics, Birkbeck College, London University; sometime University Reader and Fellow and Director of Studies in Economics, King's College Cambridge

This paper contains a constructive critique of the Penrose picture of firm, comparing it with, on the one hand the Coase–Williamson transactions–cost picture, and on the other with the author's own growth model. It concludes that the Penrose theory does not at the end of the day quantitatively determine the growth rate of the firm and that the transactions–cost story does not at the end of the day determine the size of the firm. The author's own model determines the growth rate of the firm, but not the size. If married to a stochastic process, the author's model contributes to the determination of the statistical size distribution. The form of the latter is determinate, but the quantitative parameters in any one society at any one time, are not. As there is, however, a strong case that the administrative efficiency of firms has no robust relationship to absolute size this indeterminacy may not be economically significant. In other words, the static efficiency of an economy is not sensitive to either the absolute size distribution of firms or to their growth processes. Instead, the key contribution of firms' growth processes to social welfare probably lies mainly in the realm of competitive dynamics. Interfirm competition for growth is also competition to innovation, à la Schumpeter.

I may be a couple of years younger than George Richardson but I also have reached my anecdotage. I am probably going to be somewhat more critical of *The Theory of the Growth of the Firm* (hereinafter *The Growth of the Firm*; Penrose, 1959) in the positive sense of the word, than other contributors, but claim that right in view of my special relationship with Edith and her work. I wrote the first academic review of *The Growth of the Firm*, I wrote Edith's entry in the *New Palgrave* and I wrote her obituary in the *Guardian*.

I first encountered the Penrose theory in the late 1950s when I had already written quite a large part of what eventually became *The Economic Theory of Managerial Capitalism* (Marris, 1964). My *Economic Journal* review of *The Growth of the Firm* was published in 1961. In due course I met the author and we became life-long, but argumentative friends. Given the radical character of her academic work, I found Edith surprisingly conservative in other directions. She found me unnecessarily anti-establishment and also intellectually under-baked.

Apart from teaching, I had at that time two other tasks. One was conducting a research project on the economics of multiple-shift operations, the other was acting as secretary to a departmental committee, not the appointments committee. The research work was done in an attic office in a Victorian building in Sidgwick Avenue which then housed the Department of Applied Economics: while staring out of the modest window across the playing fields of King's College Choir School, my mind wandered from the theory of capital utilisation to the general theory of the firm. During one such reverie, soon after reading newspaper accounts of a failed attempt by ICI to take over Courtaulds, I conceived what sadly now seems to be my main claim to fame, namely the take-over constraint.

More relevant to the concerns of the present collection is the fact that my administrative task was carried out after teaching hours in the department office, which was then located in Downing Street. Naturally it was my habit to read any other interesting documents concerning faculty affairs which might happen to be lying around. In 1958 one such was a full set of the galley proofs of *The Growth of the Firm*. Obviously I was immediately attracted, I took all the sheets home for the night and by the time I had surreptitiously replaced them next day, I had read every one. It turned out that the reason the proofs were in the office was that the departmental appointments committee had decided to offer Edith a Cambridge lectureship, subject to interview. The interview duly occurred; there is no record of what transpired, except that no appointment was made. A bad day for Cambridge but in my opinion a good day for Edith, who I think would have been suffocated there.

Everyone knows the readability of that book. In the light of my then interests, how could I have been other than extremely excited? But I could also have been jealous. (Of course, I was and still am—consoled partly by the knowledge that Edith was jealous of me. She sent me an autographed copy of the 1990 re-issue of her book, but she did sound a little tense when I told her of the prospective publication of *Managerial Capitalism in Retrospect* (Marris, 1998). The last time we met she asked, 'Are you behaving yourself, Robin.' When I answered, 'As a matter of fact, currently, I'm not', she replied, 'Oh well, there's still time.'). Apart from the matter of the title, the relation between our work was clearly complementary rather than competitive. I think her book subsequently outsold mine by an order of magnitude, but I don't believe my sales would have been larger, and might well have been smaller, had Edith's book never been written. My theory was lacking the institutional flesh contained in Edith's. Hers was lacking a theory of market demand, of finance and of stock-market value. It was also, in my view both then and now, for reasons to follow, lacking an *economically* interpretable account of the *motives* growth. Finally, in the words of my *Economic Journal* review (Marris, 1961), it was definitely 'woolly' on the subject of profits.

In contrast, my own idea was to create an explicit economic theory of the firm based on the literature of the managerial revolution. As every reader (of course!) will know, by the time I read *The Growth of the Firm* proofs, I had in principle already created a closed economic model, replete with suitable or unsuitable mathematics as the case might be, to determine, *inter alia*, the growth rate of the firm. As the model evolved I began to have a vision of replacing size by rate of change of size as the central variable of microeconomic theory. I assumed that once this idea had been introduced to the profession, it would instantly prevail. Half a century on we can see how wrong I was.

My embryonic theory contained an explicit 'managerial' objective function based on the quantitative relations between the firm's growth rate, profit rate and the stock-market value. But owing to a disposition to enjoy formal theory while lacking the necessary mathematical IQ required to do it competently, I did not at that time understand my own model very well—a state of affairs that persisted in the eventually published version. ('Lacks focus', Edith used to say.) So I didn't really understand whether in the model I had so far created, the growth rate was already fully determined. As a matter of fact, within the narrow confines of economic theory, you can fully determine the growth rate of a firm in a model with only two equations,[1] e.g.,

$$v = (p - g)/(r - g) \tag{1}$$

$$U = U(v, g) \tag{2}$$

where p = profit rate on physical assets, exogenous;
   r = discount rate, exogenous;
   U = utility of the firm (see below).

To obtain the optimum growth rate, maximise (2) subject to (1).

Equation (1) is based on the assumption that the stock-market value of the firm will be the sum of expected future dividends discounted to infinity at the rate r, as calculated from the growth path implied in p and g. It also assumes no external finance.[2] (It must be admitted that since the mid 1980s, as a result of the junk-bond movement and the general increase in corporate leverage, the last assumption has become increasingly unrealistic.)[3]

Equation (2) is the objective function of the firm. If the firm is managed solely in the interests of the current shareholders, $U'(v)$ will be positive and $U'(g)$ zero for all v and g. If managers pursue policies reflecting an intrinsic value, to them, of growth (independent of the effect of growth on the market value), both partials are positive for all v and g, the positive $U'(v)$ reflecting the assumption that for one reason or another (e.g. managerial stock ownership and/or the take-over threat) managers like both

---

[1] The model is a version of the one I recently presented in (Marris, 1998, pp. 22–34). Equation (1) is already a reduced form. Suppose we arbitrarily assume that after initial start-up the firm never uses any kind of external finance. Then any growth rate, g, requires a specific retention ratio, g/p. Alternatively, assume that all growth is financed by issuing new shares at the ruling market price (which is of course influenced by the diluting effect). Or assume a combination of both methods. In every case, after the equations have been simplified, one arrives at equation (1), based on the assumption that the market values the *expectation* of the stream of future dividends implied in a permanent growth rate of assets, profits and dividends all equal to g, and constant retention ratio and/or new-issue rate, to infinity and discount rate r.

[2] In the absence of external finance, the current dividend per unit of assets is $p(1 - g)$. Equation (1) also assumes mathematical equivalence of dividends and capital gains.

[3] In *The Economic Theory of Managerial Capitalism* I did allow for external debt finance in the form of a constant, moderate, leverage ratio. On that assumption, the little model continues to work without qualitative difference from the case without debt. (With regard to new-share issues, it can be shown mathematically that they have the same effect on market value as retained profits, and hence can be ignored.) In addition, for the first three-quarters of the twentieth century, the stable-leverage ratio assumption was empirically valid. But the junk-bond movement both exploded and destabilised typical corporate leverage ratios.

growth and market value. The first form of the objective function is called 'classical', the second 'managerial'.

However, there are numerous problems with this model.

(1) It is lacking Penrose. It has to have constant returns to scale, otherwise p will depend on a missing variable, namely the absolute size of the firm—but the assumption is not discussed.

(2) Why should p be independent of g?—a question which breaks down into:
   (a) will g affect p, e.g. faster growth may cause lower profits relatively to size?
   (b) will p affect g in the sense that, again relative to size, profits are needed either to finance growth directly or to attract funds to finance growth?

(3) The financial model suffers from:
   (a) mathematical-cum-metaphysical paradoxes (for example infinite v) if p, g and r approach each other;
   (b) failure to take account of a rather obvious fact that the discount rate, r, must include an element of risk, which will in turn be liable to increase with the expected growth rate, i.e., we must allow the stock market, if it so desires, to regard faster growth-rates as more uncertain than slower ones.

I. CARL KAYSEN

At the time of my discovery of Penrose, I was already trying to tackle some of these problems. I was in fact inspired by a never-published talk given in Cambridge in the mid 1950s by the Harvard economist Carl Kaysen. Assuming a single-product monopolistic firm, he drew a diagram with price on the vertical axis, but on the horizontal axis, for the traditional 'Q' he substituted the growth rate of the same. The lower the absolute level of price, he said, the faster will be the *growth rate* of demand for the product. Kaysen then drew a 'dynamic' demand curve; he also drew a 'dynamic' supply curve, based on the need for cash flow. In order to finance faster inputs of productive resources to support higher growth rates it was necessary to have higher, rather than lower, profit rates. Hey presto, where the declining demand curve intersected the rising supply curve, we had determined the growth rate of the firm.

Jumping ahead, Kaysen's model has some similarity with a model developed in the late 1960s in the French *Bureau du Plan* and also subsequently by the so-called 'post-Keynesian' school of economic thought. The purpose was to determine the firm's profit margin, rather than its growth-rate, and hence the macroeconomic distribution of income between wages and profits. Micro demand for the representative firm is assumed to grow, exogenously, from macro forces. In a general regime of price-setting competition the firm picks a point on a Kaysen-type supply curve where the indicated growth rate equals the exogenous demand for growth rate and then, on the vertical axis of Kaysen's diagram, reads off the price. This last value, via real factor structures (also exogenous), determines the profit margin, which then, via a general oligopolistic game (which with hindsight one can now see could in fact take the form of imperfect polypoly with a Nash 'competitive' conjectural general equilibrium), becomes the ruling margin in the industry and the economy. How do we know that at the micro level the price charged at a given time will clear the market? We don't! Apart from that little difficulty, I have also persistently and fiercely criticised this post-Keynesian model on other grounds which have some affinity with my gentler critique of Penrose, and can therefore for the moment be deferred.

In the meantime, reverting to Kaysen, my immediate difficulty with him was that economics did not have a theory relating the absolute selling price of a product at a given time on the one hand and the product's rate of change of quantity demanded at given prices over future time on the other. Kaysen in his talk simply assumed the relationship without discussing it. In fact it must be the case that any such theory must assume that the consumption behaviour of individual consumers can influence the utility functions of other consumers. In the 1950s we did have, however, Jim Duesenberry's (1949) rich and highly original theory based on that type of assumption, designed to explain the empirical failure of the macro propensity to save to rise with income. Later, people developed similar models to explain the growth-paths of new consumer durables, such as washing machines. For me, however, the problem with these models was that they tended to predict S-shaped paths. By contrast, I was already looking for a steady-state model of the growth of the firm and hence for exponential paths. I had already concluded that no single-product theory of the firm would meet my needs, and turned to the idea of continuous growth founded in continuous horizontal diversification. New products are successively launched. Some fail, others take off and establish good markets along S-shaped curves. When (at the top of the 'S') markets saturate, if a firm's growth is to be sustained, others must be explored. Alternatively, existing established markets can be attacked by attempting to storm the barriers to entry. If successful, market share will probably increase quite briskly as the previously established incumbent firms retreat, but eventually the marginal costs displacing incumbents becomes excessive or they have all been driven out and there is no more market to capture.

Thus one can build a 'demand side' for the growth of a firm theory, but what about the 'supply' side? Obviously, that aspect was the contribution of Penrose.

II. THE BASIC PENROSE THEOREM

The two points in Penrose's book that struck me as the strongest were: (1) the concept of the firm as an administrative organisation which can always recruit new resources to expand to any ultimate size; and (2) the famous administrative constraint on the rate of change of size. The latter rang an immediate bell in connection with a remark of my father's about his experiences in World War I. 'You cannot instantly create an effective military unit by bringing together trained people who have never worked before with each other', he said, 'They need time to bed down: they need time to learn each others' ways.'

I immediately wished to include Penrose's constraint into my own model, but found that it needed some development. Let us call the process of bedding down, 'training'. Penrose's fundamental proposition was that only trained staff can efficiently carry out the functions of planning and managing the various activities required for sustainable growth, one of which, *inter alia*, is the work of recruitment. This has crucial consequences. If a firm of size X, measured in units of trained staff, hires Y more people in a single period, and if each of them takes T periods of time to train, then T periods into the future the firm can expand its effective organisational size in a lump, proportionally Y/X. The effective administrative growth rate averaged over T is therefore $[log(X + Y) - log(Y)]/T$. But why cannot Y, and hence the growth rate, be indefinitely large. The answer must be that training requires a working contact between trainers and trainees. In turn this activity to some extent, distracts the trained staff from their other functions. If X increases relative to Y, either some trainees are temporarily ignored, as often happens in real life, or the strategic function of existing trained staff is weakened. In the first case, the training period, averaged over all the recruits, lengthens. In either case, the effective growth rate is constraining. Another possibility is that the efficiency of existing operations is also prejudiced.

To the economist the obvious way of expressing all this is in the form of a negative causal relationship from g to p. Instead of saying that the firm cannot effectively administratively expand faster than at a fixed rate, we say that the faster it hires new people the greater the general distraction (for example) of trained staff, implying reduced control over the efficiency of existing operations, i.e., tendencies to launch dud new products, tendencies to pick the wrong existing markets and attack, poor operational judgement in the execution of new-entry battles, or all or any of these. Failures cost money; so cash-flow, relatively to the size of the firm, inevitably suffers. Hence we can add a third equation to the model[1]

$$p = p(g) \tag{2.1}$$

Because new blood is invigorating, over low to moderate ranges of the growth rate, $g'(p)$ will be positive. From there, however, the Penrose effect takes over and $g'(p)$ becomes negative, and also $g''(p)$. The effect on the solution of the model is obvious. Now, not only the growth rate but also the profit rate become endogenous. The new reduced form is

$$v = v(g,r) \tag{2.2}$$

Subject only to one exogenous variable, namely r, plus some restrictions to prevent v going to infinity. Equation (2.2) is my famous v–g frontier. With growth rate on the horizontal axis, v rises to a maximum, indicating the optimum growth rate from the point of view of shareholders. In the managerial interpretation, g is maximised at a rate faster than the shareholder optimum, subject to a restraint of trade-off from v. If we call the optimum growth rate on either criterion, $g^\star$, then the final reduced form is

$$g^\star = f(r) \tag{2.3}$$

where r remains exogenous.

[1] The development of the model is a simplified version of the one presented in Marris (1998, p. 124, equations 6.7–6.10).

III. FIRST CRITICISM OF PENROSE: THE MOTIVE FOR GROWTH

The current purpose of the foregoing recap is not intended to be a mere exercise of vanity but the creation of a formal framework for my two main criticisms of Penrose already mentioned. She implied that firms grow because the process of administrative expansion is always lumpy. Therefore at any one time there are always underutilised bodies. Something should be found for them to do; why not launch a new venture? One could characterise this proposition as no more than Parkinson's Second Law (Parkinson, 1958), namely, 'Work expands to fill the time available'. But that would be unfair. Parkinson's original humorous insights later found serious foundation in 'bureaucratic' organisation theory, e.g. as found in Monsen and Downs (1965). My problem lay with the quantitative value of the resulting 'Penrose' growth rate, which may be called g^P. Presumably this rate is controlled by the ratio of the average size of a lump of new resources to the existing administrative size of the firm. There are two problems with this.

(1) If the former is an exogenous biological constant, the ratio in question will diminish as the firm grows. Hence g^P appears to diminish. I don't think Penrose would have wanted that. It is also a conclusion that appears internally contradictory. Surely the fact that lumpiness diminishes in significance as the absolute scale expands would represent a growth advantage for large firms?

(2) What happens if g^P differs from $g^\star$? This is not just a self-serving question from my side; it is quite down to earth. Suppose I am a senior manager who knows that I have underutilised subordinates. In addressing the problem I find that although it would be nice to find them more to do, anything I can think of is likely to prove more costly than leaving them alone. For example, we might set the underutilised bodies to launch a new product that is highly likely to fail. Even though the bodies, being already employed, have a marginal cost of zero, this is not the case with other resources such as sub-managerial production workers, marketing costs and intermediate inputs.

What we have here, in fact, is a classic conflict between economic theory and organisation theory. We all know bureaucracies which want to expand in order to suppress the fact that they are currently overstaffed. In Parkinson (1958) and elsewhere these stories are usually set in the public sector. But Monsen and Downs (1965) presented a similar picture for large private monopolies. Nevertheless, I still hold the view that even in an institutionalist context, Penrose's argument was and is not a very robust or convincing foundation for a theory of the growth of the firm. Bluntly, as far as I am concerned, such a theory cannot, in fact, determine the firm's growth rate.

One rejoinder to the foregoing critique could come from Oliver Williamson, more precisely from the body of work (Williamson, 1975, 1981) which can be called Williamson-II. A difference between g^P and $g^\star$ could be seen as an example of bounded rationality. We have no real idea of the value of $g^\star$. Instead, for growth decisions we employ the heuristic, 'always try to use underutilised resources, unless the profitability prospects are unreasonably negative'. But, as Herbert Simon well knows, economists have one profound difficulty with bounded rationality: what happens if, in the course of time, the satisfying solution to a problem is gradually revealed to be so adverse in its outcomes that the welfare of managers and shareholders, and hence the very viability of the organisation, is seriously threatened. The Simonite answer is, 'They make an adjustment'. Is the adjustment necessarily towards the optimum solution? One problem with Simon is to know whether he is saying that the optimum solution is so complex that we cannot find it, or whether it is so complex that it does not exist in reality. However, suppose the answer is affirmative, i.e. heuristic adjustments are in practice generally beneficial; if not, the adjustments in question would not tend to be used. Then, the economist's difficulty has always been the question of whether successive bounded rational responses will tend to draw the decision maker, willy-nilly, towards the optimum solution? Granted, he or she may not know that they have arrived there, but the implications for economics are that, ignorant or not, economic agents are in practice, implicit optimisers.

My own approach to this problem paralleled an earlier work of Williamson (1964), which may be called Williamson-I.[1] In a bureaucratic world there are numerous reasons, one of which is benefit from having subordinates (Williamson's 'expense preference'), for management to derive satisfaction not only from the size of the firm but from the rate of change of size, and I spelled this out at some length in *The Economic Theory of Managerial Capitalism*.

In short, the economist's answer to Edith is that the firm's motive for growth is that in general, in a growing economy, the optimum, i.e., utility-maximising growth rate for shareholders and/or for managers, is generally positive.[2] A simple confrontation between institutionalism and economics? Indeed. But the economist is not necessarily right. Paul Samuelson once said that because organisations had the *capacity* to grow, they did just that.

Microsoft Windows, versions 1 and 2, were badly-designed market failures. As a result Gates lost interest in the Windows project and was largely concerned with OS/2.[3] The Windows development group was run down and the small team remaining to produce Windows version 3 were not under pressure. Left alone they tried something for which they had no remit, namely, to attempt to remove a fundamental weakness which had persisted from DOS into Windows 1 and 2, that in order to run one application you had to close down others. As an erstwhile programmer, Gates believed that it could not be done. However, after a struggle with the secretaries, the heroes got a chance to demonstrate their solutions to the Leader, and one of the

---

[1] It was Williamson's 1961 PhD Dissertation which won a national prize that included sponsored publication, with a publication delay of 3 years. I finally delivered my manuscript to Macmillan in early 1962 and was published, like Williamson, in 1964.

[2] The exception is the case where the $p(g)$ function is so unfavourable that no positive value of g will set $p > i$. Then the optimum growth rate is zero and $v = p/i < 1$. Probably the firm should be wound up, but the model, being steady state, cannot encompass discontinuous events. In real life, the most likely outcome is a take-over. The physical assets can be acquired at less than their replacement value and the acquiring firm may then attempt to reform the firm in the sense of creating a more favourable $p(g)$ function.

[3] Which in the event, for PCs, has sunk without trace. IBM, post-break with Microsoft, did try to market it, but were amazingly inept.

reasons why he is where he is, is that in this and similar situations, after listening and arguing for a while, he understood what was being shown him and he eventually abruptly said 'OK, lets do it'. That didn't mean that Gates would immediately abandon IBM. He had no idea that Windows 3.1, when finally launched, would sweep the world. When it did, however, he drew the obvious conclusion and took the fateful, but by now less risky decision to finally ditch IBM. Down-the-line software developers could not know what it would actually feel like to break away from the great IBM, an organisation that had been growing continuously for the best part of a century and now employed more than a quarter of a million persons worldwide. By contrast at that time Microsoft was less than 20 years old and had a worldwide employment of no more than 60,000. But as a result of the events just described, within another 5 years Microsoft (having gone public with Gates still retaining control) had a stock-market value which on a good day could exceed that of the old giant. I have to concede that in the contemporary IT industry, the steady-state model of the growth of the firm is peculiarly unrealistic.

IV. SECOND CRITICISM OF PENROSE: PROFITS

In an often quoted passage Edith stated that profits and growth were, in effect, the same thing. What can this mean? There are various interpretations.

(1) Profits are needed for growth—true, oh Queen, but 'needed for' is not synonymous with 'equal to': how much profit is needed for how much growth?
(2) The rate of growth of real profits will tend to equal that of other physical measures of size—true in steady state, but only in steady state and subject to one rather interesting exception, namely the rate of growth of the number of administrative bodies, Edith's special concern.[1] There is a one-for-one relationship between the growth rate of profits and of other size measures—so, if a firm is to grow steadily it must maintain a constant profit rate on assets, hence, in steady state, as indicated

---

[1] Why? (a) technical progress (b) the mathematics of the bureaucratic hierarchy, as follows: Suppose the degree of loss of control is a linear function of the length of the chain of command (as embodied in the height of the bureaucratic pyramid) and that the size of the organisation is measured by the number of bodies employed at the lowest level, i.e., by the 'width' of the pyramid measured at the base. With a constant, arbitrary, span of control the height increases linearly with the log of the width, implying that, given the assumption that loss of control increases linearly with length of chain of command, the result is a law of diminishing marginal inefficiency. As organisations expand, they become increasingly inefficient, but at a diminishing rate. It follows that organisations have only to develop some additional features needing only to dampen, not eliminate, loss of control, can have constant returns to scale. But to most of the contemporary economics profession, from neo-liberal, neo-classical Right to post-Keynesian and/or Institutionalist Left, such a conclusion is heresy. A decade earlier, Oliver Williamson (Williamson, 1970), having made the same embarrassing discovery, escaped perdition by forcefully arguing that the errors arising in information-chains naturally tend to accumulate, so that loss of control, rather than rising linearly with chain of command does so at an increasing rate. He inserted a coefficient into his model to represent this effect and endowed it with the quantitative value required to guarantee diminishing returns to scale. Perfectly plausible, but hardly a general proposition, especially as potential countervailing factors (such as the use of written commands) are ignored. People who believed Williamson established a general law of diminishing returns to organisation have not noticed the underlying assumptions.

above, the relation is equality; out of steady state the relation between profits and growth can be anything.

The answers to these semantic issues can be answered with the aid of my little model.[1] In the solution both the growth rate and the profit rate are endogenous: they are not equal and never can be.

If we look behind my function $p(g)$ we can write $p = p^0 + \alpha g - \gamma g^2 \cdot p^0$ is the 'operating' profit rate the firm would earn from its existing products if it did not grow, α and γ represent the growth effects; α the beneficial effect of new blood and γ the Penrose effect. So,

$$v = p^0 \alpha (1 + g) - \gamma g^2)/(r - G) \tag{4.1}$$

$$U = U(v,g) \tag{4.2}$$

$$g^\star = g[Max\ U] = F(p^0, r) \tag{4.3}$$

It is easy to see that $g^{\star'}(p^0)$ will always be positive: obviously this must be so in relation to shareholders, but also, in the case of a growth-oriented management it can never pay to indulge in expense preference or be otherwise unnecessarily inefficient. The greater the operating profit rate, the greater the dividend that can be paid to shareholders without sacrificing growth, or alternatively the greater the growth rate possible from a given retention ratio and hence the given risk of a take-over. To correct Penrose's 'woolly' statement about profits and growth, therefore, we must say, 'other things being equal, the greater a firm's operating profit rate, the greater will tend to be its planned long-term growth rate. Galling for me and Edith, and most especially for me, is the obvious fact that in the eyes of today's top management, 'long term' means about 18 months. In MBA textbooks on business strategy, the concept of a planned long-term growth rate is hardly to be found at all.

V. ENTER (RIGHT) SIR COASE AND LORD WILLIAMSON (LEFT)

When I wrote my original book, like most other economists at that time, I had not encountered Ronald Coase's now celebrated article, published in 1937, posing the question, 'Why do firms exist?'. If the market system is so efficient, why is not all economic activity carried on by one-person businesses? I just assumed that the answer was 'teams', i.e., that firms did exist, and carried on from there. I think it can be said that the same applies to Edith Penrose. But a formidable body of scholars, e.g. Coase (1937), Williamson (1975, 1981, 1985), Hart and Grossman (1986) and Hart (1989), led and inspired by Oliver Williamson who rediscovered Coase, have shown that in doing that one leaves out a line of enquiry that has proved extraordinarily fruitful in economics. The answer Coase gave was 'transactions costs'. These are various costs of using the market to organise an economic activity in contrast to organising it within the firm. By contrast, the costs of an in-house organisation may be

---

[1] The equations are a simplified version of those given in Marris (1998, p. 125). In particular, equation 6.8 that makes r increase with g (owing to greater riskiness of faster growth), is omitted.

placed under the label of 'agency costs': as a firm expands it must employ an increasing number of people who have little or no ownership stake in the enterprise and who therefore lack incentive to be efficient. Coase's original words were as follows:

When we are considering how large a firm will be the principle of marginalism works smoothly. . . . At the margin, the costs of organising within the firm will be equal either to the costs of organising within another firm or to the costs involved in leaving the transaction to be 'organised' by the price mechanism. (Coase, 1937, p. 404).

It is evident that Coase was implicitly describing a diagram in which the absolute size of the firm was represented on the horizontal axis and either marginal agency costs or marginal transactions costs were represented on the vertical. There would be a curve of the former, which was rising, and of the latter, which was falling. Lo and behold, where they intersected we had the optimum size of the firm! If this was the representative firm in the economy, and all firms reached the optimum, then the total industrial structure was optimised: society achieves a macro-optimal balance between agency costs and transactions costs. The market was a self-organising system which optimised its own structure.

The problem with this beautiful idea is that no one has produced any strong general theory to make the marginal agency-cost inevitably slope upwards. As one takes on increasing numbers of people lacking ownership rights, why should the average non-incentive effect necessarily increase more than proportionately to their output? Why not a simple law of constant unit effect? Frankly, I do not think this question has been answered. Consequently I assert that a general hypothesis of constant or increasing returns to the scale of organisations, which is essential to the Penrose–Marris theory of the growth of the firm, continues to stand up.

A more powerful and, in my opinion, much more successful application of transactions cost theory is found in Oliver Williamson's brilliant explanation of vertical integration. This was not a topic to which Penrose–Marris gave much attention, but the more one thinks about it the more one can see that, in a sense, vertical integration is the essence of the concept of the firm. More precisely, it is the essence of the beginning of the firm. It is Adam Smith's pin factory. If an already integrated firm is considering an act of vertical disintegration, it must take account of the fact that the profitability of its sunk investment in its central equipment is now partly dependent on an outside supplier of the intermediate product. Also, the outside supplier is dependent on the final producer. In theory it should be possible for the two parties to write a long-term contract dealing with all possible eventualities in their future relations. In practice such a document would be too complex to be effective. Such contracts, in Oliver Williamson's terminology, are in practice inevitably 'incomplete'. The problem of incomplete contracts lies at the heart of the nature of transactions costs.

In determining the optimum degree of vertical disintegration these transactions costs are set against the presumed marginally increasing agency costs (but see above) of vertical expansion. A major merit of this theory is that previous explanations of vertical integration were theoretically woolly and were unable to account for the observation of fluctuating trends of integration and disintegration that have historically occurred in industries and economies.

But, but, but—what about the outstanding feature of the massive growth in the twentieth-century sizes of individual firms, namely *horizontal*, or lateral, expansion? It is my contention that at the end of the day, for all its achievements, the transactions cost theory does not succeed in providing a comprehensive analysis or explanation of the process of expansion by diversification which lay, and still lies, at the heart of Penrose–Marris. My explanation, of course, is managerial motivation combined with constant returns to scale. Edith's explanation was as discussed above.

VI. A PRELIMINARY CONCLUSION

From all the foregoing I reach what must be, for this essay collection, an embarrassing conclusion: *Coase–Williamson did not determine the size of the firm and Edith did not determine the rate of growth of the firm!*

To reduce the offence I concede that I myself, while claiming to have determined the growth rate of the firm can make no claim to have determined its size. I started my model from a situation where the firm had somehow historically acquired a given size, and was not allowed to go back. Robert Solow (Marris and Wood, 1971) subsequently severely criticised this approach. In its place he offered a typically elegant mathematical model which in effect determined the size through all points in time of all firms in a general equilibrium.

VII. ENTER (BACK) ROBERT SOLOW

Embellishment (by me) with the intellectual apparatus of subsequent developments in imperfectly competitive general equilibrium theory, the Solow model can be described as follows. The economy consists of a large number of firms each producing, under constant returns to scale with constant capital–output ratio, a different, single product. Every firm is engaged in imperfect competition with every other firm. Before introducing the theory of growth the static general equilibrium of this type of economy can be modelled as a conjectural or Nash-competitive equilibrium in which each firm sets a price on the conjecture that all other firm's prices are constant. Equilibrium exists when, on the conjecture, no firm desires to change price. Assume conjectural prices are set to maximise gross profits, i.e., set marginal cost equal to conjectural marginal revenue. The result is that the gross profit margin (share of gross profit in price) is the reciprocal of the conjectural elasticity of demand. There is a constant amount of real capital in the economy, but it is mobile. In the process of adjusting their size for the purpose of conjectural profit maximisation, firms can acquire or dispose of capital. Consequently the sizes of firms and the distribution of capital are optimised to maximise the total profit in the economy, for the given total capital stock, and thus the welfare of the shareholding class. (If the firms had identical, iso-elastic demand curves they would all be the same size.) If shares were equally distributed among households, their property income would be maximised relative to their employment income and, interestingly, so would the degree of 'excess' consumption of leisure caused by the excess of price over the value of the marginal product of labour.

Now Solow introduced a minimalist interpretation of the idea that firms, by spending money out of their profits, can influence the location of their demand curves. More precisely he assumed that by spending a constant proportion of their gross revenue on marketing, the single-product demand curve could be made to grow (as defined by changing quantities demanded for given prices) at a constant exponential rate; the higher the marketing-expenditure share, the higher the rate. Solow then applied the same type of valuation formula as used by me in equation (1) above and then solved the equations for a simultaneous solution optimising both the initial size of the firm and its subsequent growth rate. He thus optimised the size of the firm through all points in time and thus, similarly, the total value of shares.

Apart from the obvious fact that this is a normative neo-classical model (which does not, however, produce Pareto optimality) my fundamental difficulties with it are that the firms are single-product, and that because the launching of new products is a dynamically constrained 'growth' activity, the model cannot be simply adapted to a world of diversification. The marketing-expenditure decision is also conjectural, the conjecture being that other firms' prices and marketing expenditures are constant. This results in a path of point-in-time equilibria. But surely, although firms will not anticipate this in their conjecture, their marketing expenditures will tend to cancel out and the whole effect will be decay. The demand-led growth of the system will, it seems to me, then converge to some exogenously given value. In the absence of new products, it seems unlikely that the marketing expenditure will have a significant role in sustaining the growth of demand.

VIII. EXIT (TRAPDOOR) *OMNE*

Where stands the reader who is convinced by the foregoing? What determined the organisational size-structure of the modern market economy? If, in historical order, neither Coase, Penrose, Marris, Solow nor Williamson-II, can give us an answer, where do we go for honey? I have no doubt abut the reply. Under constant returns, the size of the firm cannot be determined because it is indeterminate.

However, firms do have sizes with an empirical distribution that has distinctive patterns. In natural numbers grossly skewed, the data become more or less symmetrical when semi-logarithmically transformed. In consequence the character of any particular body of data relating to this topic can be validly summarised by a convenient statistic, namely the variance of the logs of the absolute sizes of the firms, which may be called *LVS*, which in turn has a close relationship to measures of concentration, such as for example, the share of total employment, value added or gross assets accounted for by the largest hundred or largest thousand firms, as the case may be. Other things being equal, the larger the *LVS*, the larger will be the average absolute size of the largest firms. Removing multinationals, large countries tend to have larger average absolute sizes among the largest firms than small countries. Hence *LVS* tends to be higher in large countries than in small countries.

There is a large amount of literature on how this comes about (Gibrat, 1931; Hart and Prais, 1957; Simon and Bonini, 1958; Steindl, 1965), to which I myself have contributed (Marris, 1979). To cut a long story short, imagine an economy governed by a Marris–Penrose type of model as described above, running over many time periods, and as now modified below.[1]

$$p^0_t = \beta_1 + \beta_2 S_t - 1 + \eta_t \tag{8.1}$$

This is a regression equation where S represents the measure of the size of the firm and η represents stochastic shock. R^2 will be moderate. The values of the beta-coefficients are unrestricted; $\beta_2 = 0$ indicates constant returns, $\beta_2 > 0$ increasing returns, $\beta_2 < 0$, decreasing returns. 'Weak' diminishing returns, à la Mirrlees implies β_2 negative but small in absolute value. Substituting back into the model we have,

$$g^\star_t = F(S_t - 1, \eta_{t}, t) \tag{8.2}$$

With *r* continuing to be treated as exogenous, this states that the optimum growth rate of the firm in any period depends (probably weakly) on its size in the previous period plus or minus a stochastic shock.

I have elsewhere shown (Marris, 1979) that in an imaginary economy whose initial state of industrial organisation consists of a large number of firms normally distributed in size, and whose subsequent development can be approximated by assuming no entry or exit, and that in each time period each incumbent firm's actual growth rate equals its optimal growth rate, the size distribution will gradually change from normal to log-normal. With $\beta_2 = 0$ (constant returns) the variance of the distribution (*LVS*) will increase linearly at a constant rate through time, implying steady persistently increasing concentration. With increasing returns it will increase at an increasing rate, implying exploding concentration. With diminishing returns, concentration will increase at a diminishing rate converging to an asymptote.

I have also shown (Marris, 1979) that the model can be effectively embellished in two ways, i.e., (1) by including the effects of a partly stochastic merger process and (2) by permitting entry and exit. The former effect is likely to increase concentrating tendencies, the latter to dampen them. There is some evidence that in the US, after three-quarters of a century of increasing concentration, dampening may now have occurred, largely as a result of new entry. Thus the Marris–Penrose model, however heroic its assumptions, modified by stochastic shocks, does predict the type of size distribution we actually see.

---

[1] The line of argument which follows, found originally, as already indicated, in Marris (1979) was not included in Marris (1998).

IX. THE WET CONCLUSION AND ITS DENIAL

The fundamental implication of all this is that both the size of the individual firm and the organisational structure of the economy remain indeterminate because both are time- and path-dependent. The statistical characteristics of the time path, however, are 'determinate' in the sense that they are capable of prediction by theoretical modelling, such as above. What then are the normative implications? What is the specific implication for social welfare as measured, for example, by the Sen (1997) formula,

$$SW = y(1 - g),$$

where y represents average per capita income; g the Gini coefficient of inequality?

The question of how the level of business concentration affects income distribution is extremely interesting and has occasionally been touched on, I think, by writers with Marxian backgrounds, but is not one on which there has been substantial empirically-tested theorising. I personally cannot pretend any competence to contribute to it. The question of how the level of business concentration affects average per capita income, however, is quite another matter. Quite apart from the elusive theoretical aspects of the question of whether the absolute sizes of organisations systematically affect their internal efficiency, the empirical picture is also cloudy. It would be a brave analyst who can find statistical results to support a general law of diminishing returns. Consequently, constant returns remains the minimalist assumption. In that case, evidently, there is a corresponding minimalist normative conclusion, *namely that the average efficiency of an economy and hence its* y *(income per head) is statically insensitive to its organisational structure, either in detail or as represented by gross measures such as macro concentration ratios. At its highest reaches, 'IO', it seems, has little to say about SW!*

For economists, especially 'industrial' economists, that is obviously a wet and therefore distasteful conclusion. Whenever, in the past, great persons have stumbled on it, they have tended to go into denial. For example Karl Marx (who in volume I of *Das Kapital* got nearer, in the 1860s, to a realistic concentration theory than any one else for the next hundred years) wrote,

Capital grows in one place to a huge mass in a single hand, because it has in another place been lost to many. . . . The battle of competition is fought by the cheapening of commodities . . . , which in turn depends on the scale of production. Therefore the large capitals beat the small capitals. . . . Hence competition rages in direct proportion to the numbers of antagonistic capitals. It always ends up in the ruin of many small capitalists, whose capitals partly pass into the hands of their conquerors, partly vanish. (Marx, 1967)[1]

---

[1] The translated quotation is from p. 686 of the first US edition. Marx referred to the process he was describing as one of '*zentralisieren*' which in modern German would imply that it must end total monopoly. It is possible, however, that had he been writing today he would have used the term '*konzentration*', which can either mean concentrating a number of things in one place (as in 'Concentration Camp'—a description which was first used to describe the holding camps for Boer civilians detained by the British during the war in South Africa early in the twentieth century) or 'to condense a large number of things into a smaller number', as in business concentration.

About a third of a century after those words were probably written the saintly Alfred Marshall wrote,

It has always been recognised that large firms have a great advantage over their smaller rivals in their power of making experiments . . . but on the whole observation seems to show . . . that these advantages count for little in the long run. . . . There are few exceptions to the rule that large firms . . . are, in proportion to their size, inferior to businesses of more moderate size, in energy, resources . . . and inventive power. (Marshall, 1890).

Here we may take a lesson from the young trees of the forest as they struggle up through the benumbing shade of their rivals. Many succumb on the way, and only a few survive; those few becoming stronger with every year. They get a larger share of light and air with every increase in their height, and in turn they tower above their neighbours, and seem as though they would grow for ever, and ever becoming stronger as they grow. But they do not. (Marshall, 1891).

Why do they not? My feeling is that as Marshall, having written that convincing prose concluding with the resounding phrase, 'and ever becoming stronger as they grow', suddenly thought, 'Oh my God, what am I saying?' So he promptly added, 'But they do not' and then took a walk around the garden to think of a justification. What he came back with is too long to quote in full but may be synthesised as follows. The only form of organisation known for an industrial firm was the family business. One day the founder-entrepreneur must die and the only candidates for inheriting not only the ownership but also the management will be sons. If the firm in question has, by the time of the founder's death, become quite larger, the family will be rich. Therefore, in comparison with that of the founder, the childrens' upbringing will have been soft— mainly conducted by nursemaids. Therefore the sons will be no good at business and the firm will begin to decline. In later editions Marshall successively weakened the passage, in particular banishing the nursemaids, and finally in the latest editions in the early part of the twentieth century, remarked that firms which had become joint-stock companies could avoid decline, but neither did they have the energy to grow; inevitably, he said, they stagnated.

We can also see rich insights in both sets of quotations. Marx's description of the process by which competition destroys competition is as valid today as it was a century and a half ago. Marshall's description of what happens to some family businesses also even today rings a bell.[1] Nevertheless each of these razor-sharp minds had seen a ghost. For Marx it was the ghost of competition, which could, as we know, rise from the grave to destroy monopolistic structures it had itself originally brought about. For Marshall, the spectre was in effect the ghost of Marx. He had found himself elegantly describing an explicit concentration process which could destroy the very neoclassical system of thought he was himself in the act of helping to establish. (The previous liberal-classical economics did not have a theory of the firm and, in fact, implicitly assumed that returns to scale were generally constant.)

---

[1] There is clear evidence of a general tendency of nineteenth century successful British entrepreneurs to actively discourage their children from following in the family footsteps into 'trade'; instead it was preferred to raise the family in the social scale by putting its money into country estates.

X. THE FINAL TRUTH

So, what is the true reality? Most especially what is the best representation of the organisational character of the market-cum-capitalist economic system, effectively born less than two centuries back, we now carry into the next millenium? What were the various essential features, that Marx and Marshall, in their different ways, both saw on the one hand and failed or chose not to see on the other? The best starting point for an answer is quite formal. What is the shape of a log-normal distribution when the data are reconverted into natural numbers? The answer is a reversed 'J' with a right tail so long that no scale can be found to permit a feasible visual presentation. At this right extreme we have IBM, Microsoft, AOL, General Motors, AT&T, etc. To the left, the high initial peak represents an extreme density of millions of small firms. If we regard all freelance professionals, both blue-jeaned and grey-trousered, as one-person businesses, then the largest size-group in the distribution is also the smallest.

The small businesses have both high birth- and high death-rates. Some survive only to stagnate, like Marris's Ltd, of Birmingham, England, born circa 1881, died circa 1981, having grown in the meantime at an average rate of 0·5% p.a. A small minority of small firms graduate into the medium or large size groups and among these are found the origins of some eventual giants. In the past, the time taken from birth to graduation would usually be measured in decades.[1] Microsoft went from 30 employees plus maybe half a million dollars of capital to world dominance, in one decade. Of course, the reason for that extraordinary story is not just the notorious mental and physical energy of Gates, but the grotesque speed, in the IT industry of our era, of technical change. Why in fact was it possible to write software that permitted Windows 3.0 to succeed where the previous versions had failed? Answer, because in the meantime unknown engineers in Japan, in the US, in Europe or wherever, had succeeded in multiplying by a factor of five the amount of hard memory that could be put into a small metal box.

The small firms are competing to become larger firms and the large and giant firms are competing to gain or hold their places. While the shape of the statistical size distribution remains constant, the identities of the inhabitants of the various size groups are in constant flux. The group of giants of today is not the same group as yesterday, nor as that of tomorrow.

It is these competitive dynamics, rather than static optimising properties, that are the genius of the capitalist system. There is nothing much wrong with large and giant firms, provided they have competition. And there is not much new in this idea, either.

---

[1] Once upon a time, a Chicago resident of Germanic origin, with the name of Mars, invented a new kind of candy bar. He had a business-minded son with whom he quarrelled. He said, 'get thee hence' and banished him with only the consolation of the rights of produce the Mars bar anywhere in the world except in the US. Young Mars proceeded to England where he took a building on an industrial estate in Slough. Thereafter in Britain the product was a success but in the US it faltered. In due course the younger branch of the firm, desirous of being able to do business in their mother country, not only developed a whole range of confectionery products but also sought sundry outlets for its core competence by acquiring brands such as Uncle Ben's Rice, Petfoods etc. In that large-scale international diversified role they went successfully through three generations as a wholly family-owned and family-top-managed non-traded corporation.

It is essentially the picture seen by Schumpeter (1943) when he wrote *Capitalism, Socialism and Democracy*. In that book, however, he predicted that in practice socialism would triumph because capitalism, for all its dynamic merits could not deliver a morally acceptable income distribution.

Well, socialism has gone but the distribution problem has not. In this article I have run away from discussing it, not because I am not concerned about it, but because I do not have a clear opinion about how contemporary tendencies in capitalist institutions are affecting it.

There seems to be an increase in vertical disintegration ('outsourcing') which can be explained well in Williamson terms through the effects of IT on both transaction costs and agency costs. There is also an evident tendency for giant firms to continue to grow horizontally; in effect, they are becoming thinner without becoming less wide; some observers think that average width is increasing. These tendencies seem likely to support, if not encourage, the growth of general productivity, but what of their effect on distribution? One tentative, but extremely interesting answer has been suggested by Snower (1994). He argues that the old-style vertically integrated firm inherently involved strong internal specialisation. Some people are more naturally versatile than others. The old system was on the one hand accommodating to the less versatile and on the other, implicitly frustrating to the versatile. The IT revolution, however, not only encourages vertical disintegration but, in helping information-sharing, also encourages less rigid and less specialised internal methods of organisation and working. Hence in the post-IT world, people who are born and raised to possess less than average versatility would develop a comparative disadvantage. Snower suggested these people would necessarily tend to become unemployed. I have macro-theoretical difficulties with that neoclassical style of reasoning. Rather I would say that if, as unfortunately seems likely, 'versatility' becomes another name for IQ, we can see a possible cause for a new form of adverse general effect of inter-personal income distribution.

Let us finalise the discussion by reverting to Sen. Recollect, that

$$\text{Social welfare} = SW = y(1 - g)$$

where y = income per head:
$\quad g$ = Gini coefficient.

Twenty-first century, post-Penrose, post-Marris capitalism, characterised by old Schumpeterian dynamics plus new IT features, seems as likely as ever to sustain the growth of average living standards as signified by y. Unfortunately, the old conflict between efficiency and equity (i.e., between y and g) is not only still with us but possibly exacerbated. From now on, therefore, the most important task of public policy is to forestall deterioration and hopefully to improve that crucial trade-off. Otherwise, although calculations show global social welfare as having on balance increased throughout the twentieth century, that might not be the case in the future.

REFERENCES

COASE, R. (1937). The nature of the firm, *Economica*, Vol. 17, No. 4, 1–34.

DUESENBERRY, J. (1949). *Income, Saving and the Theory of Consumer Demand*, Harvard, Harvard University Press.

GIBRAT, F. (1931). *Les Inégalités Economiques*, Paris, Sirey.

HART, O. (1989). An economist's perspective on the theory of the firm, *Columbia Law Review*, Vol. 85, No. 5, 202–51.

HART, O. and GROSSMAN, S. (1986). The costs and benefits of ownership, *Journal of Political Economy*, Vol. 94, No. 4.

HART, P. and PRAIS, S. (1957). The analysis of business concentration, *Journal of the Royal Statistical Society*, Series A, 150–91.

MARRIS R. (1961). *The Theory of the Growth of the Firm*, review of Penrose, 1959, *Economic Journal*, Vol. 71, No. 1, 110–13.

MARRIS R. (1964). *The Economic Theory of 'Managerial' Capitalism*, London, Macmillan.

MARRIS, R. (1979). *The Theory and Future of the Corporate Economy*, London, North Holland.

MARRIS, R. (1998). *Managerial Capitalism Revisited*, London, Macmillan.

MARRIS, R. and WOOD, A., eds, 1971, *The Corporate Economy*, London, Macmillan.

MARSHALL, A. (1890). Address to the British Association for the Advancement of Science.

MARSHALL, A. (1891). *Principles of Economics*, London, Macmillan, p. 457.

MARX, K. (1867). *Das Kapital*, Vol. 1.

MONSEN, R. and DOWNS, A. (1965). A theory of large managerial firms, *Journal of Political Economy*, Vol. 73, No. 3, 221–36.

PARKINSON, C. (1958). *Parkinson's Law*, London, Murray.

PENROSE, E. (1959). *The Theory of the Growth of the Firm*, Oxford, Blackwell.

SCHUMPETER, J. (1943). *Capitalism, Socialism and Democracy*, London, Allen and Unwin.

SEN, A. (1997). *On Economic Inequality*, Oxford, Oxford University Press.

SIMON, H. and BONINI, J. (1958). The size distribution of business firms, *American Economic Review*, Vol. 48, 607–17.

SNOWER, D. (1994). *The Low-Skill, Bad-Job Trap*, Discussion Paper No. 999, London, Centre for Economic Policy Research.

STEINDL, J. (1965). *Random Processes and the Growth of Firms*, London, Griffin.

WILLIAMSON, O. (1964). *The Economics of Discretionary Behaviour*, Englewood Cliffs, NJ, Prentice Hall.

WILLIAMSON, O. (1970). *Corporate Control and Business Behavior*, Englewood Cliffs, NJ, Prentice Hall.

WILLIAMSON, O. (1975). *Markets and Hierarchies*, New York, Free Press.

WILLIAMSON, O. (1981). *The Economics Institutions of Capitalism*, New York, Free Press.

6

HERCULES AND PENROSE

NEIL KAY*

Economics Department, University of Strathclyde

Penrose's *The Theory of the Growth of the Firm* is one of most important modern contributions to the theory of the firm. In this paper we are concerned with her analysis of the Hercules Powder Company published separately in *Business History Review*. The Hercules paper is important because it provides a unique case analysis of Penrosean growth processes, and indeed because there is reason to believe that it may have had a strong influence on the development of the book itself. Since Hercules is still operating today, it also constitutes an interesting living laboratory that may allow us to explore whether it has continued to pursue Penrosean growth processes over the intervening years.

Penrose's *The Theory of the Growth of the Firm* (1959) is now rightly regarded by many as one of the most important modern contributions to the theory of the firm. It has become one of the most widely cited texts in economics and her contribution is being increasingly recognised in a variety of ways, including the project to which this paper is a contribution. We are concerned here with a little known aspect of potential relevance to the development of the book, her paper on the Hercules Powder Company published in *Business History Review* in 1960. The Hercules paper is important for a number of reasons.

First, it should be regarded as belonging to the book and indeed it was written with the intention that it should be included as an integral part of that work. Penrose explains her omission at the beginning of her paper.

The following analysis of the growth of the Hercules Powder Company was originally intended for inclusion in my *The Theory of the Growth of the Firm* but was omitted to keep down the size of the book. The Hercules case was designed to illustrate the arguments of that study (Penrose, 1960, p. 1).

While Penrose does give illustrative examples of other firms in her book, these tend to be limited and occasional, confined for the most part to passing references in the footnotes.[1] The Hercules paper is unique in that it provides us with a systematic and

---

* This paper draws heavily on Ingram (1998), Anon (1987) and Hercules Inc. company reports for information on the development of this company over recent years. I gratefully acknowledge the help provided by Strathclyde University library staff, especially Elaine Blaxter and Jean Webb in assisting with the University's Company Report Collection. This includes two decades of Hercules annual reports.

[1] An exception is to be found in Chapter 7 where Penrose looks at General Motors and General Mills as examples of diversification strategies. However, the analysis here is mostly limited to evidence culled from the respective company's annual reports. There is no observation of the companies as in the Hercules case.

comprehensive case study of a single firm in terms of the processes that Penrose saw as influencing the growth of firms.

Second, not only does Hercules help to illustrate the arguments of the book, it almost certainly had an influence on the development of these arguments in the first place. The book was published in 1959 (completed June 1958) and the Hercules paper was published in 1960. However the Hercules study had been made possible by a Fellowship which allowed Penrose to spend 6 weeks studying the company from within, in the summer of 1954 'with the full cooperation of all of its personnel' (Penrose, 1960, p. 2). Although the paper was completed in 1956 she made no attempt to bring it up to date for eventual publication in 1960, having been assured by the company that Hercules had not undergone major strategic changes in the intervening period (Penrose, 1960, p. 1). Given the extent to which she had been involved studying this particular company in its role of potential exemplar, it would be strange if it had not strongly coloured her eventual analysis. Yet there is no mention of it in the book.[1]

Third, Hercules is still in operation[2] and has continued to develop and grow as an independent company. Penrose analysed its growth over the first 43 years of its existence (1913–56) and it would be interesting to see what has happened to it over the last 43 years. Does it continue to exemplify and illustrate Penrose's growth theory? Have subsequent developments introduced new and different elements onto this agenda? The history of Hercules provides a direct link back to the origins of Penrose's theory and constitutes an invaluable living laboratory that may assist in probing the robustness of Penrose's contribution.

For these reasons, Hercules is a unique element in the context of the theory of the growth of the firm. After briefly summarising Penrose's overview of the growth of Hercules, we shall look at the Hercules paper through spectacles provided by the analysis of Kay (1997), before considering possible implications of the company's recent history for Penrose's theory.

I. PENROSE AND HERCULES

Penrose begins her account of Hercules by summarising some arguments from her book with relevance to the analysis of Hercules. She points out that the firm is both an administrative organisation and a pool of resources. The expansion of the firm draws on both acquired or inherited resources that must be obtained from the market. The services of the firm's existing management set a fundamental limit on the amount of expansion that can be planned or executed at any point in time.

---

[1] Given that Hercules was originally intended to be such a central element in the book, it is not clear why the company was then excised so thoroughly from it. Certainly the case study derives much of its power from the complex market and technological inter-relationships analysed by Penrose and she may have felt that partial and limited reference to the firm would not do this justice. Also, the style of illustrative examples in the book generally differs from that of the Hercules case. The book examples tend to involve secondary sources while the Hercules case is built around direct observation, interviews and discussions with Hercules personnel.

[2] It changed its name to Hercules Inc in 1968.

She points out that pools of unused productive resources (including entrepreneurial services) create opportunities to innovate, incentives to expand and a source of competitive advantage to the firm. These unused services tend to be related to existing resources and so the kind of activity that a firm diversifies into usually has some relationship to the existing technological and market 'bases' of the firm; 'a firm's productive opportunity is shaped by its ability to use what it already has' (1960, p. 3). She then argues that the history of the Hercules Powder Company illustrates the influence of the technological and market bases of a firm when it attempts to move into new areas which differ markedly from its existing areas of specialisation.

The Hercules Powder Company was formed in 1913 after a US Federal antitrust suit broke E. I. Dupont de Nemours into three parts in 1912. DuPont had been seen as close to monopolising the explosives business and the Hercules Powder Company was one of the two 'new' firms created. At the time of its formation it had 1000 employees and nine plants. It only produced explosives, black powder and dynamite.

Over the next 43 years 'this amputated piece of DuPont, like a cutting from a plant, continued to grow' (Penrose, 1960, p. 4). It grew in different directions from DuPont and the two firms found themselves in direct competition in only a few fields. By the early 1950s, Hercules had introduced a number of new products, explosives accounting for only 18% of sales by 1951. By 1956, Hercules had 11,365 employees, 22 domestic plants and total assets of nearly $170 million. No single group of products accounted for more than 16% of sales. It was 165th on the Fortune 500 list of largest US industrial companies (Penrose, 1960, pp. 4 and 19).

Penrose notes that despite the innovations and expansion in its explosives business, the industry itself was not one that would allow extensive growth and development of the firm, especially with the drastic decline in demand following the end of World War I. In particular, it provided little opportunity for the application of Hercules' experience in organic chemistry and the accumulated unused resources in these areas.

One important technological base was cellulose chemistry. Nitrocellulose is one of the most important basic raw materials in the production of explosives. Soon after World War I, Hercules developed a soluble nitrocellulose product with applications in industries other than explosives, such as lacquer, film and protective coatings. Continual development of new products in this field provided Hercules with an important technological base in cellulose chemistry that enabled it to take advantage of growing markets in artificial fibres and plastics. Over the years, Hercules developed numerous cellulose products and specialities, including a cellulose gum CMC which was found to have applications in products for a wide variety of industries, including foods, lotions, drugs, cosmetics and oil-well drilling.

Naval stores provided a second technological base for expansion opportunities. The term 'naval stores' is slightly misleading, being the old colonial name for the wood products industry that had supplied the British navy. Hercules had created an Industrial Research Department in 1919 to explore new growth opportunities. This department decided that the firm could go into naval stores (rosin, turpentine and pine oil) obtained from the left-over stumps of felled long-leaf southern pines.

Hercules already had made creative use of other waste products and this avenue promised to make use of its expertise in organic chemistry (as well as a use for its dynamite in blasting out the stumps).

The wood naval stores business actually turned out to be disappointing for many years due to an unexpected fall in demand for the main product, wood rosin. Hercules was unable to sell its existing wood rosin in commercial amounts in competition against gum naval stores, and responded by using research to convert it into novel rosin variants for which many new uses could be found. Penrose notes that Naval Stores became the equal of cellulose chemistry as a major technological base of the firm, and in 1928 it became organised as a separate department.[1] The firm continued through the inter-war years to develop new derivative products from rosin and the other joint products of the pine stump.

By 1956 the firm had diversified into a large number of areas. This now included petrochemicals which drew on the company's existing expertise in process technology in a number of contexts, and agricultural chemicals which drew on all three of Hercules major technological fields: explosives (nitrogen chemistry), nitrocellulose (chlorine chemistry) and naval stores (turpene chemistry). Hercules' expansion paths are documented in her paper in the chart reproduced in Figure 1. What is particularly interesting about this is that it anticipates similar graphical representation by Rumelt (1986, pp. 17–19) of what he described as related-linked growth by Carborundum Inc. We shall discuss related-linked growth further below. However, although Rumelt (1986) refers to Penrose's book, there is no reference to the Hercules paper. There was no reason why writers in the field of strategic management should have been aware of Penrose's 1960 paper published in the business history field, especially when Penrose herself did not refer to it in her book. Consequently, it was left to Rumelt to effectively re-discover and represent such growth some years later.[2]

Penrose looked at some specific aspects of Hercules' growth to examine how the company illustrated the type of growth processes she discussed in more detail in her book. We shall look at some of these in the next section.

II. PENROSE AND THE GROWTH OF HERCULES

We shall draw on the mapping approach and arguments developed in Kay (1997) to look briefly at some relevant issues raised by Penrose in her Hercules case. These are

[1] Penrose uses the term 'department' where today we would use 'division'. Hercules became one of the world's first multidivisional (M-form) corporations in 1928 (Anon, 1987). This was quite a remarkable organisational innovation for the company given that it had only been founded 15 years earlier as a highly specialised explosives company. Indeed, DuPont and General Motors had only just developed the M-form structure at the beginning of this decade (Chandler, 1966). Hercules still had strong informal links with DuPont even after its separation from its parent (Ingram, 1998, p. 260) so it is possible that Hercules' early adoption of the M-form structure may have been influenced by this association. It is also probable that such reorganisation helped to facilitate the growth process, although this is not a point taken up by Penrose.

[2] The Rumelt volume referred to here is the second edition, the related-linked strategy was first described in Rumelt's first (1974) edition.

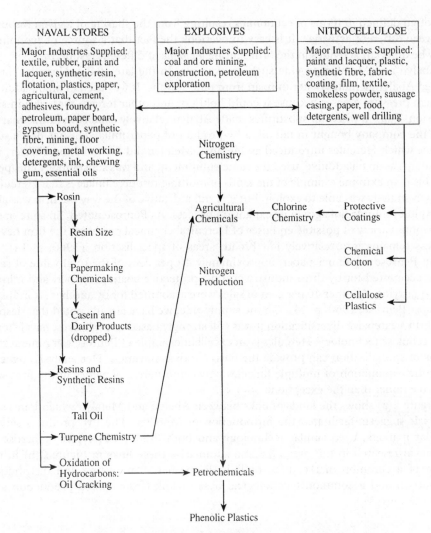

FIGURE 1. The direction of expansion of Hercules as outlined by Penrose.
(Reproduced from Penrose, 1960 with permission.)

intended for illustrative purposes, and are not a substitute for the deeper and more extensive discussion of Hercules carried out by Penrose in her paper.

II.a. Multiple resource linkages

A recurring theme in Penrose's analysis of Hercules was the role of multiple linkages between activities of the firm. Some other approaches such as transaction cost economics tend to take single assets as a basic unit of analysis, and economies of scope are often defined in terms of economies resulting from the sharing of a given resource between different activities. It is easy to overlook the fact that a source of economies may be different activities in the firm sharing a *bundle* of different resources.

Abalyn and Metalyn provide such an example. In the late 1940s, Hercules introduced Abalyn, a methyl ester of rosin from pine stumps. It was used as an oil additive in high pressure greases because it could hold grit and other foreign materials in suspension. Falls in prices of substitutes made Abalyn relatively expensive in this market and the company bought in tall oil, a by-product of paper mills. This yielded a substance which Hercules introduced as Metalyn. Metalyn did the same, or a better, job as Abalyn as an oil additive, used the same equipment, and was substantially cheaper.[1]

This is an extreme example of the kinds of multiple resource linkages that Hercules' growth strategy was able to exploit, but it is still indicative of the way the firm was able to exploit multiple linkages in its expansion process. As Penrose noted, 'in spite of the enormous variety of possible end uses of Hercules' chemical products, the firm nevertheless remains in a relatively few broad 'areas of specialisation' (1960, p. 14). So when Penrose wrote her paper, approximately 40 per cent of the total value of sales were accounted for by three industry groups; protective coatings, paper, and mining and quarrying. Another 40 per cent of sales were accounted for by another six industry groups (Penrose, 1960, p. 14). On the supply side, we have already noted that despite the firm's extensive diversification it was still strongly based in explosives, naval stores and cellulose technology. Hercules is an excellent example of Penrose's argument that areas of specialisation can provide the basis for diversification. Her analysis suggests that the exploitation of multiple linkages between different activities of the firm was the rule rather than the exception.

Figure 2(a) shows the kinds of links between Abalyn and Metalyn which Penrose's analysis suggests facilitated the introduction of Metalyn. The two products served similar markets, used similar technology and both drew on Hercules' expertise in chemical research in this area. We also summarise these links in Figure 2(b) in the form of a common market base or link (from shared marketing and distribution resources) and a common technological base or link (from shared production and R&D resources).

II.b. *Technological bases facilitating diversification*

While multiple linkages helped facilitate diversification in many cases, the linkages were frequently concentrated in the strong technological bases of explosives, wood naval stores, and cellulose technology. Some developments along the technological route led the firm into new market opportunities. An example is given below of soluble nitrocellulose. This led Hercules into the lacquer, film and protective coatings industries (Penrose, 1960, p. 6). These would entail new marketing and distribution bases

---

[1] Abalyn is still produced today by Hercules and is now used in adhesives as a softening, plasticising, and tackifying rosin, and in coatings to contribute gloss and fullness to lacquers.

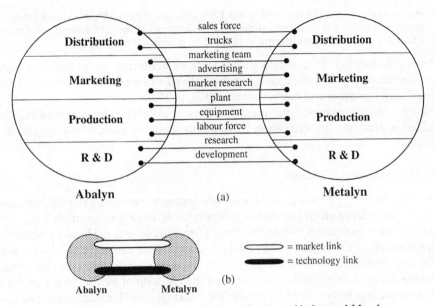

FIGURE 2. Examples of resource linkages between Abalyn and Metalyn.

FIGURE 3. Technology-based diversification.

to be set up, but the expansion was greatly facilitated in such cases by Hercules' technological strengths in nitrocellulose (see Figure 3).

II.c. *Market bases facilitating diversification*

Not only did technological bases provide a route for the exploration of diversification opportunities, so also did the firm's market bases. Penrose pointed out that celluloid[1] was the forerunner of modern plastic materials and was produced by Hercules from the very beginning. The development of cellulose acetate further committed the firm to the plastics industry. Penrose noted that

the various kinds of chemical plastics, which in a broad sense can often be regarded as the same 'product', are made by substantially different chemical processes. Hence the widening of

[1] Celluoid is essentially composed of nitrocellulose and camphor.

Hercules' position in the plastics field stems from different types of chemical technology, the development of which has itself been stimulated by the firm's attempt to maintain and improve its position as a supplier to manufacturers of plastic products. (1960, p. 17).

Penrose noted that the new plastics products were selected on the basis not only of what new markets they may create but 'how they fit in and can be developed along with existing resources and market areas' (1960, p. 17).

Figure 4 illustrates how using manufacturers of plastic products as a market base provided a platform for expansion of different kinds of plastics manufactured by Hercules.

II.d. Multiple sources of linkages

As well as noting the existence of multiple linkages between pairs of businesses, Penrose noted that a given Herculean business could itself have multiple links with a variety of other businesses within the firm. For example, when Hercules opened up petrochemicals as a new base within the firm it was able to exploit links with a number of existing businesses within the firm. The process used by Hercules to produce ammonia for explosives production involved the cracking of natural gas; some of the processes in naval stores could be used in oil cracking; also Hercules developed a new process for making phenol in the course of experimenting with terpene chemistry and naval stores products. These multiple links in production and R&D are illustrated in Figure 5.

In short, the past diversity of Hercules' activities often helped reinforce the resource base that helped to create increased future diversity.

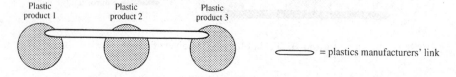

FIGURE 4. Market-based diversification.

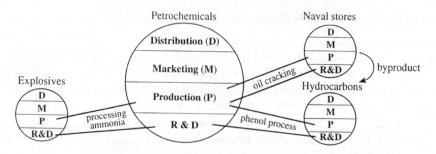

FIGURE 5. Multiple resource linkages and petrochemicals.

II.e. Related-linked expansion

One of the features of Penrose's discussion was what Rumelt (1986) would later independently rediscover and identify as related-linked diversification, in which individual products are at least indirectly linked to most other products of the firm. An example of this is shown in Figure 6 where Hercules' interests in explosives have some links with its paint and lacquer interests through the cellulose chemistry technological link. But these cellulose-based lacquers may also exploit market links in the protective coatings industry with its rosin-based paints and lacquers. In turn, these rosin-based paints and lacquers have technological links with other rosin products, such as chewing gum. Even though they are indirectly linked, dynamite and chewing gum have no strong market or technological links in common.

Penrose traced the development of these links, showing first of all how cellulose chemistry became a base for technological expansion into non-explosives markets, and protective coatings became a market base for further expansion, including rosin-based paints and lacquers (1960, pp. 7, 15–16). In turn, rosin-based technology became a platform for a number of new products, including chewing gum (1960, pp. 11–12, 18). Figure 1 charts the evolution of the major links.

The related-linked strategy is of major importance in the story of post-war diversification, and we shall return to this issue later.

II.f. Response to external threats

Another feature of Penrose's analysis of Hercules is her discussion of how the company dealt with threats to its businesses. One frequent strategy it adopted was to seek growth opportunities away from the threatened sector by expanding along a related market or technological base:

movement into new aspects of cellulose chemistry . . . was an obvious entrepreneurial response to the postwar decline of nitrocellulose markets in the explosives field' (1960, p. 9);

unable to sell rosin in its existing forms in competition with gum, Hercules learned how to modify the product . . . and thus convert it into various kinds of rosins for which many new uses could be found (1960, p. 11).

The threat to particular kinds of business could thus be localised, with Hercules being able to pursue escape routes along various market and technological links. This is

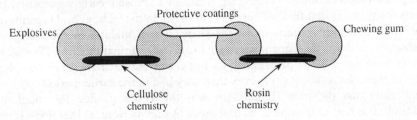

FIGURE 6. Related-linked diversification.

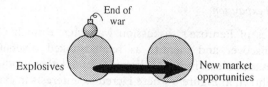

FIGURE 7. Diversifying away from external threats.

shown for Explosives in Figure 7, with the termination of war (the external threat, or 'bomb' in Figure 7) having an obvious adverse effect on this aspect of Hercules' activities.

What is particularly interesting about this is that it shows how firms could themselves respond creatively to the threats posed by the process of creative destruction (Schumpeter, 1947). Just a few years earlier, Schumpeter had written that the

competition from the new commodity, the new technology, the new source of supply, the new type of organisation . . . strikes not at the margins of the profits and the outputs of the existing firms but at their foundations and their very lives (Schumpeter, 1947, p. 84).

Schumpeter tended to see the life and death of technologies and products as naturally implying the life and death of firms; product life-cycles implying firm life-cycles. This was understandable given that even large firms tended to be highly specialised entities even up until the early post-war period. Indeed, Penrose noted in her book that 'until recent times the rise and eventual fall of individual firms in a competitive economy was often treated as almost a law of nature' (1959, p. 153). Although she did not explicitly link her discussion in this context to Schumpeter's earlier analysis, Penrose showed how diversification could enable the firm to separate its own growth and survival prospects from that of its constituent technologies and products.

III. HERCULES AFTER PENROSE

The first 43 years of Hercules' life provided a great deal of material of relevance to Penrose's theory of the growth of the firm, and it would be surprising if the second 43 years did not continue to provide examples of the kinds of growth processes she described. In fact, the company has continued to grow and diversify. Its subsequent history not only continues to illustrate many of Penrose's early arguments, it also reflects many issues that have achieved prominence in the strategy and organisation of large diversified corporations in recent years. These include merger, acquisition, collaborative activity, divestment, strategic focus and multinationalism. Penrose noted some of these issues in her earlier analysis, but each was to achieve greater emphasis at points in Hercules' subsequent history than they had in the earlier period.

One issue that did pass Hercules by was the 1960s passion for conglomerate expansion. It is true that some of its new areas of activity were, at best, loosely related to its existing activities, but it was usually possible to trace some market or technological thread linking the old to the new. At the same time, Hercules' subsequent history raises some aspects relating to the growth of firms that were not so evident in its previous history or in Penrose's analysis, and we discuss some of these below.

First, we shall summarise what happened to Hercules after Penrose's study. Hercules' sales doubled between 1955 and 1963, much of this growth being fuelled by government contracts. In 1959 it diversified into rocket fuels and propulsion systems for Polaris, Minuteman and Honest John missiles. Sales of aerospace equipment and fuels accounted for almost 10% of Hercules' sales in 1961 and 25% in 1963. Hercules derived approximately 25% of its profits during the Vietnam War from rocket fuels, anti-personnel weapons, and chemicals such as Agent Orange and napalm.

However, the mid-1970s were a difficult time for Hercules because of two unrelated problems; the energy crisis hit its petrochemical business (in 1975, 43% of its fixed assets were now in petrochemicals), and problems of over-capacity in the rosin industry led to a major slump in its sales of this product. The late-1970s led to a major rationalisation of the managerial structure with the forced retirement of 700 middle managers and three executive vice-presidents. The number of managerial levels between foreman and chief executive was reduced from 12 to 6. The 1970s also saw the managerial strategy shift from commodity to speciality chemicals. The company began to use global joint ventures to assist in this switch of strategy. R&D was refocused more towards applications and away from the basic research end of the R&D spectrum. Divestment of unprofitable lines was pursued.

In the 1980s the company emphasised growth paths in advanced materials and composites, aerospace instrumentation and communications, fragrance and food ingredients, polypropylene fibres, and high performance structural plastics. Each of these areas still tended to exploit market and/or technological links with established bases within Hercules. By 1990, Hercules had established major bases in (1) absorbent and textile products, (2) paper technology, (3) Aqualon (largely cellulose based), (4) advanced material and systems, (5) fragrance and food ingredients, (6) aerospace, (7) packaging films and (8) rosins. The first three core businesses were performing extremely well, but the last three were giving management cause for concern.

The company was now emphasising selective divestment and focus as well as continuing to pursue growth opportunities where appropriate. It divested its packaging films business in 1994, and in 1995 sold the aerospace division on the basis that is represented an awkward fit with the company's other speciality chemical businesses. In 1995 it also sold its electronics and printing division, and in 1996 it sold its composite products division. At the same time, it confirmed its intention to continue to search for acquisitions technically related to its existing activities in chemicals.

By 1997 the company was now a more clearly focused multinational chemical speciality firm. Its annual sales of $1·9 billion were split equally into what were now its two main areas: chemical specialities, and food and functional products. In turn, both these main areas had two constituent core businesses: paper technology and rosin in the chemical specialities area, and Aqualon and food gums in the food and functional products area. The company's competitors included some of the major corporations in these areas, but interestingly there was little overlap in terms of who Hercules saw as its major competitors across its major products and markets.

It is also interesting to note that Hercules' rate of new product introduction appears to have increased since Penrose's paper, and that the company intends to increase this rate still further in the future. Penrose (1960, p. 20) noted that in 1926 (13 years after the firm's creation) new product lines accounted for 35% of sales, and that by 1952, 40% of sales were attributable to products that had arisen from internal R&D over the previous 22 years. However, by 1996, 17% of sales were from products that had been on the market less than 5 years and the company's goal is to raise this to 30% by the year 2000.

In the previous section we discussed how Hercules helped to illustrate many aspects of relevance to Penrose's theory of the growth of the firm. However, the company's history also helps to illustrate how (with the wisdom of hindsight) Penrose's theory may be regarded as incomplete or unsatisfactory in certain aspects. We are now in a position to discuss how the early history of Hercules may have influenced Penrose's analysis in these respects, and how a study of Hercules' history (both pre- and post-Penrose) may help to illuminate some of these issues.

III.a. Vertical integration

Penrose's approach to vertical integration (1959, pp. 145–9) was constrained by her regarding it as 'a special form of diversification' (1959, p. 145). As such, one of the reasons she gave for it was fuller use of the firm's specialised managerial resources, for example, 'since management itself is frequently specialised, a company may well find that opportunities for using their specialised services are greatest in some form of backward integration'.

This perspective appears puzzling because it is not really apparent in the obvious and archetypal case of vertical integration in the petroleum industry, discussed by her in this section of her book. Instead (as she discusses in this section) backward integration tends to be stimulated by problems of ensuring security, quality, and continuity of supplies. None of this need involve similarity in managerial services, and indeed it is difficult to discern much in the way of managerial similarity between the various stages of extracting, refining, transporting and selling petroleum. Even if some transferability of managerial skills exist, it is difficult to believe that they could exceed those associated with horizontal moves into related products or activities. So while the specific reasons she gives for vertical integration are credible, it is difficult to see them as involving a natural extension of her earlier discussion of the role of existing managerial services as a basis for diversification. Nowadays we would say that motives, such as security, quality and continuity of supplies reflect some form of transaction or co-ordination costs in the market for intermediate products.

Hercules actually limited the extent of its vertical integration and tended to avoid activities that would take it into the final product stage. Instead, it sold to industrial or military customers and generally avoided what could be the very different skills involved in making and selling final products. Also, it was deterred from vertically integrating forwards for strategic reasons, in case its customers became reluctant to openly discuss problems and cooperate with a supplier who was also now a rival (Penrose, 1960, p. 20). When the firm did vertically integrate backwards into supplying its own chemical cotton (1960, p. 9), the reasons given by Penrose were a desire to control quality, a transactional rather than a resource-based reason.

In short, Hercules was limited in the extent to which it was vertically integrated, and this may have influenced the relative neglect of this growth strategy in her book.[1] Nor is there any strong evidence that when the firm did vertically integrate, that it did so in order to transfer the specialised productive services of the firm. This does suggest that Penrose's analysis has difficulties in being applied to vertical moves, and that other perspectives may be required in this context.[2]

III.b. Links as double-edged swords

As we discussed above, Penrose showed how the firm might be able to diversify itself out of trouble if one of its businesses was attacked. She also suggested that

in the long run the profitability, survival and growth of a firm does not depend so much on the efficiency with which it is able to organize the production of even a widely diversified range of products as it does on the ability of the firm to establish one or more wide and relatively impregnable 'bases' from which it can adapt and extend its operations in an uncertain, changing, and competitive world (1959, p. 137).

However, it could be argued that this does not go far enough and that Penrose understated the survival potential of large diversified firms pursuing a related-linked strategy. First, again with the wisdom of hindsight, it is difficult to identify any impregnable market or technological bases that any firm nowadays could feel they could safely defend to the death. Yet that in itself does not necessarily imply the death of the firm, any more than it does the death of an individual business.

Penrose's own examination of the Hercules case and the subsequent history of the company, shows how entire technological and market bases may be attacked by external threats. The company recently reported that competition from a new competitor and a new technology (spunbond technology) was threatening its existing markets based around existing polypropylene fibre. These markets included disposable diapers and adult hygiene products.

Figure 8 shows how a single technological innovation can threaten an existing technological link, the external threat or 'bomb' here being spunbond technology. The link not only provides internal economies, it provides a potential source of vulnerability to external changes that threaten to displace or weaken it. In short, the link can

---

[1] Of course, while vertical integration is an important issue in the setting of the boundaries of the firm, it has limited potential as a growth strategy. There is only so far upstream or downstream that a firm can expand before it hits the walls of basic raw materials or the final goods market.

[2] Transaction cost economics (Williamson, 1985) provides a basis for one such perspective. See Kay (1997) for discussion of transaction cost economics and other approaches to coordination problems in firms.

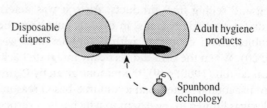

FIGURE 8. Vulnerability of links to external threats.

be a double-edged sword, providing both a source of internal economies and potential vulnerability to external threats such as Schumpeterian processes of creative destruction (Kay, 1997).[1] The external threat can be to technological or market links. For example, the energy crisis of the early 1970s threatened Hercules' technological base in petrochemicals, while the end of World War II clearly had an adverse impact on its market base in defence.

The related-linked strategy has properties that make it particularly well suited to dealing with externally generated shocks (Kay, 1997). Suppose Hercules had been a more specialised firm with its markets based around its existing polypropylene fibre technology. Although the firm would have been diversified into different markets, a single technological innovation (such as spunbond technology) could have threatened all its products and markets simultaneously along this single technological link. The firm could have responded by replacing or upgrading its existing technological base. However, to do so would have needed time and internal or external finance at a period when all of these resources would probably be in short supply.

The related-linked strategy operated by Hercules and displayed in Figure 1 has the valuable property of not only allowing the firm to diversify its *products*; the firm here diversifies its *links*. Consequently, any single external threat is more likely to have a localised impact on the firm's business. The firm's overall viability is less likely to be threatened than would have been the case had diversification been around a single market or technological link. In this respect it offers similar diversification advantages to that provided by the conglomerate or unrelated strategy. However, it has the advantage over the conglomerate that expansion and operations are facilitated by the firm being able to exploit resource linkages between old and new areas of the firm.[2]

This is illustrated in miniature in Figure 6. It is possible that competitors could develop technological or market-based competences that attack particular Herculean

[1] Hercules' reaction to this threat was interesting. It first formed a joint venture in 1997 with Danaklon, a Danish company that was one of the two major competitors to its fibres business in these areas. This allowed it to combine technical and market strengths of the two companies as well as effectively eliminating any competitive threat posed by Danaklon. It then bought out its partner in July 1998 with Hercules purchasing Danaklon's 49 per cent share of the joint venture.

[2] It would seem from this that the related-linked strategy is able to exploit the best of both worlds by being able to enjoy the same diversification benefits as the conglomerate while still being able to obtain high levels of internal economies through market and technological linkages. However, while this strategy has very attractive strategic properties, it is one of the most difficult to deal with from the perspective of design of efficient organisational structure (Kay, 1997, pp. 257–60).

bases, for example it could lose ground in cellulose or rosin technology, or a competitor might begin to attack Hercules' market base in protective coatings. However, these specific threats would be localised to attacks on particular bases and enable Hercules to isolate and limit external threats that could have more serious implications for a more specialised firm. The related-linked strategy demonstrates that *pattern* of diversification (in terms of linkages) may be more important than *degree* of diversification (in terms of numbers of product markets) in the context of corporate survival (Kay, 1997).

III.c. Shifting core

One interesting issue that follows from the previous discussion is that there is no reason why the core of the firm should remain fixed over time. As noted above, Penrose (1959, p. 137) argued that the long-term growth, profitability and survival of the firm depended on its being able to exploit one or more 'relatively impregnable bases' around which it could plan its expansion. In the Hercules case, she identifies these bases as cellulose chemistry, rosin and terpene chemistry and nitrogen chemistry (1960, p. 19), with petrochemicals as an emerging new base (1960, pp. 12–13), and with evidence that other new bases were emerging for expansion possibilities.

In fact, although petrochemicals had become Hercules' dominant base by the mid-1970s, it had liquidated all its petrochemical assets by the mid-1980s. Also, Penrose described explosives as 'the original technological base' of Hercules (1960, p. 6)[1] and it is represented as such in Figure 1. However, the company sold its commercial explosives business in 1985, ending Hercules' participation in the business on which the company was founded.[2]

So relative 'impregnability' of bases has not been fundamental to Hercules' strategy, instead the company's attitude to bases could be characterised better in terms of relative flexibility. It has not shied away from terminating bases such as explosives and petrochemicals if it feels that such a decision is warranted. The company has subsequently demonstrated even more adaptability than was evidenced at the time that Penrose wrote.

III.d. Divestment

The shifting core illustrates one aspect of divestment, a strategic issue that has become of importance in recent years. Again, this was an issue that was not really evident in the Hercules that Penrose studied, although it was to become extremely important to the company later. Penrose did not really pursue the notion of failed diversification very far, rather her emphasis in the Hercules case tended to be diversification as a

---

[1] Although here she appears to be interpreting technological base as referring to a group of products, she later interprets technological base as referring to a particular form of chemistry, such as cellulose chemistry.
[2] The company's decision to exit was based on explosives being a mature business with limited growth potential.

response to failure, rather than a cause of it.[1] While Hercules did not succumb to the fad for conglomerate diversification of the 1960s, nevertheless the company later concluded that it had moved too far into areas that were, at best, loosely or poorly related to the more focused strategy that it wished to pursue. This had some similarities to the problems encountered by the conglomerates in that the company had pursued a much more active acquisition policy than had been the case when Penrose studied it.

If mergers and acquisitions have a place in growth-oriented strategies as Penrose argues, then it might seem a natural corollary that an active divestment policy signals at least a temporary cessation to a firm's interest on growth. However, the case of Hercules indicates that this may be too simplistic an interpretation. First, like many other firms in recent years, Hercules has pursued an active mergers and acquisition policy alongside a selective divestment policy. The net effect has been to narrow the focus of the firm while still allowing it to pursue explicitly growth-oriented objectives. As Penrose might have noted, increased specialisation (of bases) may here facilitate increased diversification (of product-markets). Second, a more focused strategy may itself actively reinforce growth objectives if the result of the firm pursing more strongly related activities is increased profitability. This may enhance access to both internal and external sources of capital, and in turn enable the firm to fund internal and external growth opportunities.

In short, divestment may not necessarily signal an end to interest in growth, but may instead be an integral part of such a strategy.

III.e. Multinational enterprise

An important feature of post-war growth of many large corporations has been multinational expansion, and Hercules has been no exception. However, when Penrose wrote, she documented the company's preoccupation with domestic industrial and military markets. She also summarised the production base of the company as consisting of 22 domestic plants (1960, p. 4). There were no apparent signs that Hercules was going to become a major international player. Yet between 1962 and 1972, sales from foreign production increased sixfold, and by 1978 it was producing in 30 different countries. By 1996, almost half the company's identifiable assets were held outside the US and more than half its sales and profits from operations were attributable to non-US sources. Hercules' major sphere of operation outside the US was Europe. As we might expect from a Penrosian perspective, Hercules' international activity tended to be in areas that emphasised the company's core technological strengths.

Clearly there was no opportunity for Penrose to really observe how her theory could be applied to multinational expansion in the case of Hercules. Modern multinationals do appear to expand internationally along lines that we would expect from Penrose's

---

[1] For example, the firm over invested in extraction of rosin from pine stumps in the 1920s, but Penrose analyses how this mistake stimulated the search for new uses for rosin products.

theory, exploiting technological competences (e.g. IBM, Ford) or competences in marketing (e.g. Proctor and Gamble, Coca Cola) out of strong domestic resource bases (Kay, 1997, pp. 153–76). Yet, curiously, Penrose did look at the evolution and operation of multinational enterprise, and when she did, she tended not to analyse the phenomenon in such terms.

Penrose had in fact already published a paper (1956) which did apply her resource-based approach to the case of multinational enterprise. But rather than consider the resource-based linkages between parent and overseas subsidiary, these considerations were downplayed in the earlier analysis; 'once established a subsidiary has a life of its own, and its growth will continue in response to the development of its own internal resources and the opportunities presented in its new environment' (pp. 225–6). She recognised that the foreign subsidiary will need managerial and financial resources from the parent to set it up, but 'a foreign subsidiary, once it is established, is, with important exceptions, more appropriately treated as a separate firm' (pp. 226–7). Penrose argued that the managerial resources necessary for expansion would now tend to be largely contained within the foreign subsidiary, with the possible exception of finance (p. 226).

This is not a very satisfactory treatment of the perpetuation of multinational enterprise once the subsidiary has been created. It is not clear why the multinational firm continues to hang together if the only glue to keep it together is financial. Furthermore, Penrose (1968) actually later looked in detail at one particular form of multinational, the oil majors. Unfortunately, if there is one form of multinational to which her 1959 theory is least applicable, it is probably the petroleum industry. Multinational activity here is heavily characterised by vertical moves, and as we have seen such expansion does not necessarily sit easily with her resource-based perspective. She makes little reference to her earlier theory in her 1968 work, and makes no systematic attempt to show how it can be applied in this context.

Consequently, it could be argued that Penrose was not very Penrosean in her treatment of multinational growth, but that this was only partly due to lack of opportunity to observe multinational expansion in the Hercules case.

III.f. Collaborative activity

One issue neglected by Penrose was the evolution of collaborative activity in the firm. This was understandable given that it was not a major strategic issue for Hercules or most other firms, even by the early post-war period. Penrose does note that Hercules entered into a collaborative agreement in 1954 with the Alabama By-Products Corporation, setting up the Ketona Chemical Corporation to produce anhydrous ammonia (1960, p. 16). However, she does not explore why the firm chose to enter into a collaborative arrangement in the first place. For example, why did Hercules collaborate with Alabama rather than going it alone though internal expansion or acquisition? What did a joint venture offer that a simple contract did not?

Joint venture differs from other cooperative solutions such as licensing and subcontracting in that it involves the creation of an administrative entity (the joint subsidiary or 'child'). It tends to be appropriate to cases such as R&D and market entry where there remains considerable uncertainty concerning important strategic elements and variables, and provision has to be made to make future decisions relating to these issues. In these respects, it appears to offer similar advantages to full scale internalisation through internal expansion, merger and acquisition. By way of contrast, contractual alternatives such as licensing tend to be more appropriate to cases where technical and market features are more easily specified and incorporated in standard market agreements (Kay, 1997).

This raises an obvious question: why should collaborative arrangements such as joint venture have become important in recent years? More precisely, if they have efficiency advantages over merger and acquisition, why were they not so much in evidence in the early part of the twentieth century?

We can consider this issue by looking at Hercules' subsequent forays into collaborative arrangements. These also help to illustrate points made in Kay (1997, pp. 177–207) about the evolution of joint venture and other forms of collaborative activity in the later post-war period.

In the inter-war period, firms tended to be relatively small, specialised or only modestly diversified. Some firms like Hercules grew through a combination of internal expansion and modest merger/acquisition activity. Penrose reported that only eight small companies had been acquired by Hercules in its lifetime up until then, and collectively these accounted for less than 10 per cent of the company's net worth at the time of writing. Absorption and integration of acquisitions is less of a problem where the acquisitions are small and relatively specialised, and especially if they are in areas related to the firm's present interests. For example, Hercules' first acquisition was in 1914, the Independent Powder Company which operated a dynamite plant in Missouri. Horizontal expansion such as this fitted easily with the firms existing competences and product lines. Where expansion took place organically through internal expansion, the new areas also tended to be limited in scope (at least at first) and related to the firm's existing activities.

Even firms that did pursue major merger or acquisition opportunities still tended to become only modestly diversified as a result. If firms wished to combine forces to exploit major market or technological links, then merger was the obvious method of doing so. Figure 9(a) shows a population of relatively specialised firms that include three pairs of firms that have merged to exploit links. Cooperative options do not pose obvious advantages in this context since they tend to be associated with complex transactional problems and administrative difficulties arising from dual control and ownership.

However, Hercules' expansion in later years led it deeper into new areas and it became an even more complex and diversified system, although still based broadly in industrial chemicals. By 1978 it produced 1000 industrial chemicals with more than 80 plants world-wide. It became a more aggressive acquirer and this reinforced the growth and diversification trends on the part of the company.

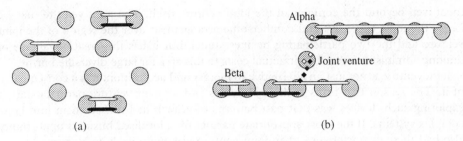

FIGURE 9. Diversification and joint venture.

How other companies grew and diversified over this period is equally important in this context. Rumelt (1986) traces the changing strategies of the Fortune 500 and estimated that 70 per cent could be classified as single or dominant category firms in 1949, the remainder could be categorised as related or unrelated diversifiers.[1] By 1969, the proportions had just about reversed, with 65 per cent categorised as related or unrelated diversifiers and only 35 per cent falling into the specialised or dominant category.

The key to analysing this issue lies in examining not only Hercules, but also the firms with whom it has collaborated. Hercules is no different from most large diversified firms in having entered into a considerable number of joint ventures in recent years. Over the last two decades its partners have included: American Petrofina, BAT International, Boots, Dainippon Ink and Chemicals, Henkel, Montedison, Orbital Sciences, Phillips Petroleum, Shell, Solvay, Sumitomo, Teijin and Thiokol. It has also entered into collaborative arrangements with Du Pont, Fluor Daniel and Rhone-Poulenc. This list includes firms that are: large, diversified, loosely related at best to Hercules' existing activities and foreign. Some of the firms score on all four of these counts.

Now consider the situation facing large diversified firms such as Hercules in the later post-war period. If there are potential opportunities to be exploited from sharing linkages, these will tend to be reflected in resource-sharing between business units, as in Figure 9(b). Here two business units of two separate and highly diversified firms are pooling their expertise in a joint venture, one partner providing market-based resources (the white dotted line) and the other technological resources (the black dotted line).

In a case such as this, the merger of the two participating firms could have created a highly complex and unfocused entity, raising problems of coordination and manage-

---

[1] Single category firms were those estimated by Rumelt to have more than 95 per cent of their revenues from a single business, dominant firms were estimated to have at least 70 per cent revenues in a single business. Related and unrelated firms were both diversified with less than 70 per cent of revenues in a single business, although related diversifiers differed from unrelated diversifiers in having at least 70 per cent of revenues in a group of businesses linked together by common resources or skills (Rumelt, 1986, pp. 29–32). The skills or resources did not have to be the same ones throughout the group, they could be linked or chained as in Hercules' strategy in Figure 1.

ment well beyond the confines of the joint venture itself. The joint venture may be *locally* the most expensive and cumbersome arrangement over the region of the joint venture and the two participating business units, but it has the great advantage of limiting administrative and contractual costs to this area for large diversified firms.

Joint venture is not just an alternative to merger and acquisition, it is a consequence of it. The reason that firms such as Hercules and its joint venture partners started exploring such devices was that past patterns of growth had turned them into large complex systems. If the most appropriate partner for a localised business opportunity also had these characteristics, then joint venture was more likely to be turned to, as Hercules recent history illustrates.[1]

IV. CONCLUSION

In some respects the Hercules paper is like a missing piece of a jigsaw. It represents an important element in the development of Penrose's theory and it almost certainly influenced the ideas and arguments set out in her book. The company has continued to grow and diversify in the years following Penrose's study of it and it effectively constitutes a link right back to the genesis of her theory.

Hercules' history provides numerous illustrations of the principles underlying Penrose's work on the growth of firms. At the same time, this same history helps illuminate areas where the theory is incomplete or is otherwise unsatisfactory. For these reasons, the Hercules paper should be regarded as occupying an important niche in the modern theory of the firm.

REFERENCES

ANON (1987). The first 75 years: the transformation of Hercules (supplement to Hercules Inc. 1987 Annual Report, celebrating Hercules' 75th anniversary).

CHANDLER, A. D. (1966). *Strategy and Structure*, New York, Doubleday.

INGRAM, F. C. (1998). Hercules Inc., in *International Directory of Company Histories*, Vol. 22, 260–3, Detroit, St James Press.

KAY, N. M. (1997). *Pattern in Corporate Evolution*, Oxford, Oxford University Press.

PENROSE, E. T. (1956). Foreign investment and the growth of the firm, *Economic Journal*, Vol. 66, 220–35.

PENROSE, E. T. (1959). *The Theory of the Growth of the Firm*, Oxford, Basil Blackwell.

PENROSE, E. T. (1960). The growth of the firm—a case study: the Hercules Powder Company, *Business History Review*, Vol. 34, 1–23.

PENROSE, E. T. (1968). *The Large International Firm in Developing Countries: the International Petroleum Industry*, London, Allen and Unwin.

RUMELT, R. P. (1986). *Strategy, Structure, and Economic Performance*, 2nd edn, Boston, Harvard Business School.

SCHUMPETER, J. A. (1947). *Capitalism, Socialism and Democracy*, London, Allen and Unwin.

WILLIAMSON, O. E. (1985). *The Economic Institutions of Capitalism*, New York, Free Press.

---

[1] See Kay (1997, pp. 177–227) for discussion of the evolution of various forms of collaborative arrangements.

7

THE GROWTH OF NEW VENTURES: ANALYSIS AFTER PENROSE

ELIZABETH GARNSEY

I. INTRODUCTION

Despite the intense interest generated by new venture activity and the growth of new firms in recent years, the entrepreneurial venture remains an under-theorised area in the economic and management literature. With the emergence of microeconomic theory in the 1930s, academic study of the entrepreneur became a marginal area, arousing little interest even among those working in the tradition of Schumpeter (Nelson, 1995).[1] In economic textbooks, the entrepreneur still receives scant mention (Baretto, 1989).[2] Industrial economists examine largely aggregate cross-sectional data on firm growth and use averaging methods which provide overviews with policy implications, but do not address the internal dynamics of firms (Jovanovic, 1982; Audretsch, 1994; Stanley et al., 1996). In the early management literature on high tech ventures, attributes of successful ventures were ascertained by Roberts, Utterback and others (Utterback et al., 1988; Roberts, 1991). Statistical associations have been examined between new venture characteristics and various independent variables (sector, strategy, etc.), in efforts to further understanding of characteristics of successful firms (Romanelli, 1989; Eisenhardt and Schoonoven, 1990; Keeble, 1998). These studies did not aim to examine the internal dynamics of firm growth. More relevant to understanding internal mechanisms of growth of new firms is the stages of growth, or organisational life cycle literature (Greiner, 1974; Churchill and Lewis, 1983; Quinn and Cameron, 1983; Kazanjian and Drazin, 1989; Churchill, 1997). However, these authors do not use proven tools of analysis nor offer a theory of the firm. They have been criticised as empirically unsound (Bhidé, 2000).

In the field of enterprise studies, the absence of a common discourse and shared body of relevant theory and evidence hampers those who do attempt conceptual thinking. Leading authorities in the field note that entrepreneurship has become: 'a legitimate field of academic inquiry in all respects except one: it lacks a substantial

---

[1] Microeconomic theory shifted in the 1930s to concern with the analysis of price and output by the firm and resource allocation by the market. Evolutionary economics is concerned with firm level behaviour and not with individual actors (Nelson, 1995).

[2] There is still no index entry for 'entrepreneur' in textbooks such as the UK Open University's *Understanding Economic Behaviour* (Simonetti et al., 1998).

theoretical foundation. A major challenge facing entrepreneurship researchers in the 1990s is to develop models and theories built on solid foundations from the social sciences' (Bygrave and Hofer, 1991, p. 13).

There is, however, an established basis on which to develop a theory of new venture formation and growth in the work of Penrose, which despite a lack of direct attention to new venture formation contains a wealth of relevant reasoning and evidence (Penrose, 1955, 1959, 1960). Moreover her approach is unique in providing conceptual links between the internal and external dynamics of firm growth. The accepted designation of Penrose's approach as the resource-base theory of the firm, which is not her terminology, diverts attention from the importance accorded in her work to the market position and external opportunities of the firm. While the perspective offered by Penrose is inevitably dated in certain respects, a conceptual scheme based on her work remains relevant, and can be used to throw light on the most recent kinds of new firm formation, including the prospects for Internet ventures.

It has been shown that an enriched Penrosean account of the firm viewed as 'an historically and path-dependent' system of activity has much to offer the study of business strategy (Spender, 1996, p. 59). Her work is equally relevant to creating a theoretically grounded research agenda for the study of new venture growth. This chapter outlines key concepts used by Penrose of special relevance to the study of new ventures. An account of new venture growth is proposed, derived from Penrose's theory of growth in an established firm, and informed by evidence on new venture growth.[1] This account is used to classify and characterise contemporary new venture growth paths and to explain growth mechanisms. In the final part of the chapter, the strength of Penrose's approach is attributed to her methodology and to her coherent analysis of the internal drivers of firm growth and the external conditions for growth.

II. PENROSE'S CONCERNS

Edith Penrose claimed that she knew little about new firms.[2] Her theory of the growth of the firm was based on studies of firms already of some size. She referred only very occasionally to the new venture in the *Theory of the Growth of the Firm*. She pointed to the 'extreme case of the prospective new firm with no resources at all other than the entrepreneur himself and what capital he can raise' (Penrose, 1959, p. 82). But her acute understanding of entrepreneurial outlook and behaviour is highly relevant to the new venture. She recognised that the entrepreneur's skills and contacts are critical: they make possible the recognition of potential resources and the detection of opportunities which together allow of productive activity. This process is shaped by the entrepreneur's perceptions of the environment: 'the environment is treated in the first instances as an "image" in the entrepreneur's mind of the possibilities and restrictions

---

[1] The author has studied over 50 Cambridge high tech firms and drawn on case projects and published case materials (Garnsey, 1998).

[2] Keynote address by Professor Penrose, Stockholm World Conference of the International Council on Small Business, June 1996.

with which he is confronted, for it is after all, such an image which in fact determines a man's behaviour; whether experience later confirms expectations is another story' (Penrose, 1959, p. 5).

Opportunities and restrictions will therefore be viewed in a variety of ways by different entrepreneurs.[1] Penrose identified linkages between constraints as objectively measured and as perceived by entrepreneurs. These have not been recognised in more recent literature on constraints on firm growth (A. Hughes, 1998; Keeble, 1998). She saw that entrepreneurs build their awareness of objective constraints in their environment into their subjective assessments of opportunity, with objective consequences (Penrose, 1959, p. 217; Garnsey, 1995).

For Penrose, it seems, the essence of enterprise was the ability to detect and connect internal resources and external opportunities.[2] In a firm with an entrepreneurial outlook, there is a continual interplay between the 'market opportunities of the firm and the productive services available from its own resources' (Penrose, 1960, p. 14).[3] Successful enterprise involves the creation of a base of internal resources from which further opportunities can be pursued (Penrose, 1959, p. 217).

A firm's internal resources include its members' problem-solving skills and experience. This raises the following question. What problems have to be overcome initially if an entrepreneur is to set up a new venture to engage in productive activity? What further problems must be solved if the venture is to become viable? These questions underlie the account of new venture growth proposed here.

III. IS NEW VENTURE DEVELOPMENT SEQUENTIAL?

A concern in the literature is whether the early development of new ventures has a sequential character. Viewing new firms' development over a ten year span, Bhidé concludes that 'ventures evolve in unpredictable, idiosyncratic ways' (Bhidé, 2000, p. 245). He rejects the notion that firm development shows any 'recurring temporal sequence' (Bhidé, 2000, p. 247).

But though there is infinite diversity among instances of firm growth in the longer term, it is argued here that processes of growth in new ventures have sequential features. All phase models raise certain problems (Utterback, 1995). The length of the phases cannot always be precisely specified and phases may overlap, recur or be bypassed.[4] This does not invalidate these models if they can illuminate mechanisms of growth which are manifest in phases with variable characteristics. But it is important

---

[1] 'It is clear that opportunity will be restricted to the extent to which a firm does not see opportunities for expansion' (Penrose, 1959, p. 31).

[2] Penrose states that enterprise is a psychological predisposition to take a chance in the hope of gain and to commit effort and resources to speculative activity (Penrose, 1959, p. 35). But the enterprise described in her book is more concerned with the recognition of opportunities, the creation of resources and ways of matching resources and opportunities to build a base for further growth.

[3] There is more in common between this perspective and the SWOT methodology than is usually recognised (Barney, 1996).

[4] This is acknowledged by Utterback, for example, in relation to product life cycles (Utterback, 1995).

that sequencing built into a model of growth have a logic and rationale that can allow consistent judgements to be made in comparing evidence from diverse cases. If this can be achieved, empirical variations in the degree of fit of actual cases with a composite account could be revealing of the early growth process. An outline account of new venture growth follows, drawing on Penrose (Garnsey, 1998).[1] The focus is on the nature of early problems, the learning involved in solving them and the further problems early solutions create. The aim is to provide a basis for comparing actual cases with the composite account to illuminate the causes of heterogeneity in the growth paths of new firms.

IV. DERIVING AN ACCOUNT OF NEW VENTURE GROWTH FROM PENROSE

In a Penrosean perspective, the route to growth of a new venture in pursuit of a certain opportunity set is shaped by the resource *requirements* of the activity undertaken and the resources the founders can muster. The vision and ambition of the entrepreneurs determine the scope and extent of their growth objectives; their acumen and luck determine whether they can build the resources required to realise opportunities for growth. In what follows we begin with the analysis of the pattern of firm growth characterised by Penrose as internal growth, opposed to external growth through some kind of merger. Penrose was interested primarily in this kind of internal dynamics. However, she recognised that certain firms obtain financial resources which allow them to acquire other firms and grow by assimilating the resource generating capability of their teams. She recognised that 'the longer the time required for a given internal expansion, the greater the cost saving if acquisition of already going concerns can be effected' (Penrose, 1959, p. 167). This time factor, together with time-dependent aspects of technological leadership, may be sufficient motivation to investors to make possible a strategy of expansion through acquisition by 'aggressive newer firms' (p. 180). We will turn to 'external growth' of this kind when examining firms' growth paths.

V. DOMINANT PROBLEMS

V.a. Preparation

The entrepreneur cannot organise production until 'the particular productive activities to be undertaken by such a firm [are] chosen from among the alternatives suitable to the abilities, finance and preferences of the entrepreneur' (Penrose, 1959, p. 82). This implies a process of prospecting or 'search', when entrepreneurs are seeking out opportunities and resources for a viable line of business. We can deduce that this may occur in some prior activity by the entrepreneurs, before the new firm is started. Some

---

[1] In this new venture growth theory, the firm is viewed as an administrative unit which functions as a system of activity open to a wider industrial ecology. Input resources are converted into internal resources which produce outputs. It is an enacted system, characterised not only by material and financial flows, but by culture and relations of power.

entrepreneurs see the opportunity to create demand before a market is in evidence. The recognition of what will catch on, how to make it catch on and a sense of timing, are among the entrepreneurial skills Penrose identified (Penrose, 1971, p. 39). The entrepreneurs' experience and knowledge will to a large extent determine what they 'see' in the external world (Penrose, 1959, p. 80). Resources may only be recognised in the light of the opportunities they open up. Opportunities may only be recognised because the entrepreneur is able to see the potential in some attainable resource. What is 'seen' depends both on prior knowledge and on the firm's interactions with others in its entourage.

In selecting the area of activity, entrepreneurs are also selecting the industrial environment within which the new venture must operate. They are setting themselves the problem of how to mobilise and generate the resources required to realise opportunities there.

V.b. *Mobilising resources for productive activity*

Between the initial prospecting that inspires choice of activity and the ability to perform productive services as an independent operation, a period of learning must take place. Resources must be secured and the firm organised internally for productive activity. This process can be termed 'resource mobilisation'. Penrose's much-cited lines are relevant to the new venture:

> The services yielded by resources are a function of the way in which they are used—exactly the same resource when used for different purposes or in different ways and in combination with different types or amounts of other resources provides a different service or set of services. . . . resources consist of a bundle of potential services and can for the most part be defined independently of their use, while services cannot be so defined, the very word service implying a function, *an activity* (Penrose, 1959, p. 25, italics added).

No entrepreneur can buy in instant productivity and output in a new venture. Resources must be obtained and skilfully combined to give rise to productive activity, and this requires a period of intensive collective learning. We could deduce, however, that in an industrial district in which experience of similar start-up problems is widespread, this process can be accelerated by common local knowledge.

Resource mobilisation can only be effective if internal resources are thereby created. 'Operations must often proceed in a certain sequence and many services required in expansion must be internally produced' (Penrose, 1971, p. 34). Penrose speaks of resources released for growth once planning and set-up requirements have been met for a phase of expansion: 'once expansion has been completed and the operation of the firm fully geared to the new level of activity, additional productive services will be both created and released to provide the basis for further expansion' (Penrose, 1971, p. 37).

Penrose is referring to a completed planning and production cycle. The initial cycle has a formative impact. Before any production cycles have been completed, everything has to be done for the first time. When the major effort of planning and setting up production for the first time is over, the firm has experience to learn from. Routines can form and learning can be embedded in procedures. The scale on which resources are mobilised initially to achieve future returns shapes further developments. The learning that has been achieved sparks recognition of further opportunities and appreciation of the value of potential resources.

Product life cycles have shortened since the time of Penrose, and new ventures may aim to set in motion a series of product launches, delaying profitability.

V.c. Early resource generation; gaining and sustaining returns

In a market economy, a new firm must solve the problems required for a resource generation process that will pay for inputs and provide an acceptable margin of profit. Setting up productive activity demonstrates the ability to mobilise and deploy resources. Then returns must be generated and sustained through repeated cycles of production or service provision. From the time when productive activity that yields returns is under way to the time when levels of profitability assure self-sufficiency, major problems centre around securing the viability of the enterprise. Where capital is obtained externally, the promise of viability must be made credible. In principle, the transition from preparation for production to earning returns on past activity is identifiable in the break-even point (Hisrich and Peters, 1989, p. 161), which marks early capacity for resource generation.[1]

Penrose assumes that for a growing firm in a market economy: 'Profits would be desired for the sake of the firm itself and in order to make more profit through expansion' (Penrose, 1959, p. 29). The profitability of the enterprise remains significant even though objectives other than the pursuit of profit are often important. These include: 'power, prestige, public approval, or the mere love of the game—it need only be recognised that the attainment of these ends more often than not is associated directly with the ability to make profits' (Penrose, 1959, p. 30).

VI. EARLY GROWTH

In brief, Penrose's analysis suggests a sequential character to early growth because entrepreneurs must (i) search for matching opportunities and resources and select their project, (ii) secure resources and set up production and (iii) sustain production to ensure viability through scale or scope of activity. These requirements have a step-like quality. The skills needed are developed sequentially. Key problems posed by these requirements must have been solved before viability can be achieved. This gives particular intensity to the problems manifest at this time, when survival is at stake. The problems persist as strategic issues in relation to changes in mission or environment. The firm must sustain the competence to solve problems of strategic direction and business viability.

---

[1] At break-even point, revenues equal variable costs plus fixed costs and profit is at zero.

Phases associated with dominant problems, their duration and the extent to which they are distinct, are important features of the early experience of firms. Variations are revealing. The processes of preparation, initial resource mobilisation and resource generation are analytically separable but overlap in practice to varying degrees. Firms that can engage in parallel problem solving may be able to compress early phases and reach the market sooner. A heuristic model can call attention to these variations and the need to explore their causes and consequences.

Sources of diversity persist. Among these, the personality and preferences of the entrepreneurs have a significant and singular impact. The nature and sector of activity are critical sources of variation between companies. The unique linkages each firm has with others shape the co-evolution of firm and industry.

VII. COORDINATION, SYCHRONISATION AND GROWTH PROCESSES

Sequential similarities are found mainly in the initial phases of growth. Once the firm has developed sufficient internal resources to provide a basis for viability, many contingencies affect further development. But even beyond these initial phases, Penrose identified a mechanism to which recurrent patterns in firm growth can be attributed: attempts at coordination of resources and the mismatch between actual and potential productive activities can provide an impetus to growth. She showed that managers' attempts to improve the synchronisation of productive activities are among the self-reinforcing growth mechanisms in the firm. Tolerance of resource mismatch can result in self-limiting growth. The working through of these forces is manifest in growth patterns examined below.

VIII. THE PLATEAU

For Penrose, endogenous growth in the firm arises because there is opportunity cost in avoiding growth. Where entrepreneurs fail to detect the opportunity costs of avoiding growth, or view these costs as acceptable, growth may tail off.[1] Their lack of vision or limited horizons keep them from reckoning the possibilities they could pursue:

it often happens that the horizon of management is extremely limited, particularly in smaller firms. Content with doing a good job in his own field, the small entrepreneur may never even consider the wider possibilities that would lie within his reach if only he raised his head to see them. If occasionally he gets a glimpse of them, he may lack the daring or the ambition to reach for them . . . Whether such a decision is taken at all is often a matter of temperament (Penrose, 1971, p. 39).

Penrose emphasised that there may be no will to grow once a firm had established a position of reasonable security, so that 'expansion was neither pressing nor particularly obvious [the firm might] continue in its existing course', avoiding the effort and

---

[1] A sigmoid growth curve is found in many complex systems (Richardson, 1991).

resource commitment entailed in even investigating possibilities for growth (Penrose, 1959, p. 33). The statistics show that stability following early growth is a common pattern among small firms (Kirchhoff, 1994; D. J. Storey, 1994). Opportunities for expansion may be forgone from lack of vision, lack of ability or through risk aversion, as where a niche ceases to grow and prospects for diversification are uncertain.

IX. FURTHER GROWTH: INCENTIVES AND CONSTRAINTS

Some firms, however, have a general entrepreneurial bias in favour of growth and the opportunity costs of failing to grow create strong inducements to growth. The nature of these incentives is a central theme of Penrose's book. 'As management tries to make the best use of the resources available, a truly dynamic interacting process occurs which encourages continuous growth but limits the rate of growth' (Penrose, 1995, p. 5).

Underloading and overloading of resources occur because resource inputs are only available in uneven multiples—the capacities of new machines and new recruits do not precisely match needs. Consequently there will be resource shortages and surpluses resulting from earlier activity. This points to a serious coordination problem. But Penrose saw that this problem and the tensions it creates could stimulate growth:

Many of the productive services created through an increase in knowledge that occurs as a result of experience gained in the operation of the firm as time passes will remain unused if the firm fails to expand. Thus they provide an internal inducement to expansion as well as new possibilities for it (Penrose, 1959, p. 54).

Knowledge by individuals and teams in the firm often moves ahead of current productive activity (Grant, 1996). There is opportunity cost in not pursuing possibilities for expansion based on new knowledge. For the entrepreneurial firm, this creates incentives to exploit unused resources, including knowledge, through further growth. This opens up further opportunities in the environment (Penrose, 1959, p. 58).

The continual adjustment between internal resources and evolving external opportunities in entrepreneurial firms would make likely the prevalence of 'improvisation' and 'adaptation' strategies followed by entrepreneurial managers (Brown and Eisenhardt, 1998, p. 33; Bhidé, 2000, p. 61).

X. THE PRESSURES OF GROWTH

However, managers of an entrepreneurial firm intent on growth face endemic problems. The rate at which expansion can be achieved is limited by the co-ordination requirements of planning and resource provision. Effective planning calls for teamwork: 'planning requires the co-operation of many individuals and this requires knowledge of each other' (Penrose, 1959, p. 47). Resources essential to growth have to be secured and cannot be rendered productive until they have been internalised. The mismatch between available resources and required resources 'limits the amount of expansion that can be undertaken at any given time' (Penrose, 1971, p. 31).

The principal constraints Penrose identifies lie in managerial services. Teamwork is essential, but so is executive authority:

The very nature of a firm as an administrative and planning organization requires that the existing responsible officials of the firm at least know and approve, even if they do not in detail control, all aspects of the plans and operations of the firm . . . The capacities of the existing managerial personnel of the firm necessarily set a limit to the expansion of that firm in any given period of time, for it is self-evident that such management cannot be hired in the marketplace (Penrose, 1959, p. 45).

If we apply these insights to the case of the rapidly growing new firm, we can foresee perverse effects of rapid growth. The growing firm must draw in new resources to support growth, but it faces planning delays and coordination problems because it is impossible to synchronise resources precisely in a dynamic system. Internal co-ordination needs set a brake on the rate at which market opportunities can be pursued. 'the rate of growth is retarded by the need for developing new bases and by the difficulties of expanding as a co-ordinated unit' (Penrose, 1971, p. 41).

If constraints on decision-making by top management limit the firm's capacity for response, rapid growth may have untoward consequences in new firms growing rapidly, but still lacking reserves. The rate at which new resources are effectively mobilised may well be insufficient to keep up with the pressures of growth. Growth may consequently stall. Given the requirements of resource synchronisation which Penrose so clearly set out, when growth stalls, key resources may be unobtainable and consequent bottlenecks may actually move growth into reverse.

The cash flow crisis endemic among growing firms is a demonstration of precisely this type of syndrome (Garnsey, 1998). Penrose was concerned with firms that had built up sufficient reserves to offset crises, and did not examine situations of this kind. But if we apply her analysis to the vulnerable new firm, we see that financial crises resulting from critical resource shortages are a likely outcome of uncontrolled early growth.

XI. MAKING SENSE OF THE STATISTICAL EVIDENCE

An analysis of this kind, grounded in the understanding of the firm provided by Penrose, is consistent with the statistical evidence on firm growth and success indicators. Detailed panel evidence needed on growth over time of cohorts of firms is still unavailable and comparisons of different data sets raise problems of consistency. Tracking and interpreting data on firm foundation, growth and closure are fraught with difficulty. Firms may change their name, activity, location and ownership.[1] But persistent patterns are in evidence. A stylised account of aggregate patterns of firm growth can be drawn from previous studies: if a cohort of 100 firms with certain common characteristics, founded in a given year, is followed through, within five or six

[1] More work is needed on 'life after death' for small enterprises to trace how the knowledge assets of firms are sustained and put to use over time even when statistical data indicate termination (Reynolds and Miller, 1988; Kirchhoff, 1994, p. 160).

years around half of those will have ceased to trade as continuous entities. Of those that survive the first five years, about half plateau and remain small, not all achieving an economic return on assets, but providing a lifestyle for their owners. The other half of the five-year-plus survivors continue for a time to grow, but in most of these, too, growth falters, reverting to the average growth rate of the industry, while many experience closure, after dramatic growth reversal or slow decline. After ten years, under 5% of the cohort have become dominant firms, providing over half of the jobs on offer by the entire cohort.[1] This is the pattern of firm growth depicted in studies by Reynolds and Miller (1988), Audretsch (1994), Kirchhoff (1994), Baldwin (1995) and Stanley et al. (1996) in the USA; and in D. J. Storey (1994), Harrison and Taylor (1996) and J. Storey et al. (1987) among others in the UK.

We can interpret these trends in relation to growth mechanisms discussed above. The difficulties of reaching self-sufficiency are to be seen in the high level of early closures. The category of firms that plateau is large because it includes those that have reached self-sufficiency and do not seek further growth, those struggling to remain viable and those lacking the capacity for further growth. The category of firms that grow and then decline is interesting in view of Penrose's analysis of the firm with the bias towards growth. UK statistical evidence shows that among firms that achieve early growth, those that sustain growth and those that subsequently fail earlier shared similar 'success' characteristics (D. J. Storey, 1994). This suggests that the pressures of growth take a toll even of promising new ventures.[2] A minority of firms become industry leaders through the operation of self-reinforcing growth processes.

New ventures are often referred to as an undifferentiated category, when in practice they are highly heterogeneous. We can use a Penrose-inspired analysis of growth processes to distinguish between some common growth patterns of new ventures. The aim is to clarify and explain differences between new ventures. This typology, like the heuristic growth model on which it is based, is not presented as empirical validation, but as enrichment of a theory which is still at the stage of an implied research agenda. Attributes of successful firms are now quite well attested. They include factors such as: motivation of founder, founding team rather than solo entrepreneur, education

---

[1] Statistical evidence shows that a large proportion of firms do not survive as identifiable units beyond their first few years, and that only a small proportion achieve significant growth (Audretsch, 1994; Kirchhoff, 1994; Storey, 1994; Baldwin, 1995). In Kirchhoff's comprehensive study of new US firms, it was the 4% of highest growth firms formed in 1977–8 that created 74% of employment growth in the whole cohort of firms six years later (Kirchhoff, 1994, p. 186). UK studies have found that under 5% of a cohort of new firms provided over 50% of all jobs in those firms ten years later (J. Storey et al., 1987, p. 152; Cosh and Hughes, 1996, p. 11).

[2] An analysis based on a large US data set indicates that the chances of achieving high growth are greater among highly innovative (mainly high tech) firms. But among these highly innovative firms, those that fail to grow sufficiently are more likely to close than those that achieve growth (Kirchhoff, 1994). This implies that an innovative firm is more likely to find a promising resource base and market position, but that those firms in this group that fail to sustain growth are likely to run into difficulties leading to closure. We suggest that failure to sustain growth may not simply be the result of resource constraints (Kirchhoff, 1994), but of growth at a rate too rapid to be sustained in relation to the resources available to the firm. Confirmation requires more detailed data, but detailed case study evidence, on which the following analysis is based, shows how this can occur.

and previous experience of founder, willingness to share equity, clear marketing plan and position, managerial team with skill mix (Utterback *et al.*, 1988; D. J. Storey, 1994). Though the reasons for the association between enterprise performance and indicators of this kind should be explored further through detailed evidence, they cannot in themselves explain growth dynamics, or why some enterprises with success characteristics fail while others succeed.

XII. PATTERNS IN NEW VENTURE GROWTH

It is not contested that new firms grow in 'unpredictable, idiosyncratic ways' as Bhidé argues, and many managerial life cycle models do not capture the extent of variation in their paths (Bhidé, 2000). However, as Penrose has shown, common processes underlie diverse, path-dependent manifestations of growth. As a brief illustration of the current relevance of an approach to new venture growth derived from Penrose, we examine patterns of new venture growth in high tech industries.[1] These ventures did not exist in their present form when Penrose was writing on the firm, but she was aware of market and technological uncertainties affecting technology-based firms (Penrose, 1960). The theory has been used to provide a unifying framework to interpret the diverse case study evidence that underlies this typology.

Many of the studies in the literature use number of employees as a growth indicator. Any one measure can be misleading; relevant growth indicators include sales, profits and asset value. The various growth indicators can move out of alignment if inputs like staff numbers grow rapidly before internal resources have been created and outputs generated. In Garnsey (1998), asset value was proposed as a measure of firm growth. Market valuations of unproven ventures reflect future prospects rather than current assets. Consistent valuations are difficult to establish, notoriously in the case of Internet ventures. Multidimensional measures can provide a basis for depicting firms' relative strengths and weaknesses (Bell and McNamara, 1991).

The types of early growth outlined below are not exhaustive, but describe a range of experience found among high tech ventures, active in new technologies and markets.

XII.a. Prospecting

Some preparatory work is universal, whether before or after the firm's legal incorporation. Whether entrepreneurs adopt a shorter term or longer term planning horizon from the start is related to their area of activity and perspective. A capital intensive start-up calling for sunk investment leaves less scope for flexibility and improvisation.

Our first category of technology-based ventures are those that engage in lengthy prospecting activities. These include the activities of inventor-entrepreneurs who do not reach beyond the preparatory phase. They are unable to turn their inventions into

---

[1] High tech industries are made up of knowledge-intensive firms producing emerging and newly diffusing technologies.

innovations. Ernst Diesel threw himself into the English Channel in frustration at the failure of the diesel engine to attract financial resources (Latour, 1987). Other inventors provide less dramatic examples of this group of would-be entrepreneurs. They may be unable to identify what business resources they need, much less to mobilise these in order to realise an opportunity.

There is increasing policy interest in technology transfer by scientist-entrepreneurs, who may be aware of the potential of their knowledge to become a commercial resource, but do not undertake realistic opportunity detection or know how to protect their intellectual property. Even the selection of a specific commercial project is often difficult for science-based start-ups aiming to exploit a technology with many possible applications. They benefit from a culture and facilities attuned to realising the potential uses of scientific research.[1] Ventures with a lengthy preparatory or search phase, uncertain over a long period as to the direction the project should take, may move rapidly into effective resource mobilisation when an entrepreneurial manager with business experience and the capacity to align resources and opportunities is put in charge and provides a sense of direction.[2]

In contrast, there are firms that go through an accelerated preparatory phase. A prospecting process may be carried out by entrepreneurs as part of a previous job, making it possible to move into resource mobilisation with a minimal preparatory phase after the firm has been incorporated. When the work involves consulting for clients or commissioned R&D, for example, entrepreneurs may shift to working on an independent basis, carrying clients with them. In a 'soft' start-up of this kind, the venture may be largely continuous with prior activity.

Alternatively, little preparation may be undertaken. Some early Internet firms speeded through the search and select phase, hitting upon an idea that enabled them to build on the timeliness of their business concept as the IT infrastructure reached a state of critical maturity and entry opportunity. The role of chance in propitious timing is considerable. However, timing of entry must be followed up by effective resource mobilisation and revenues to sustain success.

XII.b. Early resource mobilisation

A minority of firms not only resolve initial search problems early, but move very rapidly into growth, sometimes mobilising resources for productive activity even as final preparations for start-up are under way. In the 'soft' start-up, it is above all the entrepreneurs' time, contacts and expertise that are used to bring in revenues for consulting, commissioned R&D, software or other services. Resource mobilisation can be

---

[1] There is opposition to pressing premature commercialisation on scientists who could be creating more value by advancing knowledge than by attempting to reap short-term commercial returns before research has fulfilled its potential (or even reached a point where it can be effectively applied and protected). Here further investment in public domain knowledge could provide the training and competence needed for more fruitful future enterprise.

[2] Edison's reliance on an entrepreneurial manager for the take-off of electric lighting (T. Hughes, 1979) is echoed by experience in such academic spin-out companies as Cambridge Display Technology and Cambridge Antibody Technology.

speeded up in places geared to new business creation; some Silicon Valley firms are able to set up a production facility within months, by drawing on local facilities and using share options to cover costs.[1]

In many cases, rapid mobilisation takes place in firms with some prior existence in a project or unit in another organisation, in which the preparatory phase and resource mobilisation had already taken place. The project may have been incorporated as a 'company' with the capacity for generating resources already in place. Software firms built on a supervised academic software project with corporate sponsorship provide examples. Low entry barriers and few infrastructural costs, or access to a valuable and scarce resource, reduce delays.

Firms that spin out of another company with an experienced team in place may appear to omit a period of prospecting and resource mobilisation.[2] But these problems have been solved within the parent organisation. These are not what Penrose refers to as firms 'with no resources at all' but are more like the case of Hercules Powder, de-merged from Dupont (Penrose, 1960). One firm may be the source of many new spin-outs, where the inherited resource is expertise in a valuable technology with many applications.[3]

When firms incubated within a parent organisation are protected from the considerable risks of the preparatory and mobilisation phases, they are less subject to infant mortality. They must be accorded autonomy as spin-outs, however. Over-involvement by the corporate parent in the early life of a venture can make it hard for corporate ventures to achieve sustained self-sufficiency (Christensen, 1997). The impetus to flexibility and improvisation, which up-date the alignment of resources and opportunities, appears to be impaired in such cases.

For those with intellectual property in a valuable technology, a growth strategy minimising early resource requirements (other than legal protection of intellectual property) is to license the technology to other firms and use royalties as a basis for revenues. However, the technology must provide a considerable incentive to purchasers if sufficient revenue is to be generated. Many technologies require substantial resource expenditure on development to reach the point where licensing is possible, consequently early licensing often implies incubation in another organisation. Defence of intellectual property may pose considerable problems if early licensing is pursued.

XII.c. Slow resource mobilisation

Some ventures are unable to engage in productive activity until they have secured the required resource combination. This group is heterogeneous, since infrastructural, R&D and manufacturing requirements can all slow down the mobilisation of resources required for output, while resource requirements for marketing can delay access to

---

[1] Silicon Valley offers examples such as the medical instrumentation firm Fem RX, in production within nine months of foundation.

[2] For example, Domino Printing Sciences and the phone software company, Vocalis.

[3] For example, Fairchild in Silicon Valley and Cambridge Consultants.

customers and returns. These ventures are unable to expand employee numbers or otherwise grow in terms of measures of resource mobilisation.

Firms that cannot take off until they obtain funds to develop an infrastructure (e.g., in telecommunications) have to find funders with sufficiently patient capital. So too do firms founded to exploit a technology that still has to go through a long development process. This is the case with biotechnology start-ups which require venture capital or transfer income from pharmaceutical companies to fund development work over a long period while drug design and clinical trials are under way. In sectors like semiconductor fabrication, facilities are very costly to set up. These firms require extensive funding to enable them to start building internal resources capable of generating sales. Beyond a certain point of capital intensity, entrepreneurs are deterred by entry barriers and entrepreneurial ventures are rare in these sectors (Porter, 1980).

The firms in this group usually struggle to obtain initial funding. Penrose accorded great importance to entrepreneurs' perceptions of the world. How entrepreneurs are perceived by the world is equally important. This category of firms is likely to fail if resource mobilisation extends well beyond the expected period. Investment climate can make a considerable difference to their prospects. Some among this group of firms make a considerable success of growth once they are sufficiently well viewed to receive investment funding.

XII.d. Plateau firms

A fourth group of firms consists of those that mobilise enough resources to achieve output and revenues, but do not sustain their expansion, even where they originally had growth ambitions. Growth proves too difficult, or entails costs greater than personal returns to the entrepreneurs. The statistics indicate that is a very common pattern (Kirchhoff, 1994). In some cases the plateau does not represent stalled growth, but the achievement of the entrepreneur's goals for a lifestyle company (D. J. Storey, 1994). What was earlier an innovative enterprise run by an entrepreneur seeking new opportunities may become a cautious small business with an owner intent on preserving what has been achieved. The average size at which the firms in a sector plateau is important for the economy, because a firm may require critical mass to support ancillary and supplier firms in industrial clusters (Garnsey and Alford, 1996). Plateau firms may experience renewed vitality with changes in circumstance, or undergo a decline in size. Acquisition may provide resources for further growth, albeit no longer on an independent basis.

XII.e. Self-reinforcing growth with early profitability

A fifth group of firms achieves sales and profits early on. Resource mobilisation has taken place elsewhere, or resource requirements are not costly, and a customer base is rapidly established. Licensing may provide early revenues without the delays of committing resources to manufacturing. A generic technology with many applications can enable the firm to license into multiple markets which can deliver strong revenue growth in successful cases.[1] The licensing route to growth needs fewer internal resources than manufacturing, but requires a compelling technology and developed networking capability to establish partnerships with licensees.[2]

Most firms in the early profitability group start off with significant growth ambitions, though occasionally the extent of success comes as a surprise. These firms are viewed as good investment prospects. They are attractive targets for acquisition and candidates for a successful stock market launch.

In addition to the internal pressures for growth identified by Penrose, there are external pressures as funders attempt to realise their investment and insist that managers aim for growth. Exit objectives by fund managers may be imposed on them by their own investors. Customers whose demand provided for initial growth may be pressing for more products or demanding further services more quickly than these can be made available under current arrangements. Distributors may threaten to turn to competitor products unless their demands are met. Pressures of this kind may prevent stability from being an option. Among firms in an innovative industry, growth may be a requirement for survival (Kirchhoff, 1994, p. 184). Without growth of market share, competition in emerging industries may erode the firm's viability.

Firms propelled by an early spiral of (mainly) profitable growth that was self-reinforcing include Hewlett Packard (the doyen), Microsoft and Sun Microsystems, Cisco Systems and Dell, and others in a small elite of firms that dominate their industry. Even in cases of sustained growth, early growth crises which endangered survival are remembered.

XII.f. Growth before profit

In contrast to firms that sustain high growth of sales together with early profitability, there are ventures that build internal resources, but make a lengthy transition to profitability. We can distinguish between firms that grow on (certain measures) before having the capacity for generating revenues, and those that achieve some revenue-earning capacity early, but aim to extend the scale or scope of their activity and meanwhile incur greater costs than earnings.

Some ventures rapidly build up employees and an asset base, but do not yet have the internal resource combination for productive activity. Growth here is in terms of input measures: investment funds and staff numbers expand; sales and profits are for the future. Some biotechnology firms become large organisations before their technology is ready for the market place. Extensive growth of this kind, before profitability, can intensify many of the organisational problems faced by newly profitable firms. One entrepreneur whose enterprise ran out of funds before it had become profitable, stressed the dangers of dependence on external funding. Addressing his staff he said:

---

[1] There is also less likelihood of predatory acquisition by the licensee.

[2] Revenues from royalties can be greatly enhanced by providing customer services and support. Along with increasing specialisation of functions in the production chain, the licensing-with-customer-support route is gaining ground as a start-up strategy; a high profile example is the UK semiconductor firm, ARM. Success on this scale involves growth as part of new business networks and the shaping of emerging industry standards.

How do you like our new offices? . . . Now we look like a real company. But we're missing one thing . . . Revenues. We look like a company, but we are only a venture. Ventures have investors, while companies have revenues. Every month we delay a revenue stream, we have to sell off more equity to stay alive. If we delay too long, the price of the equity goes down . . . eventually no one wants to buy . . . (Kaplan, 1995, p. 93).

Other firms may achieve early sales, but are mobilising resources to extend their product range or to produce on a larger scale, incurring fixed and variable costs greater than their early sales revenues, and so delaying break-even. The break-even point identifies the stage at which the firm can become profitable because revenues have caught up with costs. But a rise in investment costs delays break-even. Ventures with sufficient capital may sacrifice the potential for earlier self-sufficiency and incur higher costs in order to increase their customer and installation base.

Traditionally it has been hard for firms to attract capital without a profitable trading record. The investment climate shifted as a critical mass of technology enterprises became highly profitable, enabling some new ventures to attract sufficient capital to invest in substantial growth. When a longer time horizon is adopted, short-term profitability may be viewed as an inadequate strategy. Extensive market share may be viewed as essential to secure the business future of companies like Amazon.com, to stave off competition as first-mover benefits wear out.

Among these firms, input measures of firm growth such as number of employees and investment funds expended move out of alignment with measures of sales and profitability. If they receive massive venture capital funding, or experience inflated share prices after an early stock market launch, these ventures are able to take on numerous employees before they have extensive sales or any profits.

The Internet firms that sped through the preparatory search and select phase, hit upon an idea that enabled them to attract funding in a propitious investment climate. They have benefited from the Internet infrastructure reaching maturity and a critical mass of users, making it easier to target specific users and specific opportunities. They have been able to mobilise resources to an unprecedented level for start-up companies. But because of substantial marketing costs and the complications of organising distribution, profitability is delayed. In cases of this kind, firms must keep funders on board with the prospect of future gains. Ventures favoured by investors may have more reserves than firms without external funding on this scale. But early learning involves errors. New ventures are exposed to continual scrutiny and subject to investor disaffection.

XII.g. Growth reversal

Many firms that grow explosively in their early years subsequently experience major setbacks.[1] These firms are a select group in that they have early achievements or are

---

[1] In a longitudinal study of UK 'supergrowth' companies, Harrison and Taylor (1996, p. 38) found none of the UK technology firms that showed initial growth and promise were still successful by the end of the period.

viewed by investors as having exceptional prospects. But they are subject to intensive pressures to which many succumb, by closing, by being acquired at a lower value than would have earlier been achieved or by retreating on to a plateau. Among this group are high growth ventures reliant on stock market valuations that are subject to severe corrections. The reasons for this type of growth path may be explicable in terms of Penrose's concepts.[1]

In firms of this kind, case study evidence shows how growth has fed further growth (Fleck and Garnsey, 1988). But this process can have perverse effects when it creates resource shortages and synchronisation problems. Penrose pointed out the difficulties of coordinating resource requirements. The cash flow crisis is a sychronisation syndrome; if cash returns do not come in fast enough to pay for outlays required by the pace of expansion, further productive activity can be halted by lack of cash. If other key resources—people, components, premises—cannot be renewed at the pace required, productivity suffers. Bottlenecks create delays with knock-on effects.[2]

Sometimes problems build up steadily and visibly—as a niche market becomes saturated, for example, the order book ceases to grow and no new sources of custom emerge. More insidiously, when rapid growth is under way, feedback may go undetected and reversals can occur unexpectedly as an essential resource is exhausted. All aspects of growth are interconnected, hence these difficulties may multiply before they are addressed. They call for a shift of strategy where opportunities and resources no longer match as expected. This is especially likely when a new industry is in transition, as occurs during the period when industry standards emerge.

The limits on the capacity of decision-makers to assimilate knowledge, coordinate and plan was identified by Penrose as the major internal constraint on growth. It also plagues attempts to turn around a company that runs into growth problems. Complexity greatly increases the knowledge those in charge have to assimilate, slowing the rate at which they can take considered decisions. Coordination and planning, together with attempts to redress errors, require intensive effort by a select group of people—founders or leading managers who have the authority to take decisions for the firm.

Informal entrepreneurial leadership cannot cope as the firm grows more complex (Kotter and Sathe, 1978). The transition to new leadership patterns is accelerated under rapid growth. Professional managers from big business lack experience of the special problems of young growing firms. People who know how to institute efficient procedures cannot be brought on board fast enough. This advantage underlies the statistics showing that relevant business experience among founders improves performance (D. J. Storey, 1994).

These firms are subject to limits to intensity of effort.[3] In the start-up period, the unmistakable impact of individual effort is highly motivating, creating the new venture

---

[1] The following account of pressures leading to growth reversal draws on Garnsey (1998).

[2] See also Garnsey (1998) on growth crises of new firms, with examples.

[3] This is another dimension of the sigmoid growth curves characterising complex systems (Richardson, 1991).

buzz. As the organisation grows more complex, communication problems set in. As routines develop, repetitive attention to detail is required. When the excitement of the start-up period fades and new procedures are experienced as contraints, motivation and commitment decline (Slatter, 1992). Overworked staff can suffer burn-out. Conflict and departures occur. This often results in competitive spin-outs of a new venture by former employees. Loss of key members of staff can be disastrous to new firms which still rely largely on tacit knowledge. Crises are particularly difficult to overcome where the whole basis of the firm's activities has to be reconsidered (Argyris and Schon, 1978). This may require the departures of founders and new forms of leadership which may take longer than circumstances allow.[1]

If firms with a strong early track record run into difficulties of this kind, firms that lack the early experience of successful sales and profit and rely on investors' belief in their future potential are still more vulnerable to investor disaffection. A slump in performance measures following the Initial Public Offering (IPO) is a common pattern. Firms enjoying a boom in share price may be able to buy into real assets sufficiently rapidly to grow through acquisition. As we see below, in some cases sufficient capital is obtained early to allow an early acquisition strategy to be pursued. Though Penrose did not address the issue directly, the management of acquisition is fraught with difficulty. Growth through acquisition of other firms is also subject to Penrosean constraints: planning delays, delays before new staff become useful, constraints on decision-makers. Methods used to motivate staff can backfire; for example share options designed to motivate staff do not act as incentives if performance targets are not met, and there is a drastic fall in share prices. At the high profile UK telecommunications venture Ionica, for example, the collapse in share prices occurred when pressures on staff were heaviest.

Synchronised growth, paced to prevent major bottlenecks and resource crises, entails less risk. If the firm has built up versatile internal resources, new opportunities should be within reach even if earlier targets are missed: 'The multiple serviceability of resources often gives firms a flexibility in a changing and uncertain environment which may become of great importance in determining the direction of growth' (Penrose, 1971, p. 33). Access to specialist market information and expertise can help these firms to identify markets where demand conditions are favourable.

The growth process of new firms is closely linked to the evolution of new industries, where pioneering firms can have a formative influence. They may be able to take on technological leadership, shape standards and pre-empt key assets. They are also subject to the hazards of uncertainty in an emerging industry without accepted standards (Porter, 1991). An understanding of industry evolution can assist firms in devising realistic strategies.

---

[1] David Potter, the founder of Psion, believes that early growth crises are a key learning experience. If the crises do not destroy the firm, the experience 'keeps entrepreneurs humble' and make them more apt to go on learning. This was also the experience at Oxford Instruments, as recalled by the founder, Martin Wood, at the second Cambridge Enterprise Conference, 1999.

XIII. PENROSE: LIMITATIONS AND STRENGTHS

Penrose's theory of the growth of the firm is of contemporary relevance, in a field where neither the mechanisms of new firm growth nor the paths of growth have been adequately theorised (Autio *et al.*, 1997). There are inevitable gaps in her analysis of the growth of the firm, some resulting from a deliberately selective approach and some from the elapse of time. Penrose made no claims to providing an exhaustive analysis. Writing before the growth of venture capital funds, she did not anticipate the impact of exit strategies designed to realise investment in a new firm. She did not contrast the independent route to growth through an IPO on the stock exchange, with a trade sale of the venture. Both routes to growth involve a valuation of the internal resources of the firm and their capacity to secure advantages in the pursuit of market opportunity. Both routes are subject to Penrosean growth constraints. The overload and stretched resources required to achieve an IPO are often followed by a slump in firm performance for reasons this analysis has briefly examined.

Penrose was aware of the potential of growth through acquisition. She saw that this could be a source of massive growth and the building of extensive business empires. But from her point of view, this type of growth did not fit her approach: 'an industrial empire built up by acquisition and merger, and carried out with little regard for administrative organization is not an industrial firm in our sense until a certain minimum of integration has been achieved'. In some cases 'we must seriously question whether the whole should be treated as a single firm' (Penrose, 1959, p. 190).

An important implication of the recent change in selection regime that has made available massive funding to new ventures is that in some cases growth through acquisition is pursued from the initiation of the company. This has enabled companies like Cisco Systems to acquire teams with multiple competences and take up opportunities in a variety of related markets. But these companies are exceptional in their effective pursuit of growth through acquisition. Acquisition is no easy route to success. It is often subject to problems of mismatched resources and executive overload at the interface, which impair 'internal' growth.

Thus many issues relating to new ventures and their growth crises were not addressed by Penrose. She did not set out formal hypotheses nor provide a rigorous statistical investigation of firm growth (Jovanovic, 1982; Romanelli, 1989; Eisenhardt and Schoonhoven, 1990; Hart and Oulton, 1996). How then can her analysis provide the basis for a coherent conceptual scheme for the study of new ventures? I will look at two main reasons.[1] The first is the way in which Penrose combined the use of theory and evidence. The second is the way in which her approach allows the analysis of the internal and external dynamics of firm growth to be integrated.

---

[1] See also Garnsey (1998). There is implicit systems thinking in Penrose; cf., the impetus provided by leads and lags in dynamic systems (Garnsey, 1995).

XIV. PENROSE'S METHOD

Industrial economists work with a strong body of theory at the industry level. However, where they use the cross-sectional, averaged data in place of observations of specific growth processes within the unit of analysis, they have to stretch their inferences to draw conclusions on growth dynamics within the firm.[1]

Penrose's work on the firm was based on close observation and documentation from individual firms. She studied case histories of companies and conducted detailed fieldwork.[2] This brought to her attention both incentives and constraints experienced by managers, and the influence of their perceptions and responses on the growth of firms. From these observations she distilled her conceptual model. Penrose reasoned on the basis of empirical observation underpinned by logic. Axiom and inference strengthen the argument from evidence. She avoided *a priori* assumptions and positivist positions.

Her theory focuses on the way behaviour is shaped by the meaning agents attribute to situations: 'the experience and knowledge of a firm's personnel . . . are the factors that will to a large extent determine the response of the firm to changes in the external world and also determine what it "sees" in the external world' (Penrose, 1959, p. 80).

An appreciative approach of this kind lacks the formalism with which economics has been invested by those seeking for it the status of science. Marris, tongue in cheek, described why Penrose's approach represented a threat to orthodox economics, in his review of *The Theory of the Growth of the Firm*:

We all agreed that static profit maximization represented a gross simplification of business behaviour, but tacitly assumed that more complex, less 'economic', motivation would be incapable of formal analysis: if the gates were opened, description would largely replace analysis, and a delightful intellectual discipline lost, possibly for ever. What would be left to teach? (Marris, 1961, p. 144).

The vindication of Penrose's work lies in its usefulness. Analysis of the kind she put forward can be used to derive causal propositions, which can be applied to further empirical evidence for support or challenge. The model can accordingly be modified, extended and refined. This iterative process has already occurred in work on the resource-based theory of the firm, notably on competence studies and on resource-based strategy issues (Grant, 1991; Peteraf, 1993; Teece and Pisano, 1994). Her approach could also provide the basis for theoretically grounded investigation of new venture growth in relation to industrial evolution.

---

[1] Lacking both theory and evidence on internal dimensions of the firm, industrial economists have fallen back on assumptions and suppositions to explain their aggregate growth data. Audretsch (1994, pp. 48, 64) assumes that learning is an individual attribute of entrepreneurs. Finding evidence of path-dependent factors in growth, Stanley *et al.* (1996) suggest disobedience as a factor explaining firm performance despite the absence of evidence on this factor.

[2] Penrose had intended to include the case study of the growth of the Hercules Powder Company, resulting from the de-merger of Dupont, in her volume on the growth of the firm, and was only dissuaded by her publishers on grounds of printing costs.

XV. FROM FIRM TO INDUSTRY: SYNCHRONISATION
AND SPECIALISATION

A second source of strength in Penrose's work is the basis she provides for a coherent analysis of the internal and external dynamics of the firm, lacking in much of the literature. Microeconomic theory provides coherence by sacrificing empirical relevance.[1] Industrial economists can draw conclusions at the level of industry more appropriately than at the level of the firm. Penrose drew insight from the way in which the entrepreneur's mental map links the internal and external dynamics of growth; they internalise their environment as a mental map of 'possibilities and restrictions'. But entrepreneurs also act on their perceptions, and in so doing can change their environment: 'Firms not only alter the environmental conditions necessary for the success of their actions, but, even more important, they know that they can alter them and that the environment is not independent of their own activities' (Penrose, 1959, p. 42).

Penrose's theory is usually referred to as resource based (Barney, 1991). But her approach could be viewed as implicit systems thinking (Garnsey, 1995). She placed continual emphasis on the firm as an open system which converts input resources into the productive services needed to produce outputs (Penrose, 1959, p. 25). The firm's output must be valued in the market environment if the firm is to survive.

Hitherto neglected aspects of the internal dimensions of the firm have received attention in work inspired by Penrose, focusing on the firm's strategic competences and capabilities (Barney, 1991; Grant, 1991; Peteraf, 1993). But the resource-based view of the firm has been judged introspective and as failing to see that resources are only valuable in so far as they 'allow firms to perform activities that create advantages in particular markets' (Porter, 1991, p. 108).

This was evident to Penrose, who recognised that to make good use of their resources, firms must develop 'increased experience and knowledge of the external world and the effect of changes in the external world' (Penrose, 1959, p. 79), as can be seen in her case study of a firm de-merged from Dupont in which she states:

> The extensive knowledge of cellulose chemistry possessed by Hercules has provided a continuous inducement to the firm to search for new ways of using it. [In this company, technically trained sales men were] . . . expected to take an active interest in the production and market problems of their customers. This permits them to acquire an intimate knowledge of the customers' businesses and not only to demonstrate the uses of their own products and to suggest to customers new ways of doing things, but also to adapt their products to customers' requirements and learn what kinds of new products can be used (Penrose, 1960, p. 13).

Here is a study of the use of competence as leverage in a strategy of moving into promising markets.[2] If the scope for expansion on the basis of the firm's existing

---

[1] Baretto (1989). Penrose's theory of the firm reveals the disparity between the value to the firm of its unique internal resources and their value on the market (Marris, 1961).

[2] Hercules Powder was able to develop new products on the basis of its competence in nitrocellulose (originally used in the production of explosives) to develop lacquers for the rapidly growing automobile industry. Its technological base enabled it to enter new markets, which in turn led the company to refine its technological knowledge.

capabilities is limited by external market opportunities, the firm can aim at 'building up an experienced managerial and technical team in new fields of activity' (Penrose, 1960, p. 14).

XVI. SYNCHRONY, COMPETITION AND CO-OPERATION

Another way in which Penrose's approach links internal and external dynamics stems from her understanding that in an interconnected system, the impetus provided by synchronisation propels change. A firm grows as attempts are made to redress asynchronies by enlarging capacity. An industry grows as firms expand their activities to exploit new opportunities.[1] The continual adjustments made by the firm are a reaction to feedback and the result of learning by trial and error. In a world of mutual adjustment between firms and their customers, competition provides the spur to remain alert, to make improvements, to avoid complacency: 'continual attention to the development of new products and new methods to meet or surpass competitive developments has been required' (Penrose, 1960, p. 7).

But symbiosis between firms also operates. The specialist division of labour is the outcome of co-operation and complementary activity and not solely of competition. So effective was the key firm studied by Penrose at achieving quality and specifications that many of their competitors simply withdrew from the field and became their customers. This suggests a form of co-evolution among firms. Many innovative firms are so specialised as to have few competitors. They find ways to position themselves to realise opportunities through innovation.

The new firm has to operate in a dynamic environment. Even innovative small firms are threatened as an industry matures and dominant players enter the scene (Penrose, 1959, p. 224). Recent thinking suggests that the firms that sustain expansion are those that shape or obtain impetus from emerging industry standards, creating partnership, which ensure that they co-evolve in an expanding network (Christensen, 1997). The requirements were anticipated by Penrose: 'Knowledge of markets, of technology being developed by other firms, and of the tastes and attitudes of consumers are of particular importance' (Penrose, 1959, p. 79).

XVII. PERCEPTIONS AND PRACTICE

Thus, Penrose's approach has numerous implications for strategy on new business development. New firms' strategies, deliberate or emergent, are shaped by 'entrepreneurial ideas about what the firm can and cannot do, that is . . . the "subjective" productive opportunity of the firm' (Penrose, 1959, p. 42).

[1] Like Smith (1933, p. 4), who saw that the division of labour is limited by the extent of the market, Penrose saw that opportunities could be limited by demand, while markets could be extended by entrepreneurs specialising, moving into new interstices and creating new demand. Both appreciated that firms respond to customers in finding new ways to create value through specialisation.

Entrepreneurs can be helped to develop informed assessments of opportunities and difficulties ahead. Mentoring and the provision of resources at critical junctures in the growth process could make a difference to survival rates, drawing more firms from micro to medium size.

A parallel challenge is to understand how the productive opportunities of entrepreneurs are viewed in the mental maps of investors. Greater public understanding is required of the nature and prospects of new firms in new industries. The ways in which entrepreneurs can set in train dynamic processes were accounted for by Penrose at a time when they were viewed as marginal players in a world dominated by big business. Structural constraints can alter unpredictably as mental horizons alter and change behaviour. Recent shifts in investment patterns affecting IT ventures are a clear demonstration of this process.

REFERENCES

ARGYRIS, C. and SCHON, D. (1978). *Organisational Learning*, Reading, Addison Wesley.

AUDRETSCH, D. (1994). *Innovation and Industry Evolution*, Cambridge, MA, MIT Press.

AUTIO, E., GARNSEY, E. and YLI-RENKO, H. (1997). 'Early Growth and External Relations in New Technology-based Firms', Proceedings of the 42nd conference of the ICSB, San Francisco, CA.

BALDWIN, J. (1995). *The Dynamics of Industrial Competition: A North American Perspective*, Cambridge, Cambridge University Press.

BARETTO, H. (1989). *The Entrepreneur in Microeconomic Theory*, London and New York, Routledge.

BARNEY, J. (1991). Firm resources and sustained competitive advantage, *Journal of Management*, Vol. 17, No. 1, 99–120.

BARNEY, J. (1996). *Gaining and Sustaining Competitive Advantage*, Reading, Addison Wesley.

BELL, C. and MCNAMARA, J. (1991). *High Tech Ventures: The Guide for Entrepreneurial Success*, Reading, Addison Wesley.

BHIDÉ, A. (2000). *The Origin and Evolution of New Businesses*, New York, Oxford University Press.

BROWN, S. L. and EISENHARDT, K. (1998). *Competing on the Edge, Strategy as Structured Chaos*, Cambridge, MA, Harvard Business School Press.

BYGRAVE, W. and HOFER, C. (1991). Theorizing about entrepreneurship, *Entrepreneurship Theory and Practice*, Vol. 16, No. 2, 13–22.

CHRISTENSEN, C. (1997). *The Innovator's Dilemma: When New Technologies Cause Great Firms to Fail*, Cambridge, MA, Harvard Business School Press.

CHURCHILL, N. (1997). The six key phases of company growth, in *Mastering Enterprise*, edited by S. BIRLEY and D. MUZYKA, London, Financial Times–Pitman.

CHURCHILL, N. and LEWIS, V. (1983). The five stages of small firm growth, *Harvard Business Review*, May–June, 30–50.

COSH, A. and HUGHES A. (1996). *Enterprise Britain*, ESRC Centre for Business Research, University of Cambridge.

EISENHARDT, K. and SCHOONHOVEN, C. (1990). Organizational growth: linking founding team, strategy environent and growth among U.S. semi-conductor ventures 1978–1988, *Administrative Science Quarterly*, Vol. 30, 504–29.

FLECK, V. and GARNSEY, E. (1988). Managing growth at Acorn Computers, *Journal of General Management*, Vol. 13, No. 3, 4–23.

GARNSEY, E. (1995). The resource-based theory of the growth of the firm: a systems perspective, in *Critical Issues in Systems Theory and Practice*, edited by K. ELLIS *et al.*, New York and London, Plenum Press.

GARNSEY, E. (1998). A theory of the early growth of the firm, *Industrial and Corporate Change*, Vol. 7, No. 3, 523–56.

GARNSEY, E. and ALFORD, H. (1996). Innovative interaction: supplier relations among new technology-based firms, in *High Technology Firms*, edited by R. OAKEY, London, Paul Chapman.

GRANT, R. (1991). The resource-based theory of competitive advantage: implications for strategy formulation, *California Management Review*, Spring, 114–35.

GRANT, R. (1996). Toward a knowledge-based theory of the firm, *Strategic Management Journal*, Vol. 17, 109–22.

GREINER, L. (1974). Evolution and revolution as organizations grow, *Harvard Business Review*, July–August, 37–46.

HARRISON, J. and TAYLOR, B. (1996). *Supergrowth Companies; Enterpreneurs in Action*, Oxford, Butterworth Heinemann.

HART, P. and OULTON, N. (1996). Growth and size of firms, *Economic Journal*, Vol. 106, 1242–52.

HISRICH, R. and PETERS, M. (1989). *Entrepreneurship: Starting, Developing and Managing a New Enterprise*, New York, McGraw Hill.

HUGHES, A. (1998). 'Growth Constraints on Small and Medium-Sized Firms', ESRC Centre for Business Research, University of Cambridge, working paper No. 107.

HUGHES, T. (1979). The electrification of America: the system builders, *Technology and Culture*, Vol. 20, 124–61.

JOVANOVIC, B. (1982). Selection and the evolution of industry, *Econometrica*, Vol. 50, No. 3, 649–70.

KAPLAN, J. (1995). *Start Up: A Silicon Valley Adventure*, London, Little Brown.

KAZANJIAN, R. and DRAZIN, R. (1989). An empirical test of a stage of growth progression model, *Management Science*, Vol. 35, No. 12, 1489–503.

KEEBLE, D. (1998). Growth objectives and constraints, in A. COSH and A. HUGHES, *Enterprise Britain*, ESRC Centre for Business Research, University of Cambridge.

KIRCHHOFF, B. (1994). *Entrepreneurship and Dynamic Capitalism*, Westport, Praeger.

KOTTER, J. and SATHE, J. (1978). Human resource problems in rapidly growing companies, *California Journal of Management*, Vol. 21, No. 2, 29–36.

LATOUR, B. (1987). *Science in Action*, Milton Keynes, Open University Press.

MARRIS, R. (1961). *The Theory of the Growth of the Firm*: book review of Penrose 1959, *Economic Journal*, Vol. 71, No. 1, 110–13.

NELSON, R. (1995). Recent evolutionary theorising about economic change, *Journal of Economic Literature*, Vol. XXIII, March, 48–90.

PENROSE, E. (1955). Limits to the size and growth of firms, *American Economic Review*, Vol. XLV, No. 2, May. [Reprinted in E. PENROSE (1971). *The Growth of Firms, Middle East Oil and Other Essays*, London, Frank Cass.]

PENROSE, E. (1959). *The Theory of the Growth of the Firm*, Oxford, Basil Blackwell.

PENROSE, E. (1960). The growth of the firm—a case study: the Hercules Powder Company, *Business History Review*, Vol. XXXIV, No. 1, 1–23.

PENROSE, E. (1971). Limits to the size and growth of firms, in *The Growth of Firms, Middle East Oil and Other Essays*, edited by E. PENROSE, London, Frank Cass.

PENROSE, E. (1995). Introduction to *The Theory of the Growth of the Firm*, 3rd edn, Oxford, Oxford University Press.

PETERAF, M. (1993). The cornerstones of competitive advantage: a resource-based view, *Strategic Management Journal*, Vol. 14, 179–91.

PORTER, M. (1980). *Competitive Strategy*, New York, Free Press.

PORTER, M. (1991). Towards a dynamic theory of strategy, *Strategic Management Journal*, Vol. 12, 95–117.

QUINN, R. and CAMERON, K. (1983). Organisational life cycles and shifting criteria of effectiveness: some preliminary evidence, *Management Science*, Vol. 29, No. 1, 33–51.

REYNOLDS, P. and MILLER, B. (1988). *Minnesota New Firms Study: An Exploration of New Firms and their Economic Contributions*, Minneapolis, MN, Centre for Urban and Regional Affairs.

RICHARDSON, G. (1991). *Feedback Thought in Social Science and Systems Theory*, Philadelphia, PA, University of Pennsylvania Press.

ROBERTS, E. B. (1991). *Entrepreneurs in High Technology, Lessons from MIT and Beyond*, New York, Oxford University Press.

ROMANELLI, E. (1989). Environments and strategies of organization start-up: effects on early survival, *Administrative Science Quarterly*, Vol. 34, 369–87.

SIMONETTI, R., MACKINTOSH, M., COSTELLO, N. *et al.* (1998). *Understanding Economic Behaviour: Households, Firms and Markets: Firms*, Milton Keynes, Open University Press.

SLATTER, S. (1992). *Gambling on Growth: How to Manage the High Tech Company*, London, Wiley.

SMITH, A. (1933). *The Wealth of Nations*, London, Everyman.

SPENDER, J. (1996). Making knowledge the basis of a dynamic theory of the firm, *Strategic Management Journal*, Vol. 17, 45–62.

STANLEY, M., AMARUL, L., BULDYREV, S. *et al.* (1996). Scaling behaviour in the growth of companies, *Nature*, Vol. 379, No. 6568, 804–6.

STOREY, D. J. (1994). *Understanding the Small Firm Sector*, London, Routledge.

STOREY, J., KEASEY, K., WATSON, R. and WYNARCZYK, P. (1987). *The Performance of Small Firms: Profits, Jobs and Growth*, London, Routledge.

TEECE, D. and PISANO, G. (1994). The dynamic capabilities of firms; an introduction, *Industrial and Corporate Change*, Vol. 3, No. 3, 537–56.

UTTERBACK, J. M. (1995). *Mastering the Dynamics of Innovation*, Cambridge, MA, Harvard Business School Press.

UTTERBACK, J. M., MEYER, M., ROBERTS, E. and REITBERGER, G. (1988). Technology and industrial innovation in Sweden: A study of technology-based firms formed between 1965 and 1980, *Research Policy*, Vol. 17, 15–26.

8

A THEORY OF THE (GROWTH OF THE) TRANSNATIONAL FIRM: A PENROSEAN PERSPECTIVE

CHRISTOS PITELIS*

Judge Institute of Management Studies and Queens' College, Cambridge

In this paper I apply a Penrosean perspective to the theory of the growth of the transnational corporation (TNC). I suggest that TNCs are the result of a dynamic interaction between endogenous factors to the firm and external opportunities and threats, as perceived by the firm's management. Trans-border geographical diversification is the result of limits to domestic expansion, failed or missing markets, perceived differential opportunities abroad, and oligopolistic interaction. The choice of institutional mode (foreign direct investment over putting out, subcontracting, licensing or exporting) is also the result of missing or failing markets, perceived differential capabilities, and oligopolistic interaction.

Transnational or multinational firms are most significant economic players in the global economic scene. The reasons behind their existence and boundaries have preoccupied theorists of economic and business studies for many years. The analysis of the theory of the firm has become one of the fastest growing areas of economics and business in the past 25 years. Being firms, transnationals have also received a lot of attention, with debates on this issue largely building on the theory of the firm. The 1990s have experienced significant developments on the theory of the firm in general and the transnational corporation (TNC) in particular. In part, this is a reflection of a dramatic increase in the scope and activities of TNCs. Despite remarkable progress, there are gaps in the existing literature. These concern, in the main, the issue of growth, to include the motivation behind it, the direction of expansion and limits to growth. Moreover, despite wide recognition of the need for dynamics, the theory of the TNC remains largely static. Progress on the Penrosean approach to the theory of the firm, has introduced welcome dynamics and can be of great use in explaining the existence, growth and boundaries of TNCs, in conjunction with existing theories and required developments.

In this paper we suggest that TNCs are the result of a dynamic interaction between

* Many people have contributed over the years to the development of my ideas on the TNC, too many to mention individually here. I am particularly grateful to Edith Penrose and Roger Sugden for discussion and comments pertaining to the contents of this paper. Errors are mine.

endogenous factors to the firm and external opportunities and threats, as perceived by the firm's management. Growth is motivated by the pursuit of long-term profits and is shaped by internal inducements, oligopolisic interaction and imperfections of, or missing, markets. Growth is constrained for reasons internal to the firm, perceived external opportunities and threats, and because of oligopolistic interaction. Diversification is the result of push and pull factors. Its direction is the result of internal resources, and external opportunity. It is predictable, but only partly; because of the potentially accidental nature of internal resource generation and differential entrepreneurial capability. Trans-border geographical diversification is the result of limits to domestic expansion, failed or missing markets, perceived differential opportunities abroad, and oligopolistic interaction. The choice of institutional mode (foreign direct investment over putting out, subcontracting, licensing or exporting) is also the result of missing or failing markets, perceived differential capabilities, and oligopolistic interaction.

I. THEORIES OF THE TNC: CRITICAL ASSESSMENT

The TNC is a firm that *controls* productive assets in countries other than (and including) its country of origin.[1] The emphasis is on control, and it is well established in the literature; see, for example, Hymer (1976), Cowling and Sugden (1987). The act of acquiring and/or creating such productive assets abroad is called foreign direct investment (FDI). FDI is distinguished from other forms of international activities, such as exports, licensing, franchising, etc., which are purely market-based transactions.[2] Reference to *countries* (as nation states) is crucial (something that arguably has received limited attention in the literature). It involves two dimensions: geographical *and* political. The same activity that counts as FDI before, for example, the merger of east and west Germany is simply domestic expansion after the merger. The separation of Czechoslovakia into two republics, and many other recent such cases, has generated 'new' FDI overnight! In the absence of nation states no FDI can exist. All we would need is a theory of the (growth of the) firm. TNCs cannot be separated from (the theory of) nation states.

The first, well known analysis of the TNC is by Stephen Hymer, to whom this theory is attributed. Arguably the most comprehensive reading of Hymer is to be gained by looking at the 1979 collection of his selected papers on the TNC, by some of his colleagues, after his death. Hymer's approach was based on a theory of the firm and a theory of industry organisation. Building on Chandler's (1962) classic work, Hymer suggested a 'law of increasing firm size'. According to this 'law', firms have grown in four stages. First, small, owner-managed and controlled firms with unlimited liability. Second, large limited liability companies he called *national*. Third, M-form

---

[1] There are also 'free standing' TNCs, without a home base, see Hennart (2000) for a discussion. Our focus here is on the international operations of national firms, so we will ignore this case.

[2] Things are more complicated in cases such as putting-out, subcontracting, joint ventures, alliances and such forms of inter-firm cooperation, for which see below.

firms, multidivisional organisations with various business units (divisions) account-able to the headquarters. Finally, multinational firms with control of productive assets (within and) without their country of origin.

In Hymer's history-based interpretational account, firm growth is achieved through internal expansion, mergers and acquisitions, vertical integration and diversification, intra-country and (then) foreign. The overall motive for firm expansion is the pursuit of profits by firms' management. Within its context both defensive and proactive factors are in operation. Both of these are in the context of ensuring control over markets, they are largely anti-competitive. For these reasons they can both be taken to belong to a broad 'market power' perspective.

Proactive factors first refer to the pursuit of 'attractive' (to use Porter's, 1980 terminology) industries in the same or different activities. These should not face a high degree of competition *ex ante*, and should be rendered (more) 'attractive' by using the usual anti-competitive conventions *ex post*: e.g. collusion, barriers to entry, anti-competitive mergers and acquisitions, vertical integration and the likes. Defensive factors are also, however, in operation. For example, to explain domestic diversifi-cation Hymer explicitly draws on the product life-cycle. Diversification is seen as an attempt to smoothen the product life-cycle, to create, in the words of strategic management, an 'all-weather company'. The M-form organisation is seen as a means of realising this aim more effectively.

The very process of making markets 'attractive' (i.e. monopolisation) generates competitive advantages, which Hymer called monopolistic advantages, and which are also known as ownership advantages in the literature. Ownership need not presuppose or imply monopoly, so the two terms are evidently not synonymous. Hymer's ter-minology aimed to emphasise the fact that, in his view, advantages were the result of a monopolistic process, not just of a natural or innocent growth process. Anything that has to do with this process—i.e. (retained) profits, brand names, marketing and distri-bution, the very M-form, being 'all-weather', and even the expanded horizons and vision afforded to these firms through their expansion process—is seen by Hymer as monopolistic advantages.

The very need for and use of 'monopolistic' advantages presupposes rivalry between and/or collusion with potentially rival firms. Oligopolistic interaction is the very back-ground, the scene of the battle. Both extinction of (some) rivals and collusion can render markets more attractive. In this struggle, all moves mentioned above can be seen as defensive and proactive at the same time. Outcompeting rivals and/or col-luding with them increases expected profits or reduces expected loss.

In this context, the TNC is a natural extension of firms' activities. 'Attractive' (monopolisable) markets may always exist; abroad too. The one difference is that there exist natural disadvantages of being foreign. These are due to language, knowledge of culture, connections, potential discriminatory behaviour by the govern-ments and others, etc. It follows that, given the incentive, differential (monopolistic) advantages are required. Differential in that foreign firms, too, are likely to have some such advantages, so that if one is foreign they need to have more of these (and/or better ones) to outcompete foreign rivals. While not emphasised by Hymer, this may presuppose some degree of differential economic 'development' as firms acquire their advantages in the process of growing, and assuming that economic development is partly facilitated by, and partly facilitates, this process of growth, the firms of earlier developed countries are more likely to acquire such advantages. Hymer was describing the process of foreign expansion of American firms, so to a large extent the micro–macro dimension was always in the background. In addition, the very process of targeting industries that are *ex ante*, reasonably attractive and with potential to be even more attractive *ex post*, involves rivalry and collusion with other TNCs (to be) as well as firms in the foreign country.

Mainstream interpretations of Hymer focus on the monopolistic advantage part (often called ownership advantages), plus the concept of pre-required differential of such advantages, on the part of the TNC-to-be. The 'reduction of rivalry' aspect of Hymer's account (see, for example, Cantwell, 1991; Yamin, 1991) is, however, part and parcel of an integrated whole.

Missing from the above account is the issue of the choice of entry modes. For example, why not licensing, exporting, franchising and other such market-oriented forms of international operations? This aspect of Hymer's theory has been overlooked. The established view appears to be that Hymer's focus was on 'monopolistic' and not 'natural' market failures (of the transaction costs type); see Dunning and Rugman (1985) and below. This might have led to an insufficient explanation of the entry mode issue. This is likely to be true, save for two observations. First, Hymer was very explicit about the market-hierarchy dichotomy, he used arguments akin to transaction costs theorising, and even had an explicit reference to Coase (1937) in one of his articles; see Yamin (1991). More to the point, Hymer's very focus on *control*, rendering industries 'attractive' through the very process of monopolisation, implies almost by definition the choice of FDI over market-based activities; FDI involves a degree of control which other modes do not. In this sense Hymer's theory is more self-contained than often believed.

An implication of Hymer's tradition is that TNCs need not be Pareto-efficient.[1] Their monopolistic advantages may facilitate a process of monopolization abroad, potentially reducing the welfare of the 'host' countries. Although proponents of the theory, for example Kindleberger (1969), express the belief that the long-run benefits of TNCs will offset any short-run costs, the Hymer tradition recognises the possibility of the existence of both efficient and inefficient aspects of TNCs' operations.[2]

The other major tradition and arguably the dominant approach to the TNC to date (see Pitelis and Sugden, 1991; Dunning, 1998; Ostry, 1998), is the transaction costs-

---

[1] It is worth noting that Caves' later work (1982, 1996) is in line with the transaction costs perspective and arguably a dynamic version of it, akin to resource-based ideas; see Pitelis and Sugden (2000).

[2] The possibility that TNCs will consciously monopolise global markets so as to obtain monopoly profits and that in so doing they will tend to behave collusively (eliminate conflict) has been developed by the 'global reach' variant of Hymer's theory (see Jenkins, 1987). Important contributions along this line of thought are Newfarmer (1985) as well as a good part of Cowling and Sugden (1987). Unlike Kindleberger, this line of thought emphasises the (Pareto) inefficiency aspects of TNCs' operations.

internalisation perspective. Affiliates of this school emphasise internalisation of market imperfections. It is suggested that TNCs internalise 'cognitive' or 'natural' market imperfections, defined as those arising out of excessive market transaction costs; see Dunning and Rugman (1985). The basic notion that the firm exists in order to reduce the costs associated with the operation of the price mechanism, dates back to Coase (1937). The forceful reintroduction and extension of Coase's insight is due primarily to Williamson (1975, 1981). For Williamson, three main factors—i.e. bounded rationality, opportunism, and most importantly, asset specificity—give rise to high market transaction costs, such as the costs of searching, contracting, negotiating and policing agreements. These costs can often be reduced if the market is superseded by a hierarchy, such as the firm. The existence of firms can thus result in reduced transaction costs.

While most proponents of this perspective would agree with the above statement, not everybody would agree with the reasons why. In Williamson (1981), for example, the driving force behind the emergence of excessive transaction costs leading to internalisation is *ex post* hold-ups due to asset specificity. Buckley and Casson (1976) on the other hand, have focused on appropriability problems of intermediate products (usually intangible assets), such as technology, know-how, managerial skills, etc. All these are claimed to exhibit 'public goods' characteristics, giving rise to excessive transaction costs of markets, thus to integration (the choice over FDI to, for example, licencing). Another view is that of Hennart (2000). Hennart in effect goes back to Coase's focus on the employment contract. He claims that internalisation requires not just market failure but also firm success. The latter is attributed to the ability of (foreign) hierarchies to coordinate, more efficiently than indigenous firms, employees in the 'host' country. In addition, Hennart's approach focuses on internalisation of markets, not just advantages, in apparent contrast to other transaction costs-related perspectives.

An eclectic synthesis of the Hymer and the internalisation/transactions costs tradition has been offered by Dunning's (1981, 1988) 'eclectic theory' and, more recently, the 'Ownership, Location, Internalisation' (OLI) paradigm (Dunning, 1991, 1998, 1999). In Dunning's theory, ownership advantages and internalisation of market transactions are reasons for TNCs as well as 'locational factors'. Dunning's (1958) early work on the TNC explained US TNCs' activities in Europe in terms of such locational factors; see also Fieldhouse (1986). The role of such factors had received little attention from other authors, partly owing to the belief that locational differences between developed countries are of no importance; see Gray (1985). This has now changed radically (see next section).

In a more radical vein, Marglin (1974) focused attention on the labour market. Starting from the classical dichotomy between capital and labour, Marglin suggested that the emergence of the factory system from the putting-out system was more likely to have been due to the desire of capitalists to increase their control over the work force than to any alleged technological superiority of the factory system. Accordingly, the emergence of the more hierarchical structure from the more decentralised market-based one need not necessarily be associated with exclusively (Pareto) efficiency attributes. Labour, for example, may not perceive the change of institutional form as an improvement.

Sugden (1991) has extended Marglin's analysis to the TNC. According to Sugden, the TNC allows capital to increase their control over global labour, which thus allows them to pursue 'divide and rule' policies in order to reduce further the power of trade unions by playing off country-specific worker groups against each other. The potential mobility of TNC operations, as compared to the inherent immobility of country-specific labour groups, increases TNCs' bargaining power, allowing them to derive distributional gains from labour. This redistributional aspect of TNCs implies Pareto-inefficiency by definition; see also Peoples and Sugden (2000).

To summarise, existing theories of the TNC tend to emphasise the exploitation of monopolistic advantages by firms (the Hymer tradition), the internalisation of market transaction costs, an eclectic synthesis cum locational factors (the OLI paradigm), or the increased power over labour markets ('divide and rule' theory). In principle, all these theories can be integrated within the general concept of internalisation. TNCs can be argued to arise in order to reduce 'natural' and structural market failures (as suggested by Rugman, 1986) as well as what they perceive as labour market imperfections. The problem for the internalisation theory, resulting from this integration, is that the efficiency-only property fades away. The internalisation of monopolistic advantages may result in the restriction of competition, which need not be Pareto-efficient or socially beneficial despite its private gains to the TNC itself.[1] This is clearer in the case of the internalisation of 'labour market inefficiencies'.[2]

There are various problems with all the aforementioned theories. Williamson's almost exclusive focus on hold-up problems due to specific assets has been criticised widely; see in particular, Kay (1991), Demsetz (1995) and, more recently, Holmström and Roberts (1998). Kay was among the first to observe that TNCs internalise both specific and non-specific assets, indeed that the focus of 'other' internalisation theorists (i.e. Buckley and Casson) was mostly on non-specific assets. On the other hand, Buckley and Casson's focus on the alleged 'public goods' nature of the TNC has been criticised by Kogut and Zander (1993). Building on work by Teece (1977), Kogut and Zander observed that much of knowledge-related (intangible) assets involve tacit knowledge, and thus are hard to transfer across markets. The differential ability of firms to transfer these assets internally may explain the TNC. This need not require market failure; differential ability vis-à-vis the market and other firms is sufficient. In addition to the above, it is now widely recognised that the internalisation perspective is a comparative static one, basically underplaying dynamics. This old criticism (see, for example, Pitelis 1991) is acknowledged by Buckley and Casson (1998B) who call for a dynamic theory, based on the concept of (and need for) flexibility.

[1] It is possible, for example, that the social costs from internalisation may exceed the private benefits; see Dunning (1992).
[2] Dunning (1989) traces the origin of internalisation of Coase-type costs as well as the internalisation of labour markets to Marx (1959), first published in 1867! A similar claim is made by Bowles (1985).

There are additional and related concerns. First, the issue of motivation and inducements for growth; second, the nature and acquisition of advantages; lastly, the direction of expansion. In all theories the profit maximisation concept is maintained, or at least not challenged. In Hymer, this is pursued by, among others, monopolising markets; in the internalisation school, by reducing transaction costs. In both cases, the inducement to expansion is external opportunities (monopoly or transaction cost reductions, respectively). No internal inducements to growth are recognised, whilst it remains unclear whether reference is made to short-term, or long-term profit maximisation.

Coming to the advantages (including intangible assets), it is not always clear how they derive. In Hymer's work taken as a whole (see the collection of his papers in Hymer, 1979), advantages derive from a process of expansion motivated by external market opportunity. In the internalisation school, advantages are taken to be either inherited, or the result of the very action of internalisation. In neither case are advantages derived through a process of internal inducements to growth.

Concerning the direction of expansion, this is unclear. From the monopolistic advantage perspective one could suggest that diversification would take place where 'attractive markets' exist, domestically and/or abroad. This would be very much in line with Porter's (1980) five-forces analysis, whereby attractive industries are chosen on the basis of the lack of competitive forces within them. A push factor in Hymer is the product life-cycle, which induces domestic diversification so as to create what is known in the strategic management literature as the 'all weather company'. However, this is not sufficient to distinguish between domestic diversification and expansion abroad. So, why the latter? Does the profit motive á la Hymer or Coase suffice?

To summarise, existing work fails to address adequately the issues of motivation, nature and acquisition of advantages, internal inducements, and direction of expansion. In view of the wide recognition of the need for dynamics, this is the issue to be addressed. In addition, there is scope for refining the distinction between domestic diversification and expansion abroad.

II. EDITH PENROSE AND THE THEORY OF THE GROWTH OF THE (MULTINATIONAL) FIRM

Edith Penrose's classic *The Theory of the Growth of the Firm* was published in 1959, albeit the main arguments behind the book were already in print in Penrose (1955). In her book, Penrose defines the firm as a bundle of resources under administrative coordination, producing for profit realised through sale in the market. Among the tangible and intangible resources most important are the latter, notably the human resources and, amongst these, management. The cohesive shell of the administrative structure called a firm gives rise to knowledge through teamwork, learning by doing, learning to work with others, etc. Learning, indivisibilities of resources and unused resources more generally, give rise to endogenous growth. The availability of (firm specific), in particular, managerial resources puts a limit on the rate of growth but not on the size of the firm. Unused resources provide an incentive for expansion, but also determine in part the direction of expansion (where such resources are most suitably of use). They also provide an incentive for innovations. There is a dynamic interaction between the internal and external environment, the latter being an 'image' of management's perceptions. This dynamic interaction results in endogenous and external growth, horizontal, vertical, and diversified. Diversification is the natural result of the process of growth. This applies to both intra-national and inter-national diversification. In this sense, the TNC is the outcome of an (endogenous) growth process.

Crucially, competitive advantages are generated within the firm. They, notably managerial skills, are often unavailable in the spot market. This generates a limit to the rate of growth, albeit not to the size of the firm *per se*.

We do not intend to provide a full account of the Penrosean contribution here; for further details see Pitelis and Wahl (1998) and Pitelis and Pseiridis (1999). It will be sufficient to observe that within her theory we have an explanation for internal growth which supplements existing theories. Moreover it explains the process of generation of internal competitive—firm specific advantages, which is in accord with, and supplements, Hymer's view. Last but not least, Penrose's theory is the only one that provides a hint as to the *direction* of expansion. This should, in the first instance, be in activities where firms have already developed a resource base, or what in today's parlance is called a competence; see, for example, Prahalad and Hamel (1990). Importantly—as Penrose shows in what is arguably the best case study of all times, the Hercules Powder Company (Penrose, 1960)—the intra-firm innovation process may well be the result of quasi-chance. To that extent the direction of diversification is only partially predictable. Importantly, the internally generated resources may well be firm specific and an ownership or monopolistic advantage, but need not be specific assets in the Williamsonian sense. Some may be, some others may not. In the latter case the fundamental issue returns: why not sell these?[1] There are two generic reasons: inability and/or unwillingness. Inability could be the result of failing and/or missing markets. Unwillingness can be the result of expectations for higher profits, and/or lower losses. The former case is for example that of differential efficiency, the latter can be the avoidance of creating a competitor. There can simply not be a general answer to all cases. Each case should be reviewed individually by the actors involved.

Penrose's focus on endogenous growth and the role of knowledge have currently given rise to an industry of writings in the theory of the firm. Resource, competence and knowledge-based theories of the firm are arguably now dominant, in the strategic management literature. While not always drawing explicitly on Penrose, her motherhood of such theories seems to be very widely acknowledged; see for example, Foss (1997) for extensive coverage. However, with the notable exception of Kogut and Zander (1993), this extensive literature has found very little application in the theory of the TNC.

One can hazard various reasons why the Penrosean perspective, while arguably

---

[1] Penrose (1955) herself raised this question but, intriguingly, finished the article without returning to it.

dominant on the issue of diversification (see Foss, 1997), has failed to make inroads so far on the theory of the TNC. While Penrose has spent a lot of time and effort analysing TNCs (see Penrose 1956, 1968, 1985, 1987, 1995, 1996), she did not make much effort in showing the links between her analysis of the theory of the growth of the firm and the TNC itself. In part, this is due to her view that once up and running a subsidiary could usefully be regarded as a separate entity (Penrose, 1956).[1] In addition to the above, Kay (2000) observed that Penrose's choice of (the oil) industry was not the best case study of her theory of growth, exhibiting hardly any (internal resource-related) diversification strategies.

There are additional reasons which might explain the failure of the Penrosean approach to make inroads into the theory of the TNC. Arguably, her theory is more amenable in explaining the direction of expansion rather than the mode (see Kay, 2000).[2] When it comes to understanding the mode, transaction costs-related arguments may well be indispensable. Another reason may be the result of strength not weaknesses. The Penrosean approach is arguably fully compatible with the Hymer story. It complements the latter by providing an endogenous growth theory cum (relatedly) an explanation of the direction of diversification. Other than these, the resource-based related firm specific advantages can be easily translated as monopolistic or ownership advantages.

In contrast to Hymer, however, Penrose clearly distinguished between monopolistic and non-monopolistic advantages. She recognised the existence of both. She suggested that larger and older firms have a 'competitive advantage' over smaller firms in terms of non-monopolistic advantages, such as size, experience, access to funds, etc., and also due to sheer 'monopolistic power' (1956, p. 64).[3]

The affinity between Hymer and Penrose is not really surprising, given Penrose's own recognition of the close links between her work and that of Chandler (1962) (see Penrose, 1995), and the fact that Hymer's own account of the historical evolution of firms and the related acquisition of advantages drew explicitly on Chandler's work (Hymer, 1972).[4]

A final, but very important reason for failure to recognise the Penrosean contri-

---

[1] Another possible reason is her tendency not to go back to issues she had already covered before.

[2] The need to incorporate transaction costs in her theory is acknowledged by Penrose (1996).

[3] In discussing monopoly and competition in the international petroleum industry, Penrose (1964) suggested that the firms'

. . . superior efficiency in production and distribution, in inventions and technological advance, could not account for the dominant position they achieved. Their record in finding, producing and distributing oil and its products is indeed impressive, but efficiency in this respect would not have been enough to secure their dominance. Hence the story of the rise of the great companies deals as much with financial power, commercial and political negotiations and intrigue, with cartel agreements, marketing alliances, price maintenance arrangements, price wars and armistices, mergers and combination, actions to avoid taxes, and the national and international political interests of governments, as it does with the economics of production and distribution. This statement does not necessarily imply any condemnation of the companies (1964, p. 155).

[4] The link between Hymer's work and that of Penrose is also noted in Buckley and Casson (1998B). Sanna Randaccio integrated Penrose's work with oligopolistic interaction, see Cantwell (1991), who also noted the links between the Penrosean perspective and the technological accumulation approach to TNCs.

bution to the issue of the TNC is Edith's own choice. With the only exception of her very late writings, Penrose refused to attribute much significance to the issue of multinationality *per se*, versus expansion in general.

Her earliest work on the topic was in 1956 for the *Economic Journal*, but she kept revisiting the issue, importantly enough, authoring the *New Palgrave* entry to multinationals. In most of these writings, Penrose views the multinational as the natural outcome of the process of expansion,

In her 1987 entry to the *New Palgrave*, she points out that

There are differences between national and international firms but the differences are not such as to require a theoretical distinction between the two types of organization, only a recognition that national boundaries make an empirical difference to their opportunities and costs (1987, p. 563).

In the introduction to the 1995 edition of *The Theory of the Growth of the Firm*, she suggests that

. . . it is easy to envisage a process of expansion of international firms within the theoretical framework of the growth of firms . . . It is only necessary to make some subsidiary 'empirical' assumptions to analyze the kind of opportunities for the profitable operations of foreign firms that are not available to firms confining their activities to one country as well as some of the special obstacles (1995, p. xv).

It is in her very last published paper, the 1996 entry in the *International Encyclopaedia of Business and Management*, that she concedes some importance to the issue of international borders. These, she suggests, make enough difference to

justify separate treatment of international firms. The differences arise from the additional obstacles (or advantages) relating to culture, language and similar considerations (which may not apply nationally within ethnically diverse countries), to different currencies, border controls or other types of physical or financial regulations, political attitudes of foreign or home governments, size of protected markets, the configurations of firm cultures or associations, the type of technology involved, and so on (1996, p. 1720).

Arguably, this late recognition of the significance of national borders in requiring a separate treatment of the TNC explains the earlier de-emphasis on the theoretical *raison d'être* of the TNC, despite the strong interest on the topic generally and the impact of TNCs' actions in particular, especially for developing countries.[1] Penrose's reluctance to acknowledge any major need for separating national from international expansion stresses the point mentioned in our introduction, that the *raison d'être* of the TNC should be looked at in the (sometimes artificial) differences that emerge from the existence of nation states. In the absence of these, FDI and TNCs would simply not exist, and similarly all we would need is a theory of the growth of the firm, not the TNC. In this context, it is interesting to note Penrose's observation-critique in her

[1] In this latter context Penrose's work was pioneering. She was one of the first people to discuss the issue of transfer pricing, and to recommend the need for restrictions on the repatriation of profits in certain cases (Penrose, 1962). She introduced the concept of 'infant firms' (like infant industries) as a reason for protectionism in some cases (Penrose, 1973) and observed the endemic nature of dumping in a system of big-business competition (Penrose, 1990); see Penrose and Pitelis (1999).

New Palgrave entry, of Coase (1937) and Hymer (1970)-based theories which, in her view, fail to 'distinguish the multinational corporation from domestic firms' (1987, p. 562).

Having said this, it is arguable that Penrose herself did not fully provide a satisfactory reason to distinguish between national firms and TNCs. To this end, developments on the issue of inter-firm co-operation and locational factors can be of use. In addition, we claim here that demand-side, business cycle-type differentials across countries may provide some answers to our concern.

III. INTER-FIRM RELATIONSHIPS, LOCATION AND THE NATIONAL 'BUSINESS CYCLE': TOWARDS AN 'ALL-WEATHER COMPANY'

We start with inter-firm relations, a major development of the 1980s and 1990s. This links with Penrose's contribution, but arguably owes more to the work of George Richardson (1960, 1972). In Richardson's now classic article in the *Economic Journal*, the conventional distinction between market and hierarchy was put to task, given the 'dense network of *co-operation* and affiliation by which firms are inter-related' (1972, p. 884).

Echoing Penrose, Richardson attributed such inter-relationships to production-side related capabilities, as well as informational interdependence; see Richardson (1999). Co-operation provides an alternative to integration and it is pursued when there exist complementary but dissimilar activities. When activities are complementary but similar, integration is more likely, while weakly complementary activities are more amenable to market coordination.[1]

A particularly significant aspect of inter-firm relations are clusters, networks, webs and/or industrial districts.[2] These represent inter-firm linkages, often with a territorial dimension. They have been hailed in recent years as the major alternative to integration, in that they can achieve unit cost economies, normally associated with large size, through co-operation. At the same time, further such economies can be achieved due to the development of trust (which reduces transaction costs) and 'external economies', which further reduce unit costs. Clusters are also arguably more

[1] In her latest writings Penrose (1995, 1996) paid a lot of attention to 'networks'. She saw a growing tendency in the 1980s and 1990s for a number of firms 'to join together in corporate alliances or co-operative agreements, not necessarily with monopolistic intent but as a means of gaining mutual access to such resources as technology, regional markets and information services' (1966, p. 1722). She went on to observe that in such networks 'the individual companies do not lose their "independent" identity; but the administrative boundaries of each of the linked firms may become increasingly amorphous and the effective extent to which any individual firm exercises control not at all clear' (1996, p. 1723).

[2] It is beyond the scope of this paper to review developments in this field; for surveys see, among others You (1995), and more recently Holmström and Roberts (1998). Such developments as they impact on the theory of the TNC are covered in Dunning (1991). For Dunning, the significance of such inter-firm relations today are large enough to have led to what he calls 'alliance capitalism', see also Dunning (1998). Dunning attributes much of these relationships to the need to acquire complementary capabilities, in line with Richardson.

'bottom-up', a seedbed of larger firms and a source of innovation and competition for large firms. Members of a cluster also compete, often fiercely, while maintaining co-operation for joint inputs, marketing and innovation; see for example Best (1990), Porter (1990) and, more recently, You (1995) and Porter (1998).

There is an interesting aspect of clusters which pertains to the theory of the TNC. For many years, Dunning was among the very few (arguably lonely) champions of locational factors in explaining the TNC. The resurgence of interest in clusters has greatly facilitated the reappearance of geography and location in economics. Porter (1990) and Krugman (1991), among others, have led this resurgence. Models of the TNC, building explicitly on geographical issues, have also been developed; see for example Markuzen (1995). As Dunning (1998), however, observes, such models still fail to account for multinationalism as opposed to multi-dimensionality; they help to explain why a different location, but not necessarily why a different nation.

What could potentially explain TNCs in this context is arguably differences among countries concerning the agglomeration (clusters) of certain activities. It is well known that Silicon Valley is in the US, that the best ceramic tiles are in Italy and so on. Given such (however emergent) inter-national differences, TNCs interested in such activities could (and should actually be advised to) populate such clusters. In this sense, location, geography and inter-firm co-operation can coexist and provide a *raison d'être* for the existence of the TNC.

While locational factors, particularly in the role of clusters, could provide a missing link in the theory of the TNC, it is arguably not enough. All other considerations mentioned above, and more, could well be of use. Despite repeated claims concerning the difficulties and for some the impossibility of constructing a general theory of the TNC (Cantwell, 1999; Buckley and Casson, 1998B), and almost in contradiction to this claim, the 1990s have experienced a (welcome in our view) trend for pluralism and synthesis. Besides Dunning, major contributors in the field have recognised the need for synthesis, e.g. between transaction costs and monopolistic advantages (Buckley, 1990); ownership advantages and capabilities (Kogut and Zander, 1993); resources, capabilities and transaction costs (Buckley and Casson, 1998B); location and ownership advantages (Dunning, 1998); oligopoly and resources (Cantwell, 1999); transaction costs and oligopolistic interaction (Buckley and Casson, 1998A); location and resources (Ostry, 1998). Indeed much like the 1982 edition, Caves' 1996 edition of his classic book on the TNC reads like a synthesis of resource-based and transaction costs arguments, despite Caves' own phrasing in terms of transaction costs. Last, but not least, and in recognition of the static character of transaction costs economics, Buckley and Casson (1998B) identified 'flexibility' as the hallmark of recent modelling on the TNC. Flexibility is taken to be the ability to reallocate resources quickly and smoothly in response to change.[1]

---

[1] Interestingly, flexibility is claimed to discourage integration, thus partially explaining subcontracting and other neo-market-based firm strategies, notably export over FDI and licensing over internalisation. Flexible firms should exploit the relative advantage of market, hierarchy and co-operation. This focus on syntheses and dynamics is most welcome.

The discussion above affords more scope for synthesis and integration.[1] Penrose supplements Hymer in explaining endogenous growth, the internal generation of advantages and, partially, the direction of expansion. Hymer discusses pull-and-push factors for diversification. The various versions of transaction costs analysis explain the choice of mode. Dunning's OLI addresses the issue of location. Throughout, the pursuit of long-term profit, through innovation and/or monopoly restrictions and oligopolistic interaction, motivate and shape decisions and choices. Networks and the boundaries of the firms are a result of a dynamic interplay between all these and exogenous factors, such as government policies. Geographical agglomeration of economic activities taking place across boundaries can help to explain the TNC. In addition to the aforementioned, this synthesis should consider labour, in terms of intra-firm competition between management and labour,[2] widely acknowledged as the agency issue—see Alchian and Demsetz (1972), Jensen and Meckling (1976), Grossman and Hart (1986)—and in terms of the Marglinian and Sugden's focus on divide and rule strategies to reduce labour costs. A synthesis along the above lines is not only possible but also suggested by the widely used assumption of 'profit maximisation'. Once this is accepted as an aim, it becomes difficult to explain why a firm should stop short of redeploying its whole arsenal to achieve it. This can include the reduction of market transaction costs and also of labour and other costs, and/or the increase of prices through, for example, the monopolisation of markets.[3] The ability of a TNC to achieve all the above may, in fact, be viewed as a supply-side reason *per se* why firms choose to become TNCs; see Dunning (1992).

Still missing from this synthesis is a consideration of national demand-side and/or business cycle-related factors differentiating domestic from foreign diversification. As already noted, in Hymer, a push factor for domestic diversification is the product life-cycle. Extending this reasoning, we suggest here that another (push-and-pull) factor contributing to foreign diversification can be (is) the 'business cycle' and more specifically differential effective demand-growth related (locational) factors among countries. This need not be the only such factor, but we claim it to be an important one.

While supply-side considerations are important in motivating international produc-

---

[1] The need for synthesising transaction costs to the resource-based view is acknowledged by Penrose (1996) and by Coase (personal correspondence). In Penrose's words 'the firm that expands internationally does so for much the same reasons as the firm that expands nationally. The relevant considerations in both cases are as follows: (1) the cumulative growth of knowledge combined with the perceived advantages of expansion for both the firm and the individuals concerned with decision making; (2) the internalization of activities; (3) the fact that size, scope, market position and existing integration give rise to firm-specific assets of which better use can be made internationally than nationally; (4) the subsequent cumulatively increasing production of knowledge with respect to international possibilities' (1996, p. 1720).

[2] While widely recognised in the 'agency' literature, in the Marxist tradition this intra-firm struggle provides an inducement for endogenous technological developments (of the labour saving type). This is akin to, and complements, the Penrosean endogenous growth theory, and is to our knowledge the only (along with Penrose's) theory of endogenous innovation.

[3] In our framework 'profit maximisation' is seen as the pursuit of maximum *possible*, long-term profitability. It is questionable whether the profit maximisation assumption of neoclassical price theory is compatible with bounded rationality; see Kay (1984).

tion, it does not follow that demand-side issues cannot play a role. Demand-side factors can be of the push type or of the pull type, or both. On the push side, declining domestic effective demand, low expected rates of profitability and growth can be of importance. From the pull side, faster expected growth rates and/or profits abroad can be seen as locational advantages; see Dunning (1998). When the two are combined, with demand declining at home alongside accumulated retained profits and other ownership or monopolisic advantages, there remains almost no other choice. It follows that a demand-side argument can legitimately be made. Interestingly, and despite the fact that such factors are often underplayed in the literature, demand-side considerations are often implicit in other theories, notably in Dunning, and they usefully complement Hymer's explanation of product life-cycle-based domestic diversification. Demand-side questions, from the point of view of the firm, also enter Vernon's (1966) 'product cycle' hypothesis.

In Vernon's approach, products are seen to have a life-cycle with three main phases: introduction and growth, maturity, and decline. In the first phase production takes place at home for various reasons, such as the need for careful control and monitoring of the market. In the second phase the product becomes standardised and given that it is already somewhat known abroad through exports, DFI is contemplated. In the third phase, DFI becomes inevitable, as tariffs tend to constrain further exports. Scale economies tend to be exhausted at home and servicing foreign markets becomes very difficult.

Vernon's theory has been attacked as an inadequate account of post World War II TNC activities, notably by Buckley and Casson (1976), on grounds such as the difficulty the theory has to explain non-export substituting investments, the appearance of non-standardised products being produced abroad and the case of carefully differentiated products to suit the local market. Another potential criticism of the theory is that the decline in the rate of growth of demand in the maturity phase can be avoided through unrelated diversification to products at a different phase of their cycle. In this sense, conglomeration can be seen as an answer to the vagaries of the product cycle, as in Hymer (1979).

All the aforementioned criticisms derive from the focus of Vernon's approach to demand for the individual product, not aggregate demand. The aggregate demand deficiency argument applies to all firms, at all phases of their products, albeit to different degrees of severity, depending, among others, on the phase of the product cycle. DFI, in our framework, can be seen as firms' reply to the vagaries of the 'business cycle', thus giving rise to a genuinely 'all weather company'. In this sense, the aggregate demand argument goes beyond and survives the criticisms of the product cycle theory while, however, retaining the important role of demand considerations.

To summarise, demand-side, business cycle-related considerations build on the tradition and insights of Penrose and Hymer and provide, in our view, an important distinguishing factors between domestic and international diversification. We believe that the focus on the demand-side is also welcome, in that it counterbalances the almost exclusive focus of existing theories of supply-side factors. To the extent international location matters, it is almost self-evident that international macro-economic, demand-related conditions will be of importance.[1] However, it is not claimed here that demand-side, business cycle-related considerations provide a theory of the TNC. To do this, one needs to bring together all issues raised in this paper; failing or missing markets, differential firm advantages, locational factors and oligopolistic interaction.

In the context of this paper, and building on Penrose, the TNC is just a firm. Its explanation requires a theory of the growth of the firm and limits to growth, plus a justification for the special importance of national state boundaries. The theory of (limits to) growth is therefore, the Penrosean one. It is compatible with, and fully complements the Hymer story. Both, however, need to be complemented with transaction costs-related arguments, in order to explain more adequately the choice of institutional mode. Locational factors can be one factor that separates national from international diversification. Geographical agglomeration can constitute a (potentially international) locational advantage, thus a reason for the TNC. Demand-side, business cycle-related considerations are also of importance. They can be seen as yet another international location (dis)advantage, a pull-and-push factor for FDI, in firms' pursuit of being 'all weather companies'. Endogenous constraints as in Penrose, alongside oligopolistic interaction, explain the limit to the growth of the TNC.

In this story, Penrose's contribution has been seminal. Hers is the endogenous (inducements to) growth theory, the explanation of the limits to growth, and of the direction of expansion. Her theory, moreover, is both compatible and complements the Hymer story in explaining expansion but also the direction of advantages (in her case both monopolistic and non-monopolistic). Last but not least, hers is the observation that to explain TNCs we need only to look at how and why national boundaries can make a difference. By grafting dynamic transaction costs onto the Penrose–Hymer evolutionary tale and by recognising the importance of international clustering and demand-side, business cycle-related reasons we can add onto the Penrose–Hymer story and come up with most elements of the puzzle. These add up, not to a general theory of the TNC, but provide the elements of an integrative framework and the tools to be applied in specific cases so as to explain and/or predict the specific actions of specific firms.

IV. CONCLUSION

Penrose's theory of the growth of the firm can provide the overall framework, additional insights to, and the glue that can bind together existing supply-side theories of the TNC. Alongside demand-side considerations, inter-firm competition and locational factors, the emergent integration can provide an overall framework (albeit not a general theory) for explaining the TNC.

---

[1] For some econometric support of the demand-side hypothesis for the case of the UK outward investment, see Pitelis (1996).

REFERENCES

ALCHIAN, A. and DEMSETZ, H. (1972). Production, information costs and economic organization, *American Economic Review*, Vol. 62, No. 5, 777–95.

BEST, M. (1990). *The New Competition: Institutions for Industrial Restructuring*, Oxford, Polity Press.

BOWLES, J. (1985). The production process in a competitive economy, *American Economic Review*, 75, 16–36.

BUCKLEY, P. (1990). Problems and developments in the core theory of international business, *Journal of International Business Studies*, Vol. 21, No. 4, 657–65.

BUCKLEY, P. J. and CASSON, M. C. (1976). *The Future of Multinational Enterprise*, London, Macmillan.

BUCKLEY, P. J. and CASSON, M. C. (1998A). Analyzing foreign market entry strategies: extending the internalization approach, *Journal of International Business Studies*, Vol. 29, No. 3, 539–62.

BUCKLEY, P. J. and CASSON, M. C. (1998B). Models of the multinational enterprise, *Journal of International Business Studies*, Vol. 29, No. 1, 21–44.

CANTWELL, J. (1991). Theories of international production in *The Nature of the Transnational Firm*, edited by C. N. PITELIS and R. SUGDEN, London, Routledge [2nd edn, 2000].

CAVES, R. E. (1982). *Multinational Enterprise and Economic Analysis*, Cambridge, Cambridge University Press [2nd edn, 1996].

CHANDLER, A. D. (1962). *Strategy and Structure*, Cambridge, MA, MIT Press.

COASE, R. H. (1937). The nature of the firm, *Economica*, Vol. 3, 386–405.

COWLING, K. and SUGDEN, R. (1987). *Transnational Monopoly Capitalism*, Hemel Hempstead, Wheatsheaf.

DEMSETZ, H. (1995). *The Economics of the Business Firm: Seven Critical Commentaries*, Cambridge, Cambridge University Press.

DUNNING, J. H. (1958). *American Investment in British Manufacturing Industry*, London, Allen and Unwin.

DUNNING, J. H. (1981). *International Production and Multinational Enterprise*, London, Allen and Unwin.

DUNNING, J. H. (1988). The eclectic paradigm of international production, *Journal of International Business Studies*, Vol. 19, 1–31.

DUNNING, J. H. (1989). *Explaining International Production*, London, Unwin Hyman.

DUNNING, J. H. (1991). The eclectic paradigm in international production, in *The Nature of the Transnational Firm*, edited by C. N. PITELIS and R. SUGDEN, London, Routledge, 116–36 [2nd edn, 2000].

DUNNING, J. H. (1992). The competitive advantage of countries and the activities of transnational corporations, *Transnational Corporations*, Vol. 1, No. 2, 135–68.

DUNNING, J. H. (1998). Location and the multinational enterprise: A neglected factor? *Journal of International Business Studies*, Vol. 29, No. 1, 45–66.

DUNNING, J. H. and RUGMAN, A. (1985). The influence of Hymer's dissertation on the theory of foreign direct investment, *American Economic Review*, Vol. 75, 228–39.

FIELDHOUSE, D. (1986). The multinational: a critique of a concept, in *Multinational Enterprise in Historical Perspective*, edited by A. TEIHOVA, M. LEVY-LEBOYER and H. NUSSSBAUM, Cambridge, Cambridge University Press.

FOSS, N. J. (ed.) (1997). *Resources, Firms and Strategies*, Oxford, Oxford University Press.

GRAY, H. P. (1985). Macroeconomic theories of foreign direct investment: an assessment, in *New Theories of Multinational Enterprise*, edited by A. RUGMAN, London, Croom Helm.

GROSSMAN, S. and HART, O. (1986). The costs and benefits of ownership: a theory of lateral and vertical integration, *Journal of Political Economy*, Vol. 94, 691–719.

HENNART, J.-F. (2000). Transaction costs theory and the multinational enterprise, in *The Nature of the Transnational Firm*, edited by C. PITELIS and R. SUGDEN, 2nd edn, London, Routledge.

HOLMSTROÖM, B. and ROBERTS, J. (1998). The boundaries of the firm revisited, *Journal of Economic Perspectives*, Vol 12, No. 4, 73–94.

HYMER, S. H. (1970). The efficiency (contradictions) of multinational corporations, *American Economic Review Papers and Proceedings*, Vol. 60, 441–8.

HYMER, S. H. (1972). The multinational corporation and the law of uneven development, *Economics and World Order*, edited by J. N. BHAGWATI, London, Macmillan.

HYMER, S. H. (1976). *The International Operations of National Firms. A Study of Foreign Direct Investment*, Cambridge, MA, MIT Press.

HYMER, S. H. (1979). in *The Multinational Corporation*, edited by R. B. COHEN et al., Cambridge, Cambridge University Press.

JENKINS, R. (1987). *Transnational Corporations and Uneven Development*, London, Methuen.

JENSEN, M. C. and MECKLING, W. (1976). Theory of the firm: managerial behaviour, agency costs and ownership structure, *Journal of Financial Economics*, Vol. 3, 304–60.

KAY, N. M. (1984). *The Emergent Firm: Knowledge, Ignorance and Surprise in Economic Organization*, London, Macmillan.

KAY, N. M. (1991). Multinational enterprise as strategic choice: some transaction cost perspectives, in *The Nature of the Transnational Firm*, edited by C. PITELIS and R. SUGDEN, London, Routledge.

KAY, N. M. (2000). The resource-based approach to multinational enterprise, in *The Nature of the Transnational Firm*, edited by C. PITELIS and R. SUGDEN, 2nd edn, London, Routledge.

KINDLEBERGER, C. P. (1969). *International Business Abroad*, New Haven, CT, Yale University Press.

KOGUT, B. and ZANDER, U. (1993). Knowledge of the firm and the evolutionary theory of the multinational corporation, *Journal of International Business Studies*, Vol. 24, 625–45.

KRUGMAN, P. R. (1991). *Geography and Trade*, Cambridge, MA, MIT Press.

MARGLIN, S. (1974). What do bosses do? The origins and functions of hierarchy in capitalist production, *Review of Radical Political Economics*, Vol. 6, 60–112.

MARKUSEN, J. R. (1995). The boundaries of multinational enterprise and the theory of international trade, *Journal of Economic Perspectives*, Vol. 9, No. 2, 169–89.

MARX, K. (1959). *Capital*, Vol. I, London, Lawrence and Wishart.

NEWFARMER, R. (ed.) (1985). *Profits, Progress and Poverty: Case Studies of International Industries in Latin America*, Notre Dame, IN, Notre Dame University Press.

OSTRY, S. (1998). Technology, productivity and the multinational enterprise, *Journal of International Business Studies*, Vol. 29, No. 1, 85–99.

PENROSE, E. T. (1955). Research on the business firms: limits to growth and size of firms, *American Economic Review*, Vol. XLV, No. 2, 531–43.

PENROSE, E. T. (1956). Foreign investment and the growth of the firm, *Economic Journal*, Vol. LXVI, 220–35.

PENROSE, E. T. (1959). *The Theory of the Growth of the Firm*, Oxford, Oxford University Press [3rd edn, 1995].

PENROSE, E. T. (1960). The growth of the firm—a case study: The Hercules Powder Company, *Business History Review*, vol XXXIV, 1–23.

PENROSE, E. T. (1962). Some problems of policy towards direct private foreign investment in developing countries, *Middle East Economic Papers*, Lebanon, American Research Bureau, American University of Beirut, 121–39.

PENROSE, E. T. (1964). Monopoly and competition in the international petroleum industry, in *The Yearbook of World Affairs*, Vol. 18, London, Stevens.

PENROSE, E. T. (1968). Problems associated with the growth of international firms, in *Tijdschrift voor Vennootschappen, Vereiningingen en Stichtingen*, Vol. II, No. 9, 219–26.

PENROSE, E. T. (1973). The changing role of multinational corporations in developing countries. Paper submitted to the United Nations Group of Eminent Persons to Study the Impact of Multinational Corporations on Development and on International Relations, Geneva.

PENROSE, E. T. (1985). The theory of the growth of the firm: twenty five years after, *Acta Universitatis Upsaliensis: Studia Oeconomicae Negotiorum*, (Uppsala Lectures in Business) Vol. 20, 1–16.

PENROSE, E. T. (1987). Multinational corporations, in *The New Palgrave: A Dictionary of Economics*, London, Macmillan, 562–4.

PENROSE, E. T. (1990). Dumping, 'unfair' competition and multinational corporations, *Japan and the Word Economy*, Vol. 1, 181–7.

PENROSE, E. T. (1996). Growth of the firm and networking, in *International Encyclopaedia of Business and Management*, London, Routledge, 1716–24.

PEOPLES, J. and SUGDEN, R. (2000). Divide and rule by transnational corporations, in *The Nature of the Transnational Firm* edited by C. PITELIS and R. SUGDEN, 2nd edn, London, Routledge.

PITELIS, C. N. (1991). *Market and Non-Market Hierarchies: Theory of Institutional Failure*, Oxford, Basil Blackwell.

PITELIS, C. N. (1996). Effective demand, outward investment and in the (theory of the) transnational corporation: an empirical investigation, *Scottish Journal of Political Economy*, Vol. 43, No. 2, 192–206.

PITELIS, C. N. and PSEIRIDIS, N. A. (1999). Transaction costs versus resource value?, *Journal of Economic Studies*, Vol. 26, 221–40.

PITELIS, C. N. and SUGDEN, R. (eds) (1991). *The Nature of the Transnational Firm*, London, Routledge.

PITELIS, C. N. and SUGDEN, R. (2000). The (theory of the) transnational firm: the 1990s and beyond, in *The Nature of the Transnational Firm*, edited by C. PITELIS and R. SUGDEN, 2nd edn, London, Routledge.

PITELIS, C. N. and WAHL, M. (1998). Edith Penrose: A pioneer of stakeholder theory, *Long Range Planning*, Vol. 31, 252–61.

PORTER M. E. (1980). *Competitive Strategy*, New York, NY, Free Press.

PORTER M. E. (1990). *The Competitive Advantage of Nations*, Basingstoke, Macmillan.

PORTER M. E. (1998). Clusters and the new economics of competition, *Harvard Business Review*, November/December, 77–90.

PRAHALAD, C. K. and HAMEL, G. (1990). The core competence of the corporation, *Harvard Business Review*, Vol. 90, No. 3, 79–91.

RICHARDSON, G. (1960). *Information and Investment: A Study in the Working of the Competitive Economy*, Oxford, Clarendon Press.

RICHARDSON, G. (1972). The organisation of industry, *Economic Journal*, Vol. 82, 883–96.

RICHARDSON, G. (1999). Mrs Penrose and neoclassical theory. *Contributions to Political Economy*, Vol. 18, 23–30.

RUGMAN, A. M. (1986). New theories of the multinational enterprise: an assessment of internalization theory, *Bulletin of Economic Research*, Vol. 38, 101–18.

SUGDEN, R. (1991). The importance of distributional considerations, in *The Nature of the Transnational Firm*, edited by C. PITELIS and R. SUGDEN, London, Routledge, 168–93.

TEECE, D. J. (1977). Technology transfer by multinational firms: the resource costs of transferring technological know-how, *Economic Journal*, Vol. 87, 242–61.

VERNON, R. (1966). International investment and international trade in the product cycle, *Quarterly Journal of Economics*, Vol. 80, 190–207.

WILLIAMSON, O. E. (1975). *Markets and Hierarchies*, New York, NY, Free Press.

WILLIAMSON, O. E. (1981). The modern corporation: origins, evolution, attributes, *Journal of Economic Literature*, Vol. 19, No. 4, 1537–68.

YAMIN, M. (1991). A reassessment of Hymer's contribution to the theory of the transnational corporation, in *The Nature of the Transnational Firm*, edited by C. PITELIS and R. SUGDEN, London, Routledge.

YOU, J.-I. (1995). Small firms in economic theory, *Cambridge Journal of Economics*, Vol. 19, 441–62.

9

EDITH PENROSE, ECONOMICS AND STRATEGIC MANAGEMENT

NICOLAI J. FOSS

Department of Industrial Economics and Strategy, Copenhagen Business School

Edith Penrose's work on *The Theory of the Growth of the Firm* is often seen as a classic in strategic management and a precursor of today's resource-based view of the firm. However, it is argued in this paper that this is, at best, imprecise and rather a misrepresentation of Penrose: her interest was not in whether there can be rents in equilibrium, but in firm growth as a disequilibrium phenomenon. Moreover, her basic vision of the competitive process is very different from that which animates modern economic approaches to strategy, such as the resource-based approach and new industrial organisation economics. This paper briefly discusses some implications for modern strategic thinking of Penrose's vision.

Strategic management does not have many classics. Granted, it is hard to have established and recognised classics in a field that is not only very young (reaching back only to the 1950s) but also very fragmented (Mintzberg, 1990). But there is a handful of books that most strategy scholars would recognise as truly classic contributions to the field. Among these is Edith Penrose's *The Theory of the Growth of the Firm*. When the book was reissued by Oxford University Press in 1995, the back of the new edition contained endorsing statements by two strategy scholars (Sumantra Ghoshal and Ikujiro Nonaka) and one heterodox economist (Richard Nelson). In fact, this rather accurately reflects the composition of the contemporary audience of this book, which originally was clearly aimed at an audience consisting of economists. Mainstream economists, even those working in industrial organisation, are unlikely to have encountered the book.[1]

Penrose's influence in strategic management has largely come about as a result of her being considered the founding matriarch of the so-called 'resource-based perspective' (henceforth, RBP). This perspective has become increasingly influential during the last 15 years (since Wernerfelt, 1984) and is arguably the dominant approach in contemporary strategic management. The RBP took over from the once dominant industry analysis approach that is commonly associated with Michael Porter (1980) and which was developed on the basis of the Bain–Mason–Scherer structure-

---

[1] Best and Garnsey (1999, pp.187–8) are surely going too far, when they argue that Penrose (1959) is 'recognized by, if not familiar to, every economist'.

conduct–performance (SCP) paradigm in industrial organisation (Scherer and Ross, 1990).[1] Thus, many discussions in strategic management have taken place under the heading of 'Penrose versus Porter' and the RBP is widely thought as a restatement, even rediscovery, of positions originally associated with Penrose (1959) (e.g. Werner-felt, 1984; Mahoney and Pandian, 1992; Peteraf, 1993; Williams, 1994; Kor and Mahoney, 2000).

I think this reading is largely incorrect, and shall explain why in this paper. In fact, the influence from Penrose to the RBP is virtually non-existent. Instead, the RBP is founded on ideas from basic neoclassical price theory, particularly of the type developed by economists associated with the University of Chicago and UCLA. Thus, many of the differences, and many of the controversies, between resource-based scholars and adherents to industry analysis or new industrial organisation approaches to strategy (Tirole, 1988) is in actuality—and *mutatis mutandis*—a replay of the classic public policy debates between Harvard and Chicago (Goldschmid, Mann and Weston, 1974).

In contrast, Penrose's work is, in the crucial dimensions, at variance with economic orthodoxy—whether Chicago, Harvard or the new industrial organisation style.[2] It should rather be thought of as a contribution to economic heterodoxy. Penrose's basic, and too often overlooked, themes are flexibility in an uncertain world, organ-isational learning as an evolutionary discovery process, the vision of the management team, entrepreneurship, etc. Her work on growth of the firm through diversification is an application of those themes.

It is therefore not surprising that only a few dimensions of her richly faceted work can be found in the RBP. Her interest was not in whether there may be efficiency rents in equilibrium (hence, sustainability of competitive advantage). Instead, her emphasis was on learning, disequilibrium, differential cognition, organisation, and the creation of real options. Her work points towards a distinctly non-orthodox approach to not only strategy but also firm organisation. It is in many respects closer to behavioural perspectives on firms (e.g. March and Simon, 1958; March, 1988; Beach, 1993; Weick, 1995; Shapira, 1997) than to more economics-based approaches.

In the following pages, I shall seek to substantiate these claims. Thus, my focus is partly exegetical, partly reconstructive. The focus will be almost exclusively on Penrose's 1959 work. This is certainly not because it is the only interesting work she wrote. Rather, it is undeniably her masterpiece; a masterpiece, however, whose fundamental message has been insufficiently appreciated. Moreover, much of her

---

[1] However, industrial organization perspectives have made a strong come-back recently (Chemawat, 1991, 1997; Brandenburger and Nalebuff, 1996).

[2] Admittedly, this is a somewhat delicate issue. To be sure, there is critique of neoclassical economics in Penrose (1959), but this has mostly to do with the inability of the neoclassical theory of the firm to explain firm growth (pp. 1, 15). Arguably, Penrose adopts something like the position later expounded by Machlup (1967) that while the neoclassical theory of the firm is indeed inappropriate for some purposes, it is clearly appropriate for other purposes (such as comparative-statics). However, the foreword to the new and third edition of Penrose (1959) is explicitly critical of economic orthodoxy. See also Pitelis and Wahl (1998) for a reading of Penrose that stresses her differences from more mainstream approaches.

other work is not related to the arguments in this book, and is, thus, not relevant to bring into the discussion.[1]

I. ECONOMIC APPROACHES TO STRATEGY

I.a. Economics and strategy

There can be little doubt that economics has become a dominant voice in the contemporary conversation of strategy scholars, at least since 1980 (Porter, 1980) and perhaps even earlier. There are many reasons for this. Among the obvious ones are that economics provides a relatively clear-cut language and a number of useful established insights, allows strategy scholars to interpret (long-lived) performance differences between firms (e.g. by means of entry, exit and mobility barriers), and that in general a number of the basic strategic issues lend themselves rather naturally to an economic treatment. Thus, increasingly, concepts such as 'subgame perfect equilibrium', 'asset specificity' and 'real options' have become part of the discourse in strategy, and even strategy scholars who otherwise have difficulties with economic approaches to strategy accept that the essence of strategy is the pursuit of economic rent.

Arguably, there are two overall approaches—each existing in different versions—that dominate contemporary economic approaches to strategy research.[2] The first one is the *industry analysis approach* originally developed by Michael Porter (1980) and later updated in the light of breakthroughs in game-theoretical, new industrial organisation economics (Tirole, 1988; Shapiro, 1989) by Brandenburger and Nalebuff (1996), Chemawat (1997) and others. The other dominant contender is the *resource-based view*, launched by a number of scholars in the mid-1980s (Teece, 1982; Wernerfelt, 1984; Barney, 1986). I shall briefly discuss these two overall approaches. The point that I shall seek to substantiate is that the much-discussed differences between these two approaches vanish if compared to their differences to the vision that animates Penrose's work.

I.b. Industrial organisation approaches

In the mid-1970s, economists-turned-strategy scholars, such as Richard Caves and Michael Porter, realised that the Bain–Mason structuralist approach in industrial organisation (IO) could be very usefully applied to the study of firm strategies and also for deriving practical recommendations. To Caves and Porter, basic IO concepts such as entry barriers and the collusion such barriers may foster offered an explanation of,

---

[1] For example, surprisingly Penrose's work on multinationals (Penrose, 1968, 1971) does not draw on her 1959 book. Thanks to Neil Kay for pointing this out to me.

[2] Transaction cost economics (Williamson, 1996) and agency theory (Holmström, 1982; Holmström and Milgrom, 1991) may also be mentioned as strong voices, but they are more narrowly concerned with the firm's boundary choices and internal organisation, respectively. Thus, neither has anything to say about key strategy issue of the firm's positioning in the product market. For a review and discussion of alternative approaches to the 'strategic theory of the firm', see Foss (1999C).

for example, the observed persistence of above-normal profit. However, it was not entirely unproblematic to rely on IO in strategy research; the would-be importer of IO to the strategy field confronted a basic translation problem, as Porter (1981) clearly recognised.[1] Thus, IO was fundamentally static, it did not seriously consider the diversified firm, it saw the firm as a unitary decision-maker, it had an industry—rather than a firm—focus, it operated with perfect competition as the ultimate yardstick for purposes of welfare comparisons, etc. (Porter, 1981; Scherer and Ross, 1990). This was much in contrast to the mainstream of management literature, which saw strategy as a matter of entrepreneurial action in an uncertain and hard-to-predict environment (Ansoff, 1965), did not neglect the large, diversified corporation (Chandler, 1962), and was very much concerned with the internal workings of the firm (Bower, 1970).[2]

Although Porter was well aware of the problems this raised for an application of IO to the strategy discipline (Porter, 1981), crucial characteristics of IO did in fact carry over to his industry analysis approach (Porter, 1980). An unfortunate example is the black-box conceptualisation of the firm that is characteristic of older IO and which is clearly present in *Competitive Strategy* (Porter, 1980). Another one is in the focus on non-cooperative equilibria where firms earn rents from their market-power because of their ability to engage in tactics designed to build and maintain mobility and entry barriers.

All this was verbally stated and the equilibrium focus in Porter's work was rather implicit. The upsurge in work on the new IO that took place at the beginning of the 1980s changed this. The study of firm strategy came to be understood as founded on game-theoretical studies of behaviour and performance in imperfectly competitive markets (Tirole, 1988; Schmalensee and Willig, 1989; Shapiro, 1989; Saloner 1994). According to the prominent new IO scholar, Carl Shapiro (1989), recent work in new IO can virtually be identified with 'the theory of business strategy'. Indeed, he goes as far as asserting that '[a]t this time, game theory provides the only coherent way of logically analyzing strategic behavior' (1989, p. 125). 'Strategic behavior', in this approach, means engaging in behaviour that by influencing rivals' expectations of one's future behaviour is able to influence significantly the behaviour of those rivals to the benefit of the strategising firm.

Although the Porter industry analysis framework is not identical to the new IO, they have a common ancestor in older IO, and share many of the same assumptions and concerns.[3] In some ways, however, the new IO represents an advance relative to the Porter framework. For example, firms in the new IO are not homogenous. Thus, they may differ not only in terms of their cost structures but also in terms of, for example, their reputations (Tirole, 1988, p. 256). Moreover, the notions of factor/resource

---

[1] For example, Bain (1959) explicitly excluded from the focus of IO any '. . . internal approach, more appropriate to the field of management science, such as could inquire how enterprises do and should behave in ordering their internal operations and would attempt to instruct them accordingly' (pp. vii–viii).

[2] We note in passing that Penrose's basic vision is congenial to the mainstream of management and strategy literature, particularly the type of work associated with Christensen and Andrews (e.g. Andrews 1971).

[3] However, Porter's (1980) analysis of issues such as signalling, is strikingly modern.

indivisibility and immobility become central, primarily because these notions play a key role in understanding entry-deterrence and, more generally, the notions of credible threats and commitments.

In this view, strategy becomes primarily a matter of deploying given resources to a product-market, and utilising them in sophisticated plays and counter-plays. Strategy becomes a matter of extracting maximum monopoly rents out of 'fixed factors over the planning horizon' (Caves, 1984, p. 128). Thus, firms in the new IO are clearly different, but the sources of heterogeneity are given and fixed; firms do not themselves create their own opportunity set. To some extent, this is because the agents that populate new IO models are incredibly smart. Here, a strategy involves anticipating any and all actions that other players might take in all future stages of the game, and calculating the optimal response. Since all players are able to do this, the equilibrium position is essentially given from the beginning. Players cannot be surprised by unexpected events, there is never any difference between the competence of players and the difficulty of decision problems, and although agents may formally learn in Bayesian games, their learning *functions* never change.

I.c. The resource-based perspective

Whereas the origins of the Porter industry analysis approach were very clearly in Bain–Mason–Scherer SCP theory, the resource-based view may be seen as drawing inspiration from the antagonist of the Bain–Mason view, namely the Chicago approach to IO.[1] In the Chicago view, entry barriers are informational, concentration is a result of efficiency, and high returns are returns to efficient underlying assets rather than monopoly profits stemming from restriction of supply (e.g. Demsetz, 1973, 1982). Such returns may be long-lived because of the complexity of the assets that cause them (Demsetz, 1973). Moreover, assets are not necessarily priced according to their value, because of informational asymmetries (*idem*). Thus, the Chicago view of IO is one that stresses efficiency in a world constrained by informational scarcity; in contrast, SCP theory stressed market power and gave little attention to issues of information.

The Chicago influence can be clearly seen in the resource-based analysis of the conditions for sustained competitive advantage (Peteraf, 1993). Thus, according to this analysis, resources yield a sustained competitive advantage when they meet the criteria of:

(1) *heterogeneity*—this implies efficiency differences across resources and therefore the existence of rents;
(2) *ex ante limits to competition*—i.e. resources have to be acquired at a price below their discounted net present value in order to yield rents. Otherwise future rents will be fully absorbed in the price paid for the resource (Demsetz, 1973; Barney, 1986);

---

[1] See Foss (1999A) for a more detailed analysis of the relation between the Chicago view and the RBP.

(3) *ex post limits to competition*—i.e. it should be difficult or impossible for competitors to imitate or substitute rent-yielding resources (Lippman and Rumelt, 1982);[1] and

(4) *imperfect mobility*—i.e. the resource should be relatively specific to the firm. Otherwise, the superior bargaining position that is obtained from not being tied to a firm can be utilised by the resource (or the resource's owner) to appropriate the rent (or, at least a large portion of the rent) that the resource helps create.

Clearly, this rather basic exercise explicitly draws on simple, economic equilibrium price theory as set out in any standard textbook on the subject.[2] It is easy, indeed, to discern the role of equilibrium assumptions in the RBP. For example, Peteraf (1993) develops the concept of Ricardian rent using efficiency differences across firms under competitive equilibrium as a benchmark. And Barney (1986) utilises the finance theory concepts of strong and weak efficiency to elucidate the reasoning behind the concepts of perfect factor markets and factor market imperfections. Indeed, the very concept of sustained competitive advantage is often defined in equilibrium terms: it is that advantage which lasts after all attempts at imitation have ceased. So, (zero imitation) equilibrium is utilised as a yardstick to define and understand (sustained) competitive advantage.

I.d. Spanners in the works

On the face of it, the RBP and (Bain–Mason or new) industrial organisation approaches to strategy are widely different, and this is indeed how they have normally been portrayed. Thus, whereas the former stresses an efficiency perspective, focuses on the firm's resources (to the neglect of the external environment) and adopts a factor market perspective, the latter stresses a market power perspective and focuses on product market interaction (e.g. Barney, 1991). Or so the conventional wisdom has it.

However, these differences may be more apparent than real. First of all, note that both approaches conceptualise competitive advantage as rents that can be sustained in equilibrium. Second, the 'trick' with which these rents are generated is in both cases the time-honoured one of throwing one or two spanners into the works of an otherwise perfect world. In the case of the RBP, it is assumed, for example, that successful firms' technologies are not observable by less successful firms (Lippman and Rumelt, 1982), and in the case of the industrial organisation approach, it is assumed that, for example, firms may establish Stackelberg commitments by committing large sunk costs to certain markets or may build reputations for 'irresponsibly' aggressive pricing behaviour in the case of entry (Tirole, 1988). Note that these spanners in the works are essentially

---

[1] As Dierickx and Cool (1989) clarify, there are a number of mechanisms at work that often make it hard for competitors to copy the sources of competitive advantage of a successful firm. For example, there may be 'causal ambiguity', which means that competitors confront difficulties ascertaining precisely how a bundle of resources contributes to success.

[2] For a very interesting Austrian critique of the RBP that also stresses the orthodox nature of the RBP, see Lewin and Phelan (1999).

informational in nature. Moreover, in all these cases, the emphasis is, in actuality, on 'sticky' resources.

Thus, in spite of their different intellectual backgrounds and ways of expressing basic insights, both of these two approaches stress that product market advantages are dependent on factor market imperfections. And both start out from an otherwise perfect neoclassical world and introduce selected imperfections in order to generate the desired results, namely the existence of rents in equilibrium. In fact, in neither the resource-based nor the industrial organisation view are innovation, learning and differential cognition prominently placed. Essentially, this is because both views start out from the same neoclassical basis in which perfection rules in all dimensions, except those few that the analyst singles out as relevant spanners in the works. This, as we shall see, is very different from Penrose's approach. However, we first need to consider how mainstream economists have tried to cope with Penrose's analysis.

II. THE RECEIVED VIEW

II.a. The main argument of The Theory of the Growth of the Firm

The conventional outline of the main argument of Penrose's 1959 book is well-known and shall be only briefly summarised: firms are collections of productive resources that are organised in an administrative framework which partly determines the amount and type of services that the resources yield. As they go along with their productive operations, firms—or, more precisely, the management team—obtain increased knowledge of the services that may be obtained from resource.

The results of such learning processes is, first, the expansion of the firm's 'productive opportunity set' (the opportunities that the firm's management team can see and can take advantage of) and, second, the release of managerial excess resources that can be put to use in other, mostly related, business areas. Since the opportunity costs of unused, excess resources are zero, there will be a strong internal incentive for such diversification which in turn causes the firm to grow—an idea that according to Penrose destroys the notion of the firm's optimum size. However, the managerial resources inherited from the past set a limit to the firm's rate of growth. In Penrose, this is rationalised by pointing to the difficulties of socializing new managers that are needed for the expansion of the firm. Thus, the main theme of the book concerns the growth of successful firms through related diversification, and this main theme has indeed been well represented in much of the diversification research since Rumelt (1974).[1] In contrast, Penrose is not taken up with what is arguably the main aim of the RBP, namely to analyse the circumstances under which resources may provide long-lived rent streams.

---

[1] In fairness, it must be observed that an important research theme in the resource-based perspective has in fact been diversification, and here Penrose's insights have arguably been more adequately represented, although much of this literature is also cast within the maximisation/equilibrium framework of mainstream economics (e.g. Montgomery and Wernerfelt, 1988).

II.b. Mainstream interpretations of Penrose

A number of mainstream writers who took an interest in Penrose's work (Baumol, 1962; Marris, 1964) imposed the Penrose effect exogenously (Gander, 1991), but eventually it became subordinated under supposedly more general theories of adjustment costs in the theory of investment of the firm (Treadway, 1970). Given this, and using a dynamic control theory approach, the Penrose effect arises naturally as the profit-maximising firm calculates its optimal time-profile of outputs. A steady-state equilibrium growth pattern is shown to exist, in which firm size and management size grow at the same rate (Marris, 1964; Slater, 1980; Gander, 1991). The high point of this literature is Rubin's (1972) attempt to reconstruct Penrose (1959) in terms of finding the solution to the dynamic optimisation problem of balancing the development of new resources (using existing resources) and the use of existing resources directly in production.

Thus, Penrose's ideas became absorbed in mainstream economics by, first, arguing that her purportedly most important point, 'the Penrose effect', was just a minor detail in the neoclassical analysis of optimal investment, and, second, by demonstrating that Penrose's critique of equilibrium economics should not really be taken seriously, as her ideas were fully compatible with extended notions of equilibrium. The third wave of inclusion of Penrosean ideas in mainstream economics is marked by the emergence of the resource-based perspective. The attempt to interpret Penrose's ideas in terms of mainstream economics is perhaps not so surprising, if we take these ideas to merely be that (1) there is a constraint on the growth of the firm stemming from the difficulties of expanding the management team, and (2) firm heterogeneity is the source of differences among revealed competitive advantages. Both of these points lend themselves to formal modelling without too much difficulty. However, here I will argue that there is more to *The Theory of the Growth of the Firm* than these points.

Moreover, we should not forget that although mainstream treatments of Penrose's insights may be valuable and justified, they run counter to Penrose's own critique of (parts of) neoclassical economics. As she saw it herself, her theory constituted a powerful critique against certain aspects of the neoclassical theory of the firm. In the neoclassical theory of the firm, she says, there is '. . . no notion of an *internal* process of *development* leading to cumulative movements in any one direction' (1959, p. 1), a notion that is absolutely crucial for understanding firm development. Rather, growth is simply a matter of adjusting to the equilibrium size of the firm. But if services are produced endogenously (and continuously) through various intra-firm learning processes involving increased knowledge of resources, 'new combinations of resources' (1959, p. 85), and an expanding productive opportunity set, there is no equilibrium size.[1]

---

[1] Moreover, because of the difficulties of managing new resources and services, and of assimilating new managers in the firm, firm growth is not smooth or 'balanced' (as in Marris, 1964; Slater, 1980). On the contrary, growth rates in succeeding periods will typically be negatively serially correlated, so that high growth in one period is followed by low growth and vica versa. In fact, *this* is the true 'Penrose effect', and a first indication that even on this fundamental level, Penrose has been partly misrepresented in the literature.

III. MISSED ASPECTS OF PENROSE'S WORK

Penrose's book is very rich indeed, and it is perhaps not surprising that important points in it have been missed, both in economics and in strategic management. Most importantly, many of the basic ideas that Penrose used to build the main argument about growth through related diversification have been missed. Arguably, this has to do with Penrose's belonging to economic heterodoxy, primarily the post-Marshallian tradition.[1] Moreover, there are also ideas in her work that have a striking resemblance to Veblen's work (Foss, 1998). I briefly discuss the heterodox nature of Penrose's work in the following.

III.a. The Theory of the Growth of the Firm *as a part of economic heterodoxy*

As a starting point, one may think of Penrose as re-stating, refining, and sometimes radicalising, the basic conceptualisation of the firm that can be found in the work of Marshall and his later followers. Specifically, like Marshall, and later writers in his tradition, Penrose emphasised that not only is the firm a repository of productive knowledge, it is also an institution that develops and manages this knowledge. Moreover, the two processes of developing and managing knowledge may be hard to separate, both in practice and conceptually.

As stated earlier there is also a Veblenian (the emphasis on cumulative causation and group-based knowledge assets) flavour to Penrose's overall argument. Although she does not refer, even a single time, to Veblen in *The Theory of the Growth of the Firm*, her later (1995) nutshell conceptualisation of the main message of the 1959 book is straight out of Veblen: 'One of the primary assumptions of the theory of the growth of firms is that "history matters"; growth is essentially an evolutionary process and based on the cumulative growth of collective knowledge, in the context of a purposive firm' (1959[1995], p. xiii). As a third intellectual allied, one may single out Schumpeter. The Schumpeterian flavour of Penrose's work is more than a matter of spicing up the arguments with the standard quotations from Schumpeter; more fundamentally, Penrose's basic vision of the competitive process in general, and of the firm in particular, is disequilibrium-oriented and subjectivist (or cognitivist). Thus, important parts of Penrose's overall argument are *cognition, learning,* and *coordination*. All of these are neglected, or interpreted very narrowly, in that part of contemporary strategy thinking that draws on (mainstream) economics. Consider each in turn.

III.b. Cognition

The firm's productive opportunity, arguably the key concept of *The Theory of the Growth of the Firm* (cf. also Fransman, 1994, p. 744), is '. . . the productive possibilities that its "entrepreneurs" see and can take advantage of. A theory of the growth

---

[1] Which is represented by Andrews (1949), Downie (1958), Malmgren (1961), Richardson (1972), Loasby (1991), Langlois (1992), Earl (1996), and Kay (1997). For a splendid discussion of the post-Marshallian stream in economics, see Finch (1999).

of the firm is essentially an examination of the changing productive opportunity of firms' (1959, pp. 31–2). By implication, differential firm growth (as in Nelson and Winter, 1982) can be explained, at least in part, in terms of different productive opportunities. Thus, the notion of productive opportunity is quite a central notion in *The Theory of the Growth of the Firm*. This is in contrast to both the RBP and new IO inspired approaches to strategy, where decision-makers are assumed not to differ in cognitive terms, and where differences in competitive advantages or the growth rates of firms, therefore, cannot be explained in such terms.

The notion of productive opportunity is clearly a subjective (or, as some may prefer, 'constructivist') category.[1] In terms of modern organisation theory (notably, Weick, 1995) Penrose is here clearly talking about the 'enactment' of the environment that the management team performs, so that '. . . the relevant environment is not an objective fact discoverable before the event' (Penrose, 1959, p. 41).[2] Penrose's subjectivism is particularly apparent in her adoption of Boulding's (1956) concept of 'the image': '. . . the environment is treated . . . as an "image" in the entrepreneur's mind of the possibilities and restrictions with which he is confronted, for it is, after all, such an "image" which in fact determines a man's behaviour' (Penrose, 1959, p. 5). It should be noted that the idea of the image is not just another version of the ideas of bounded rationality and tacit knowledge. It explicitly recognises that agents have to make sense of their world, that agents' cognitive development is moulded in social processes, and it implies that tacitness is an aspect of virtually all acts of interpretation and meaning attribution, as modern 'image theory' (Beach, 1993) has clarified.

III.c. Entrepreneurship and learning

It is seldom recognised that Penrose's book stresses entrepreneurship and learning in a world characterised by change and uncertainty. Instead there has been a tendency to associate her 1959 analysis with path-dependence and rigidity effects, at least to an extent that this analysis is interpreted as an endorsement of only extremely narrow diversification. However, this view is wrong. 'In the long run', Penrose explains,

. . . the profitability, survival and growth of a firm does not depend so much on the efficiency with which it is able to organize the production of even a widely diversified range of products as it does on the ability of the firm to establish one or more wide and relatively impregnable 'bases' from which it can adapt and extend its operations in an uncertain, changing and competitive world (Penrose, 1959, p. 137).

[1] Penrose elaborates: '. . . for an analysis of the growth of the firm it is appropriate to start from analysis of the firm rather than the environment and then proceed to a discussion of the effect of certain types of environmental conditions. If we can discover what determines entrepreneurial ideas about what the firm can and cannot do, that is what determines the nature and extent of the "subjective" productive opportunity of the firm, we can at least know where to look if we want to explain or predict the actions of particular firms' (Penrose, 1959, p. 42).
[2] Elaborating the cognitive content of the notion of the firm's productive opportunity, Penrose explains that there '. . . is a close relation between the various kinds of resources with which the firm works and the development of the ideas, experience and knowledge of its managers and entrepreneurs' (Penrose, 1959, p. 85).

Thus, seemingly paradoxically, flexibility and adaptation are really just as much a message of the analysis as specialisation is. The paradox vanishes on realising that specialisation in Penrose's analysis means specialisation in terms of the underlying base of resources and competencies (rather than products) and that such specialisation may be fully consistent with seizing new business opportunities, for example, in the form of diversifying to new product markets that are, at least in terms of products, 'unrelated' relative to the firm's existing product portfolio.

In fact, as Penrose makes clear, there may be a considerable option value associated even with a specialised base of resources and services.

A firm is basically a collection of resources. Consequently, if we can assume that businessmen believe there is more to know about the resources they are working with than they do know at any given time, and that more knowledge would be likely to improve the efficiency and profitability of their firm, then unknown and unused productive services immediately become of considerable importance, not only because the belief that they exist acts as an incentive to acquire new knowledge, but also because they shape the scope and direction of the search for knowledge (Penrose, 1959, p. 77).

In other words, 'businessmen' may have a rational expectation that the resources and services that they control may yield more options than are immediately apparent and that further learning about them may reveal these options—a striking anticipation of real-options thinking that has only made its way in economics and the firm strategy field during the last 5–10 years. Thus, firm development is essentially an evolutionary and cumulative process of 'resource learning' (Mahoney, 1995), in which increased knowledge of the firm's resources help both to create options for further expansion and to increase absorptive capacity (Cohen and Levinthal, 1990), or, to use Penrose's terminology an expanding '*productive opportunity*'. All this, however, needs to be co-ordinated.

III.d. Coordination

A major focus of *The Theory of the Growth of the Firm* lies in administrative coordination.[1] Among other things, Penrose uses the notion of administrative coordination to define the boundaries of the firm (1959, p. 20). Her primary interest, however, lies in understanding the connection between administrative coordination and the production of services. In this connection, a crucial argument is that many different services may often be yielded by the same resources, depending on the uses to which the management team decide to put them and on the knowledge that management has of these resources.

These notions of the managerial task are much in contrast to modern organisational economics (Holmström and Milgrom, 1991; Milgrom and Roberts, 1992; Hart, 1995; Williamson, 1996) where management is essentially assumed to always know

---

[1] Pitelis and Wahl (1998, p. 259) argue that 'Edith Penrose is arguably the first economist to enter "the black box", the firm'. Considering how little Coase (1937) actually said about internal organisation, there is much truth in this view.

the best uses of resources (even though incentive problems may make first-best utilisation impossible). Thus, the strong dichotomy between exchange and production, which is characteristic of organisational economics, is not present in Penrose's work.

Relatedly, the RBP has neglected the actual process of coordinating and deploying resources to alternative uses as well as the fact that the actual use may determine what services can be obtained from a resource. Instead, resource-based theorists, who only consider the issues of the terms at which resources were acquired (Barney, 1986) and/or whether they are protected (Peteraf, 1993), forget that it is the actual application in production, and not the mere possession, of resources that create revenue. The same may be said of the application of new IO in strategy. In both cases, it is implicitly assumed that managers know what are the best uses of resources right from the beginning, and there is, therefore, little need for learning, and no need for experimentation. In contrast, Penrose was clearly concerned with managerial coordination and one might even construct an argument that her theorising leads us to what would in modern management studies be called 'strategic human resource management' where the focus is on the development of the firm's pool of talents with particular goals in mind.

IV. IMPLICATIONS FOR THEORISING ON THE FIRM AND STRATEGY

In *The Theory of the Growth of the Firm*, Penrose was both critical and constructive; critical of the limitations of the neo-classical theory of the firm of her day, and clearly constructive by putting forward a new theory of the firm, based on knowledge, learning and cognition. The character of this new theory was not appreciated when the book was published and it still awaits its full development. If one wishes to develop such a theory, a pertinent place to begin is where Penrose began: with managerial cognition and decision-making. In the view taken by Penrose, managers exercise judgment based on their imagination (Shackle, 1972; Beach, 1993). Imagination, in turn, is partly rooted in (imperfect) cognition and knowledge. As I very briefly sketch in the following, this has important implications for how we think about strategy and organisation.

IV.a. Management and strategy

In a cognitive view, the essence of strategic decision-making is not choice among given alternatives, but the process by which the strategist understands his environment (including, as Penrose stressed, 'his' firm's resources), defines which variables are relevant, attaches meaning to information and produces problem-solving heuristics.[1] All our knowledge of this sort of problem-solving behaviour in complex environments

[1] There is a surprising lack of work on this in the firm strategy field. However, see Stimpert (1989) for an early contribution.

indicates that, first, it is not best represented by an optimisation calculus (Dosi and Marengo, 1994), and, second, it is of considerably broader scope than finding the transaction cost minimising organisational form or the entry-deterring level of capacity investment. If (strategic) managers are judged by their ability to make the right choices among a set which they themselves partly generate, they are also judged by their ability to break with existing practice, to mediate and to engage in sense-making (Weick, 1995). Thus, leadership and the provision of cognitive frames enter the picture.

This also implies that concepts such as adaptation, coordination, management, etc. become intimately associated with learning, rather than with picking alternatives out of an already known set.[1] Specifically, management learning must involve more than Bayesian revision of priors into posteriors; it must also involve setting up new interpretive frameworks—new 'images', in Penrose's (1959) terms—for handling new types of problems. Managerial services are based on well-validated constructs for selecting and interpreting data and for making timely and effective decisions (Loasby, 1983). More generally, in this perspective, the essence of economic behaviour would seem to lie in understanding the environment, defining what are the relevant variables in that environment, making sense of incoming information, generating procedures which can help solving problems, and, finally, actually taking action (ibid.; Dosi and Marengo, 1994; Marengo, 1995).

Clearly, in such a cognitive view, managers cannot be presumed to undertake these mental processes in the same way and they cannot be presumed to reach the same conclusions. But this implies that we have identified a whole set of determinants of differential competitive advantages that have been missed so far, at least in economic approaches to strategy. For if managers hold a different view of the world, work with different heuristics and decision-making procedures, etc., then they are also likely to hold widely different views of which resources will be valuable in the future, which sort of positioning is optimal in the same industry, or evaluate competitive threats in different ways. All this will be determinative of revealed competitive advantages on a par with the firm's endowment of resources.

IV.b. Organisation

In a Penrosean cognitive perspective, where it cannot be presumed that division X understands the same by the message 'the state of the world is Z' as division Y does, or the divisions do not understand the message at all, the over-riding organisational design objective is creating a shared knowledge-base and getting everybody on the same wavelength. In firms that primarily grow through mergers and acquisitions this may be an extremely time-consuming and costly process, and this has often been singled out as one of the important reasons why mergers may break up again. Agents that are engaged in productive activities often spontaneously develop shared mental

---

[1] This will certainly not come as a surprise to modern organisation theorists; see, for example, March (1988).

construct, or, if you like, 'corporate cultures', that help coordinating distributed knowledge by infusing employees with firm-specific shared knowledge.

As Marengo (1995) argues, if agents entering the firm held the completely same habits of thought/models of the world, the only obstacle to efficient co-ordination of their actions would be precisely the sort of incentive problems that preoccupy modern organisational economists. However, in a world in which agents do not share exactly the same models and do not know each others' models, a collective knowledge base is required for coordination. As simulation work, built on the theory of classifier systems, demonstrates, such a knowledge base—a Penrosean image—may develop as a result of organisational learning under rather general assumptions (ibid.). Moreover, because of the role of change and lock-in, firms will develop different knowledge bases for coordinating their stocks of distributed knowledge. This helps to account for firm heterogeneity—or, if you like, differential capabilities—and, to the extent that collective knowledge bases influence productive and transactional efficiency (which is more than likely), also helps to account for differences in revealed competitive advantages.

However, while this puts some more conceptual meat on the skeleton of 'knowledge-based assets' often encountered in the resource-based literature on the firm, it does not say anything directly about the firm–market boundary. In other words, why is it that sometimes markets can coordinate distributed knowledge and sometimes firm organisation is necessary? The answer may turn on differences in the histories of the emergence of collective knowledge bases which help us to understand the phenomenon of differential capabilities. Thus, if capabilities are 'dissimilar', in the Penrose-inspired terminology of George Richardson (1972), it is because they are supported by different underlying collective knowledge bases, including managerial images.

In Richardson's terminology, production can be broken down into various stages or *activities*. Some activities are *similar*, in that they draw on the same general capabilities. Activities can also be *complementary* in that they are connected in the chain of production and therefore need to be coordinated with one another. Juxtaposing different degrees of similarity against different degrees of complementarity produces a matrix that maps different types of economic organisation. For example, closely complementary and similar activities may be best undertaken under unified governance.

Richardson's insight is a simple but extremely profound one. For it suggests that, as a quite general matter, capabilities are determinants of the boundaries of the firm. Problems of economic organisation may crucially reflect the possibility that a firm may control production knowledge that is, in important dimensions, strongly different from what others control. Thus, the management team of one firm may quite literally not understand what the management team of another firm wants from them (for example, in supplier contracts) or is offering them (for example, in license contracts). Their respective Penrosean images are not overlapping, as it were. In this setting, the costs of making contacts with potential partners, of educating potential licensees and franchisees, of teaching suppliers what it is one needs from them, etc., become very real factors determining where the boundaries of firms will be placed.[1]

The upshot of all this is that an analysis of organisational cognition, for example, along the lines pioneered by Edith Penrose and founded on the idea of the image, may provide an alternative to existing contractual theories of the boundaries of the firm (Alchian and Demsetz, 1972; Williamson, 1985, 1996; Hart, 1995). Moreover, it may strengthen the capabilities theory on this subject (Foss, 1993; Langlois and Robertson, 1995; Langlois and Foss, 1999), which has merely started from the empirical generalisation that firms control different production and organisation knowledge, without fundamentally enquiring into why this should be the case.

V. CONCLUSION

The purpose of this paper has been to argue that Penrose's crucial ideas are far from perfectly absorbed in contemporary thinking in strategy (or in the theory of the firm, for that matter). The dominant economic approaches to strategy—the RBP and the industrial organisation—both follow the usual explanatory approach in mainstream economics of assuming a perfect world and then introducing a few spanners in the works. Both stress that competitive advantage should be understood in terms of rents existing in equilibrium, and both trace the sources of these rents to imperfections in the factor market.

Penrose's view is very far from this. Not only is she not interested in whether there may be rents/competitive advantage in equilibrium *per se*, she also has a completely different basic outlook. Her world is one characterised by disequilibrium, learning, experimentation, and change; it is a world in which people adopt or become socialised into cognitive frameworks that allow them to make some sense out of the world, and where the coordination of their actions is dependent upon the sharing of such frameworks. Strategy research still has a long way to go before it will catch up with Edith Penrose's four decades old insights. As Pitelis and Wahl (1998) argue, much the same may be said with respect to organisational economics.

REFERENCES

ALCHIAN, A. A. and DEMSETZ H. (1972). Production, information costs, and economic organization, in *Economic Forces at Work*, edited by, A. A. ALCHIAN. Indianapolis, Liberty Press.
ANDREWS, K. (1971). *The Concept of Corporate Strategy*, Homewood, IL, Richard D. Irwin, [1980].
ANDREWS, P. W. S. (1949). *Manufacturing Business*, London, Macmillan.
ANSOFF, I. (1965). *Corporate Strategy*, New York, Wiley.
BAIN, J. S. (1959). *Industrial Organization*, New York, Wiley.

---

[1] One is reminded of the, possibly apocryphal, story about the Japanese supplier firm, committed to total quality, zero defects managements, that unable to make sense of a requirement from its American buyer of 95 per cent defect free deliveries sent a separately boxed batch of 5 per cent deliberately broken parts and a note saying 'We don't know why you want these'.

BARNEY, J. B. (1985). Strategic factor markets, *Management Science*, Vol. 32, 1231–41.
BARNEY, J. B. (1991). Firm resources and sustained competitive advantage, *Journal of Management*, Vol. 17, 99–120.
BAUMOL, W. J. (1962). On the theory of expansion of the firm, *American Economic Review*, Vol. 52, 1078–87.
BEACH, L. R. (1993). *Image Theory: Decision Making in Personal and Organizational Contexts*, Chichester, Wiley.
BEST, M. H. and GARNSEY, E. (1999). Edith Penrose 1914–1996, *Economic Journal*, Vol. 109, F187–F201.
BOULDING, K. E. (1956). *The Image*, Ann Arbor, Michigan University Press.
BOWER, J. (1970). *Managing the Resource Allocation Process*, Boston, Harvard Graduate School of Business Administration.
BRANDENBURGER, A. and NALEBUFF, B. (1996). *Co-opetition*, New York, Doubleday.
CAVES, R. (1984). Economic analysis and the quest for competitive advantage, *American Economic Review, Papers and Proceedings*, Vol. 74, 127–32.
CHANDLER, A. D. (1962). *Strategy and Structure*, Cambridge, MA, MIT Press.
COASE, R. H. (1937). The nature of the firm, *Economica*, Vol. 4, 386–405.
COHEN, W. M. and LEVINTHAL, D. (1990). Absorptive capacity: a new perspective on learning and innovation, *Administrative Science Quarterly*, Vol. 25, 128–52.
DEMSETZ, H. (1973). Industrial structure, market rivalry, and public policy, *Journal of Law and Economics*, Vol. 16, 1–10.
DEMSETZ, H. (1982). Barriers to entry, in *Efficiency, Competition, and Policy*, Oxford, Basil Blackwell.
DIERICKX, I. and COOL, K. (1989). Asset stock accumulation and the sustainability of competitive advantage, *Management Science*, Vol. 35, 1504–11.
DOSI, G. and MARENGO, L. (1994). Some elements of an evolutionary theory of organizational competences, in *Evolutionary Concepts in Contemporary Economics*, edited by R. W. ENGLANDER, Ann Arbor, The University of Michigan Press.
DOWNIE, J. (1958). *The Competitive Process*, London, Duckworth.
EARL, P. E. (1996). Shackle, entrepreneurship, and the theory of the firm, in *Interactions in Political Economy*, edited by S. PRESSMAN, London, Routledge.
FINCH, J. (1999). The methodological implications of post-Marshallian economics, in *Economic Knowledge and Economic Coordination: Essays in Honour of Brian Loasby*, edited by P. EARL and S. DOW, Aldershot, Edward Elgar.
FOSS, N. J. (1993). Theories of the firm: contractual and competence perspectives, *Journal of Evolutionary Economics*, Vol. 3, 127–44.
FOSS, N. J. (1998). The competence-based approach: Veblenian ideas in the contemporary theory of the firm, *Cambridge Journal of Economics*, Vol. 22, 479–96.
FOSS, N. J. (1999A). Equilibrium versus evolution in the resource-based perspective: Demsetz vs Penrose, *Resources, Technology and Strategy*, edited by N. J. FOSS and P. L. ROBERTSON, London, Routledge.
FOSS, N, J. (1999B). Edith Penrose and the Penrosians—or, why there is still so much to learn from *The Theory of the Growth of the Firm*, in *Cahiers de l'ISMEA*.
FOSS, N. J. (1999C). Research on the strategic theory of the firm: 'isolationism' and 'integrationism', *Journal of Management Studies*, Vol. 36, 725–55.
FRANSMAN, M. (1994). Information, knowledge, vision, and theories of the firm, *Industrial and Corporate Change*, Vol. 3, 713–57.
GANDER, J. P. (1991). Managerial intensity, firm size, and growth, *Managerial and Decision Economics*, Vol. 12, 261–6.
GHEMAWAT, P. (1991). *Commitment*, New York, The Free Press.
GHEMAWAT, P. (1997). *Games Businesses Play: Cases and Models*, Cambridge, MA, MIT Press.

GOLDSCHMID, H. J., MANN, J. H. and WESTON, J. F., eds (1974). *Industrial Concentration: The New Learning*, Boston, Little Brown.

HART, O. D. (1995). *Firms, Contracts, and Financial Structure*, Oxford, Clarendon Press.

HOLMSTRÖM, B. (1982). Moral hazard in teams, *Bell Journal of Economics*, Vol. 13, 324–40.

HOLMSTRÖM, B. and MILGROM, P. (1991). Multitask principal-agent analyses: incentive contracts, asset ownership, and job design, *Journal of Law, Economics, and Organization*, Vol. 7, 24–52.

KAY, N. (1997). *Pattern in Corporate Evolution*, Oxford, Oxford University Press.

KOR, Y. Y. and MAHONEY, J. T. (2000). Penrose's resource-based approach: the process and product of research creativity, *Journal of Management Studies*, Vol. 37, No. 1, 99–139.

LANGLOIS, R. N. (1992). Transaction costs in real time, *Industrial and Corporate Change*, Vol. 1, 99–117.

LANGLOIS, R. N. and ROBERTSON, P. L. (1995). *Firms, Markets and Economic Change*, London, Routledge.

LANGLOIS, R. N. and FOSS, N. J. (1999). Capabilities and governance: the rebirth of production in the theory of economic organization, *KYKLOS*, Vol. 52, 201–18.

LIPPMAN, S. A. and RUMELT, R. P. (1982). Uncertain immutability: an analysis of inter-firm differences in efficiency under competition, *Bell Journal of Economics*, Vol. 13, 418–38.

LEWIN, P. and PHELAN, S. E. (1999). Rents and resources: a market process perspective, mimeo, School of Management, University of Texas at Dallas.

LOASBY, B. J. (1983). Knowledge, learning and enterprise, in *Beyond Positive Economics*, edited by J. WISEMAN, London, Macmillan.

LOASBY, B. J. (1991). *Equilibrium and Evolution*, Manchester, Manchester University Press.

MACHLUP, F. (1967). Theories of the firm: marginalist, behavioral, managerial, in *The Methodology of Economics and Other Social Sciences*, New York, Wiley.

MAHONEY, J. T. (1995). The management of resources and the resource of management, *Journal of Business Research*, Vol. 33, 91–101.

MAHONEY, J. T. and PANDIAN, J. R. (1992). The resource-based view within the conversation of strategic management, *Strategic Management Journal*, Vol. 13, 363–80.

MALMGREN, H. B. (1961). Information, expectations, and the theory of the firm, *Quarterly Journal of Economics*, Vol. 75, 399–421.

MARCH, J. G. (1988). *Decisions and Organizations*, Oxford, Basil Blackwell.

MARCH, J. G. and SIMON, H. H. (1958). *Organizations*, New York, Wiley.

MARENGO, L. (1995). Structure, competence, and learning in organizations, *Wirtschaftspolitische Blätter*, Vol. 6, 454–64.

MARRIS, R. (1964). *The Economic Theory of Managerial Capitalism*, London, Macmillan.

MILGROM, P. J. and ROBERTS, J. D. (1992). *Economics, Organization, and Management*, New York, Prentice Hall.

MINTZBERG, H. (1990). Strategy formation: schools of thought, in *Perspectives on Strategic Management*, edited by J. W. FREDERICKSSON, New York.

MONTGOMERY, C. A. and WERNERFELT, B. (1988). Diversification, Ricardian Rents, and Tobin's q, *Rand Journal of Economics*, Vol. 19, 623–32.

NELSON, R. R. and WINTER, S. G. (1982). *An Evolutionary Theory of Economic Change*, Cambridge, Bellknap Press.

PENROSE, E. T. (1959). *The Theory of the Growth of the Firm*, Oxford, Oxford University Press [1995].

PENROSE, E. T. (1968). *Large International Firms in Developing Countries*, London, George Allen & Unwin.

PENROSE, E. T. (1971). *The Growth of Firms, Middle East Oil, and Other Essays*, London, Frank Cass & Co.

PETERAF, M. A. (1993). The cornerstones of competitive advantage: a resource-based view, *Strategic Management Journal*, Vol. 14, 179–91.

PITELIS, C. N. and WAHL, M. W. (1998). Edith Penrose: pioneer of stakeholder theory, *Long Range Planning*, Vol. 31, 252–61.

PORTER, M. E. (1980). *Competitive Strategy*, New York, Free Press.

PORTER, M. E. (1981). The contributions of industrial organization to strategic management, *Academy of Management Review*, Vol. 6, 609–20.

RICHARDSON, G. B. (1972).The organisation of industry, *Economic Journal*, Vol. 82, 883–96.

RUBIN, P. H. (1972). The expansion of firms, *Journal of Political Economy*, Vol. 81, 936–49.

RUMELT, R. P. (1974). *Strategy, Structure and Economic Performance*, Cambridge, MA, Harvard Business School Press.

SALONER, G. (1994). Game theory and strategic management: contributions, applications, and limitations, in *Fundamental Issues in Strategy: A Research Agenda*, edited by R. P. RUMELT, D. E. SCHENDEL and D. J. TEECE, Boston, Harvard Business School Press.

SCHERER, F. M. and ROSS, D. (1990). *Industrial Market Structure and Economic Performance*, Boston, Houghton-Mifflin.

SCHMALENSEE, R. and WILLIG, R. D., eds (1989). *Handbook of Industrial Organization*, Amsterdam, Noth-Holland.

SHACKLE, G. L. S. (1972). *Epistemics and Economics*, Cambridge, Cambridge University Press.

SHAPIRA, Z. (ed.) (1997). *Organizational Decision Making*, Cambridge, Cambridge University Press.

SHAPIRO, C. (1989). The theory of business strategy, *RAND Journal of Economics*, Vol. 20, 125–37.

SLATER, M. (1980). The managerial limitation to a firm's rate of growth, *Economic Journal*, Vol. 90, 520–8.

STIMPERT, C. I. (1989). Managerial cognition: a missing link in strategic management research, *Journal of Management Studies*, Vol. 26, 325–47.

TEECE, D. J. (1982). Toward an economic theory of the multiproduct firm, *Journal of Economic Behaviour and Organization*, Vol. 3, 39–63.

TIROLE, J. (1988). *The Theory of Industrial Organization*, Cambridge, MA, MIT Press.

TREADWAY, A. B. (1970). Adjustment costs and variable inputs in the theory of the competitive firm, *Journal of Economic Theory*, Vol. 2, 329–47.

WEICK, K. (1995). *Sensemaking in Organizations*, London, Sage.

WERNERFELT, B. (1984). A resource-based view of the firm, *Strategic Management Journal*, Vol. 5, 171–80.

WILLIAMS, J. R. (1994). Strategy and the search for rents: the evolution of diversity among firms, in *Fundamental Issues in Strategy Research*, edited by R. P. RUMELT, D. SCHENDEL and D. J. TEECE, Boston, Harvard Business School Press.

WILLIAMSON, O. E. (1985). *The Economic Institutions of Capitalism*, New York, Free Press.

WILLIAMSON, O. E. (1996). *The Mechanisms of Governance*, Oxford, Oxford University Press.

10

EDITH T. PENROSE AND RONALD H. COASE ON THE NATURE OF THE FIRM AND THE NATURE OF INDUSTRY

JOËL THOMAS RAVIX*

IDEFI – LATAPSES, CNRS, University of Nice Sophia Antipolis

Although Edith Penrose's book entitled *The Theory of the Growth of the Firm* was published about 20 years after Ronald Coase's famous article on *The Nature of the Firm*, these two authors share the same object of study, namely the 'real-world' firm, and both emphasise the disadvantages arising from the lack of a genuine economic theory of the firm. Nevertheless they ignored each other for a long time, and it is only recently that a minor connection between them has appeared. In the new preface of the third edition of her book, Penrose (1995) recalls the publication of Coase's article without any comment. In the same way, Coase (1988) only quotes Martin Slater's foreword to the second edition of Penrose's book. This motion in parallel directions can be explained by the incompatible features of the analytical framework developed by each of these authors. Coase differentiates the firm from the market by using the concept of transaction cost and gives the bases for a contractual approach, while Penrose takes the firm itself as a point of departure to study its growth from an evolutionary perspective. This opposition can more easily be defended as today contractual and evolutionary approaches are clearly differentiated (Foss, 1993; Fransman, 1994).

However, such an interpretation is surprising when we know Penrose's scepticism to any biological analogy. She writes indeed that 'in seeking the fundamental explanations of economic and social phenomena in human affairs the economist, and the social scientist in general would be well advised to attack his problem directly and in their own terms rather than indirectly by imposing sweeping biological models upon them' (Penrose, 1952, p. 819). The same scepticism appears in the work of Coase when he asserts that 'although economists claim to study the working of the market, in modern economic theory the market itself has an even more shadowy role than the firm' (Coase, 1988, p. 7). This general sentiment is reinforced by the fact that both authors do not limit themselves to building a theory of the firm, they also propose a more general framework to analyse the organisation of industry.

* A previous French version of this article was published in *Economies et Sociétés*, Vol. 8, No. 29 (1999), 165–85.

The purpose of this chapter is to emphasise the originality of Penrose's methodological approach as opposed to that of Coase. In the first section we examine the
foundations of this opposition and its implications for the theory of the firm. In the
second section we compare the two opposite perspectives proposed by Coase and
Penrose in the field of the organisation of industry. Finally it will be suggested that
Penrose's approach offers an open system compared with Coase's approach, which is
basically a closed system.

I. THE NATURE OF THE FIRM AND THE NATURE OF THE INFORMATION

The general approach developed by Penrose can be interpreted as symmetrical with
Coase's approach. To justify such an interpretation it is necessary to clarify first the
difference between the perspectives adopted by Coase and Penrose on the question of
the firm, and to show next that Penrose's point of view is based on a specific distinction between information and knowledge.

I.a. The nature and the function of the firm

It is possible to consider that Coase analyses the question of the firm from the large
end of the telescope because 'in searching for a definition of a firm', he first considers
'the economic system as it is normally treated by the economist' (Coase, 1937, p. 387).
Therefore he admits that 'an economist thinks of the economic system as being
co-ordinated by the price mechanism' and that 'the economic system "works itself"'
(*ibid.*). From this standpoint, Coase thinks that 'outside the firm, price movements
direct production, which is co-ordinated through a series of exchange transactions on
the market', while 'within a firm, these market transactions are eliminated and in place
of the complicated market structure with exchange transactions is substituted the
entrepreneur-co-ordinator, who directs production' (*ibid.*, p. 388). Although the firm
and the market are 'alternative methods of co-ordinating production' (*ibid.*), they
serve the same function, that is to coordinate production.

Whereas Coase's aim is to show that 'a definition of the firm may be obtained which
is . . . realistic in that it corresponds to what is meant by a firm in the real world' (*ibid.*,
p. 386), he takes as a starting point an economic system dominated by a 'mercantile'
division of labour in which firms do not exist. This ambiguity is reinforced by the fact
that paradoxically Coase is not interested in the nature of the firm, but in its function.
Actually he attempts to define the role of the firm in the whole economic system and
not its characteristics as an economic entity.

Penrose's analysis is radically opposed to Coase's. She approaches the question of
the firm from the small end of the telescope. Her starting point is not the working of
the economic system in general, but the nature of the firm in itself, that is to say the
characteristics of the firm in general. Thus, 'in a private enterprise industrial economy
the business firm is the basic unit for the organization of production' (Penrose, 1959,

p. 9). As a result, the nature of the firm does not depend on the working of the economic system. By contrast, 'the very nature of the economy is to some extent defined in terms of the kind of firms that compose it, their size, the way in which they are established and grow, their methods of doing business, and the relationships between them' (*ibid.*). This statement allows Penrose to expand on her initial definition of the firm. She points out that the firm 'is a complex institution, impinging on economic and social life in many directions, comprising numerous and diverse activities, making a large variety of significant decisions, influenced by miscellaneous and unpredictable human whims, yet generally directed in the light of human reason' (*ibid.*).

These different characteristics explain why the firm can be analysed through several contexts: economic, sociological, technological, organisational, etc. However, Penrose specifies that 'much confusion can arise from the careless assumption that when the term "firm" is used in different contexts it always means the same thing' (*ibid.*, p. 10).

After she has stated precisely the nature of the firm, Penrose takes up the question of its role in the economic system. According to her, 'the primary economic function of an industrial firm is to make use of productive resources for the purpose of supplying goods and services to the economy in accordance with plans developed and put into effect within the firm' (*ibid.*, p. 15). On the one hand, her analysis is partly similar to Coase's because she considers that 'the essential difference between economic activity inside the firm and economic activity in the "market" is that the former is carried on within an administrative organization, while the latter is not' (*ibid.*). On the other hand, and unlike Coase, Edith Penrose does not restrict her analysis to the notion of *direction*, which characterises the administrative organisation. Although this concept establishes a sharp contrast between organisation and market, it is not only peculiar to firms; it also characterises other forms of socio-economic organisation. On that basis, Penrose insists on the fact that the fundamental role of the firm is to organise production, and that there is no production outside of firms. Thus, for Penrose, 'a firm is more than an administrative unit; it is also a collection of productive resources the disposal of which between different uses and over time is determined by administrative decision' (*ibid.*, p. 24).

The role of the firm in the organisation of production is also present in Coase's analysis. But by supposing that production obeys a logic of resource allocation, he makes no distinction between organisation and coordination, and assumes that the market can coordinate production. In Coase's analysis, the logic of production is the same as the logic of exchange. On the contrary, Penrose bases her analysis on a conception of production in which the time dimension of production is central. She establishes a sharp distinction between physical or human productive resources and services that the resources can generate, and this distinction is close to N. Georgescu-Roegen's (1971) analysis in terms of *flows* and *funds*. Thus, 'strictly speaking, it is never *resources* themselves that are the "inputs" in the production process, but only the *services* that the resources can render. The services yielded by resources are a function of the way in which they are used—exactly the same resources when used for different purposes or in different ways and in combination with different types or amounts of other resources provide a different service or set of services. The important distinction between resources and services is not their relative durability; rather it lies in the fact that resources consist of a bundle of potential services and can, for the most part, be defined independently of their use, while services cannot be so defined, the very word "service" implying a function, an activity' (*ibid.*, p. 25).

Penrose avoids the use of the term 'factor of production' because 'it makes no distinction between resources and services' (*ibid.*, p. 25, note 1), and that is why she prefers the notion of 'productive opportunity' of a firm, which comprises 'all of the productive possibilities that its "entrepreneurs" see and can take advantage of' (*ibid.*, p. 31).

This notion of productive opportunity is fundamental because 'a theory of growth of firms is essentially an examination of the changing productive opportunity of firms' (*ibid.*, pp. 31–2). The productive opportunity of the firm refers to 'a psychological predisposition on the part of individuals to take a chance in the hope of gain, and, in particular, to commit effort and resources to speculative activity' (*ibid.*, p. 33). More precisely it does not depend on the entrepreneur's 'sober calculations', but corresponds to 'the decision to make some calculations' (*ibid.*). As the productive opportunity is closely linked to the entrepreneur's behaviour, she stresses 'the effects of risk and uncertainty on, and of expectations in, the growth of firms' (*ibid.*, p. 41).

Contrarily to Coase who reduces uncertainty to a lack of information necessary to transaction fulfilment, Penrose considers that uncertainty is intrinsic to entrepreneurs' expectations and to the state of their knowledge. Her statement is directly related to Alfred Marshall's approach (Loasby, 1991, p. 59), which consists in saying that 'knowledge is our most powerful engine of production' and that 'organization aids knowledge' (Marshall, 1920, p. 115).

Penrose's analysis originally differs from Coase's in the determination of the entrepreneurial behaviour. At the very beginning of her book, she emphasises 'the productive services available to a firm from its own resources, particularly the productive services available from management with experience within the firm' (Penrose, 1959, p. 5). Actually for Penrose 'the experience of management will affect the productive services that all its other resources are capable of rendering' (*ibid.*). As the management team tries to make the better use of available resources, 'a truly "dynamic" interacting process occurs which encourages continuous growth but limits the rate of growth' (*ibid.*). But the accumulated knowledge of the firm does not correspond to a better knowledge of the objective environment. On the contrary, the environment of the firm is treated by Penrose as 'an "image" in the entrepreneur's mind of the possibilities and restrictions with which he is confronted'. She states that, after all, it is 'such an "image" which in fact determines a man's behaviour' (*ibid.*). This 'image' concept is explicitly borrowed by Penrose from Kenneth Boulding's book entitled *The Image* (1956), and she notes ' "image" is so apt a word for my purpose that I promptly appropriated it' (*ibid.*, p. 5, note 1). The importance of this concept in Penrose's theory sends us back to Boulding's analysis in order to specify the meaning of the *image* concept, as well as Penrose's utilisation of it.

I.b. The 'image' and the productive opportunity

Boulding's project is to elaborate 'an organic theory of knowledge' (Boulding, 1956, p. 6) that goes beyond a mere theory of the behaviour of firms (Samuels, 1997). But it is interesting to recall his starting point. Boulding thinks first that 'knowledge is what somebody or something knows, and that without a knower, knowledge is an absurdity' (Boulding, 1956, p. 16), and second, that there is only subjective knowledge because 'for any individual organism or organization, there are no such things as "facts". There are only messages filtered through a changeable value system' (*ibid.*, p. 14).

Boulding's approach brings about two main consequences. The first is that there is no objective environment which could be directly understood by individuals, but only individual images built up on personal experiences. The image of each person comes from the vision of his environment so that 'the image is built up as a result of all past experience of the possessor of the image', and that 'part of the image is the history of the image itself' (*ibid.*, p. 6). The second consequence is that the individual behaviour does not directly depend on the environment but on the image he or she builds up. As indicated by Boulding, 'every time a message reaches him his image is likely to be changed in some degree by it, and as his image is changed his behavior patterns will be changed likewise' (*ibid.*, p. 7). He introduces a distinction between the image itself and the messages that proceed from the environment. As 'the messages consist of *information* in the sense that they are structured experiences' (*ibid.*), objective messages do not exist. There are only subjective messages that may be different for different persons because 'the meaning of a message is the change which it produces in the image' (*ibid.*).

Therefore it is possible to conceive with Boulding that one message can induce a variety of influences on an individual image. First, the image may remain unaffected, if the message is going straight without hitting it. Boulding thinks, 'the great majority of messages are of this kind' (*ibid.*). We are receiving messages all the time, but generally we ignore them. Secondly, the message may change our image, adding some information to our knowledge, without fundamentally revising our own image. Thirdly, the message can change the image in a quite radical way. This phenomenon of reorganisation of the image takes the form of a complete conversion, and, according to Boulding, 'the sudden and dramatic nature of these reorganizations is perhaps a result of the fact that our image is in itself resistant to change' (*ibid.*, p. 8). When we receive messages that conflict with our image, our first impulse is to reject them as in some sense untrue. However, 'as we continue to receive messages which contradict our image, we begin to have doubts, and then one day we receive a message which overthrows our previous image and we revise it completely' (*ibid.*, p. 9).

It is on this conception of information and knowledge that Penrose bases her analysis of the process of the growth of the firm. According to her, 'the 'expectations' of a firm—the way in which it interprets its 'environment'—are as much a function of the internal resources and operations of a firm as of the personal qualities of the entrepreneur' (Penrose, 1959, p. 41). She establishes an accurate distinction between the 'objective' productive opportunity, namely 'what the firm is able to accomplish', or its *competences*, and its 'subjective' productive opportunity, namely 'what it thinks it can accomplish' (*ibid.*). If the objective productive opportunity comprises all the internal resources of the firm, its subjective productive opportunity indicates the way in which the firm interprets its environment.

But Penrose advocates that '"expectations" and not "objective facts" are the imme-diate determinants of a firm's behaviour' (*ibid.*, p. 41). She adopts Boulding's idea in which there are images which only subjectively build up the interpretation of the information proceeding from the environment. Therefore the behaviour of the firm depends on its subjective productive opportunity, i.e., on its capability to build up images, not on its ability to process information. In Penrose's argument, there is no objective information because information is part of the image created by the firm. As Martin Fransman points out: 'Penrose makes it clear that she sees the firm's planners as "image creators" rather than as "information processors". In other words, rather than beginning with the objective environment of the firm, and the information that this environment generates—in the form, for example, of market prices, market demands, the activities of competitors etc.—Penrose starts with the mental world of the planners who are situated within the context of their own firm and its specific productive services' (Fransman, 1994, p. 743).

More generally, the originality of Penrose's approach does not lie only in the fact that the firm perceives its environment in a purely subjective manner. She recognises that 'in the last analysis the "environment" rejects or confirms the soundness of the judgement about it', and she adds that 'the relevant environment is not an objective fact discoverable before the event' (Penrose, 1959, p. 41). The only phenomenon that might be objective is a past event on which the firm does not have any possibility of action on account of its irreversibility. In most cases, the expectations of the firm depend on events that have not yet occurred, and are also basically conditioned by the behaviour of other firms because 'firms not only alter the environmental conditions necessary for the success of their actions, but, even more important, they know that they can alter them and that the environment is not independent of their own activities' (*ibid.*, p. 42).

In other words, Penrose does not ignore the gaps that may occur between the image created by the firm and the 'reality' of its environment, but, as this reality is partly the consequence of the actions of firms, it necessarily results from the different images created by firms. Consequently, the environment does not correspond to a fixed and given reality on which the firm's expectations could be evaluated. On the contrary, the reality is always changing and the future is radically uncertain. That is why Penrose thinks that 'the factors affecting the relation between the "image" and "reality" are not being ignored, but for an analysis of the growth of firms it is appropriate to start from an analysis of the firm rather than of the types of environmental conditions' (*ibid.*, p. 42). So, she adopts a conception of the effect of uncertainty and risk, which is different from that of Frank Knight (1921). Penrose considers, on the one hand, that '"uncertainty" refers to the entrepreneur's confidence in his estimates or expectations'; while '"risk", on the other hand, refers to the possible outcomes of action, specifically to the loss that might be incurred if a given action is taken' (*ibid.*, p. 56). This opposition proceeds from the idea that estimation on which action is finally taken does not depend on several eventualities that could be realised, but on the entre-preneur's interpretation of such eventualities. Risk is not subordinate to the prob-ability of a random event. Rather it depends on resources and capabilities of the firm to take it on, insofar as when 'a firm expands its investment, the risk to it of a given chance of loss becomes more serious with each increment of investment—its "wealth" position becomes endangered if it operates on borrowed money; its liquidity, or ability to meet unexpected demands for cash, becomes precarious as it depletes its own reserves and if its ability to raise money is affected by its heavy illiquid investment' (*ibid.*, p. 57). This is the principle of 'increasing risk' which as a result submits the possibilities of the firm's action to its resources and capabilities and not directly to the risks peculiar to its environment.

In the long run, the Penrosean 'subjective uncertainty' is conditioned by the 'productive opportunity' of the firm. Therefore, 'uncertainty resulting from the feeling that one has too little information leads to a lack of confidence in the soundness of the judgement that lies behind any given plan of action. Hence one of the most important ways of reducing subjective uncertainty about the future course of events is surely to obtain more information about the factors that might be expected to affect it; and it is reasonable to suppose that one of the most important tasks of a firm in an uncertain world will be that of obtaining as much information as is practicable about the possible course of future events' (*ibid.*, p. 59). However, such an activity of information research needs capabilities as well as resources. The result is that uncertainty cannot be totally discarded, and that uncertainty constrains the firm possibility of action. Thus, 'the point will be reached where a firm believes it is either impossible or too expensive to attempt to obtain further information. At this point the firm must decide how far, if at all, it should commit resources to the activity in question in the face of the irreducible uncertainty and of estimates of risk' (*ibid.*, pp. 60–1).

In Penrose's analysis, the decision process is a particular sequential trial and error process since it is fundamentally associated with the environment image of the firm. It is linked with the acquisition process of competences resulting from the increasing experiences of the firm. This 'increasing experience shows itself in two ways—changes in knowledge acquired and changes in the ability to use knowledge' (*ibid.*, p. 53). It is impossible to distinguish in concrete terms these two modalities which act jointly on the knowledge process of the firm and on its action possibilities because 'this increase in knowledge not only causes the productive opportunity of a firm to change in ways unrelated to changes in the environment, but also contributes to the "uniqueness" of the opportunity of each individual firm' (*ibid.*, pp. 52–3). It is this endogenous knowledge process that explains the growth of the firm, as well as the limit of its rate of growth. As a consequence, 'once it is recognized that the very processes of operation and of expansion are intimately associated with a process by which knowledge is increased, then it becomes immediately clear that the productive opportunity of a firm

will change even in the absence of any change in external circumstances or in fundamental technological knowledge' (*ibid.*, p. 56).

By linking the growth process of the firm to the process of knowledge, Penrose's
analysis stands out from that of Coase, which basically treats of information and
knowledge as synonymous terms. When Coase considers that 'it seems improbable
that a firm would emerge without the existence of uncertainty' (Coase, 1937, p. 392),
he sees uncertainty much more in Herbert Simon's sense than in Boulding's. In
Coase's analysis uncertainty is attached to 'what the relevant prices are' (*ibid.*, p. 390).
It does not affect the knowledge process, but only depends on imperfection of the
information necessary to make transactions.

Nevertheless, this imperfect information assumption involves new issues that Coase
does not tackle in his 1937 article. How does the market work, if the price system does
not convey all the necessary information? In a situation of imperfect information, is
there a need to reconsider the 'nature of the market'?

II. THE NATURE OF THE MARKET AND THE ORGANISATION OF INDUSTRY

The question of the nature of the market could be considered as the reverse side of the
problem of boundaries of the firm. As soon as the firm emerges as a production coordinating device alternative to the market, it is necessary to explain how the coordination task is divided up between the firm and the market. This question inevitably leads to the question of the 'nature of industry', since it is necessary to explain
how firms coordinate with one another. The nature of the market problem fits into the
more general and more complex problem of the organisation of industry.

II.a. From the nature of the firm to the nature of the market

Coase's point of view on the nature of the market is synthesised in his 1988 book
entitled *The Firm, the Market, and the Law*. He notes: 'That part of economic theory
which deals with firms, industries, and markets . . . is now usually termed price theory
or micro-economics', and he specifies that 'the elaboration of the analysis should not
hide from us its essential character: it is an analysis of choice' (Coase, 1988, p. 6).
However, the reduction of economic theory to an analysis of choice has produced
'serious adverse effects on economics itself' and a 'divorce of the theory from its
subject matter', from which the results are that 'the entities whose decisions economists are engaged in analysing have not been made the subject of study and in consequence lack any substance' (*ibid.*, p. 3). According to Coase, the consequences are
disastrous because this approach leads the economic theory to analyse economic
entities without any empirical reality: 'Exchange takes place without any specification
of its institutional setting. We have consumers without humanity, firms without
organization, and even exchange without markets' (*ibid.*).

Coase attributes this deadlock to the fact that economic theory is concerned chiefly with price determination and does not worry about the market in itself. When economists study the market, they only deal with the number of firms and the determination of prices, and this 'has nothing to do with the market as an institution' (*ibid.*, p. 8). On the contrary, Coase's viewpoint is that 'markets are institutions that exist to facilitate exchange, that is, they exist in order to reduce the cost of carrying out exchange transactions. In an economic theory which assumes that transaction costs are non existent, markets have no function to perform' (*ibid.*, p. 7).

Coase justifies an institutional approach of the market structure by the fact that markets have always been organised. The origin of this phenomenon has to be sought in the role of the entrepreneur: 'the provision of markets is an entrepreneurial activity and has a long history' (*ibid.*, p. 8). To illustrate his argument, Coase speaks about fairs and markets of the medieval period, which were organised by merchants, but also, for modern times, commodity exchanges and stock exchanges which are normally organised by a group of traders. These traders, 'regulate in great detail the activities of those who trade in these markets (the times at which transactions can be made, what can be traded, the responsibilities of the parties, the terms of settlement, etc.), and they all provide machinery for settlement of disputes and impose sanctions against those who infringe the rules of exchange' (*ibid.*, p. 9).

Such an institutional conception of the nature of the market is partly similar to Penrose's idea, according to which 'markets and firms are interacting institutions, each being necessary to the existence of the other' (Penrose, 1959, p. 197). However, she specifies that 'the organization of production within the administrative framework of the individual firm is substantially different from the organization of production brought about through the operations of the open market' (*ibid.*). But Penrose also develops another vision based on the idea that 'the really enterprising entrepreneur has not often, so far as we can see, taken demand as "given" but rather as something he ought to be able to do something about' (*ibid.*, p. 80). For Penrose indeed, demand is an integral part of the knowledge process of the firm, that leads her to assume 'the essentially subjective nature of demand from the point of view of the firm' (*ibid.*).

In other words, the definition of the market appears as a specific part of the firm activities. It is identified with selling programmes or commercial activities of the firm since 'the "demand" with which an entrepreneur is concerned when he makes his production plans is nothing more nor less than his own ideas about what he can sell at various prices with varying degrees of selling effort' (*ibid.*, p. 81). More accurately, Penrose considers that 'at all times a firm has a foothold in certain types of production and in certain types of market', which are called 'areas of specialization of the firm' (*ibid.*, p.109). In order to build its areas of specialisation, each firm has to determine its own 'market areas', including the customer group which the firm hopes to influence by the same sales programme, 'regardless of the number of products sold to that group' (*ibid.*, p. 110). The result is that 'the appropriate criteria for the delimitation of market areas are different for different firms; the significance of the boundaries lies in the fact that a movement into a new market area requires the devotion of resources to the development of a new type of selling programme and a competence in meeting a different type of competitive pressure' (*ibid.*). This idea according to which the firm builds its own market is the consequence of how the firm estimates its potential outlets; 'for demand from the point of view of the firm is highly subjective—the opinion of the firm's entrepreneur' (*ibid.*, p. 85).

It appears then that Coase's and Penrose's points of view do not really differ even if they do not emphasise exactly the same phenomena. Coase deals with organised markets like stock markets or commodity markets, while Penrose is concerned with industrial product markets, which have not much in common with one another. Nevertheless, in any case, the market conceived as the encounter of an offer and a demand is ruled out in favour of a more complex approach pointing out the diversity of market institutional arrangements. The traditional dichotomy between the firm and the market becomes irrelevant because of the important issue of the nature of the market which inevitably leads to acknowledgement of the existence of an 'institutional structure of production' (Coase, 1992). Therefore, the organisation of production implies that different 'business institutions' (Langlois and Robertson, 1995) must be established.

II.b. *From the nature of the market to the organisation of the industry*

It is to such an institutional structure of production that Coase makes reference when he writes that 'we all know what is meant by the organization of industry. It describes the way in which the activities undertaken within the economic system are divided up between firms. As we know, some firms embrace many different activities; while for others, the range is narrowly circumscribed. Some firms are large; others, small. Some firms are vertically integrated; others are not. This is the organization of industry or— as it used to be called—the structure of industry' (Coase, 1972, p. 60). Although he deplores that these problems are not correctly analysed in economics, he suggests that his transaction cost theory can be a starting point to a theory of industrial organisation. To explain the organisation of industry, Coase extends his analysis of the size of the firm. He considers that 'a firm will tend to expand until the costs of organising an extra transaction within the firm become equal to the costs of carrying out the same transaction by means of an exchange on the open market or the costs of organising in another firm' (Coase, 1937, p. 395). Coase's formulation admits the existence of a type of transaction that is organised neither by exchange in a market, nor by the direction within a single firm, but results from particular relations between two firms. To explain this phenomenon, 'he then extends the principles of decreasing returns to management to these inter-firm transactions, by asserting, first, that they take place between firms that share different stages of production process, and, second, that the additional costs of organising an extra stage for one firm will be greater than the other firm's cost of performing the same activity' (Ravix, 1998, p. 74).

But this solution given by Coase is unsatisfying for two reasons. On the one hand, 'the definition of these new forms of transactions in addition to pure market transactions and internal direction remains necessarily ambiguous: these transactions are neither of one type nor of the other, but they are artificially reduced to one or to the other, depending on how the process is divided'. On the other hand, Coase uses 'a notion of the production process organised in a series of stages bound together whose logic of organisation has little to do with an analysis in terms of substitution' (*ibid.*). Coase's solution is an illusion, or an eyewash solution, because the transactional logic cannot explain the division of productive activities between firms, but only market relations.

Penrose's theory avoids this problem because in her analysis the growth process of the firm is independent of the market. She considers indeed that 'internal inducements to expansion arise largely from the existence of a pool of unused productive services, resources, and special knowledge, all of which will always be found within any firm' (Penrose, 1959, p. 66). These internal inducements have two important consequences. The first consequence is to prevent the achievement of the firm's equilibrium, in a Coasean sense because of 'three significant obstacles'. These obstacles are 'those arising from the familiar difficulties posed by the indivisibility of resources; those arising from the fact that the same resources can be used differently under different circumstances, and in particular, in a "specialized" manner; and those arising because in the ordinary processes of operation and expansion new productive services are continually being created' (*ibid.*, p. 68). Because there are continuous incentives to grow, and because there is no limit to the size of the firm, there is no room for a notion of optimal size of the firm within Penrose's analysis. The second consequence is that the internal inducements, which lead the firm to build its own areas of specialisation, are a direct function of the productive opportunity of the firm. The market areas of the firms are closely connected to their production bases. For Penrose, 'each type of productive activity that uses machines, process, skills, and raw materials that are all complementary and closely associated in the process of production we shall call a "production base" or "technological base" of the firm, regardless of the number or type of products produced' (*ibid.*, p. 109). This notion of productive base, dissociated from the notion of product, is very important in Penrose's analysis; it comprises the whole technological and managerial knowledge that the firm must mobilise to produce. According to Penrose, firms differentiate from one another because of their productive opportunity, which induce different production bases and different market areas to develop, so that they have different areas of specialisation. As Penrose emphasises, 'a movement into a new base requires a firm to achieve competence in some significantly different areas of technology' (*ibid.*, p. 110), which means that the firm must acquire or develop new knowledge to create new competences.

It is eventually the process of knowledge that leads to the expansion of the firm and which pushes the firm to adopt a logic of diversification. However, Penrose takes the precaution to stress that diversification is not to be considered in its static dimension because 'not only is a comparison of the "extent of diversification" of different firms likely to be meaningless in itself, but statistical studies of the number of different "products" produced by firms are also of very limited usefulness' (*ibid.*, p. 107). Two reasons explain this point of view: 'the "number of products" produced by a firm has no general significance', and 'the definition of a product is at best arbitrary' (*ibid.*). Penrose's analysis is perfectly compatible with Richardson's approach to the organisation of industry. As Richardson puts it, 'Mrs Penrose has provided us with excellent accounts of how companies grow in directions set by their capabilities and how these capabilities themselves slowly expand and alter' (Richardson, 1972, p. 888). Like Penrose, Richardson refers to *activities* rather than products, and to *capabilities* rather than productive factors. Activities are distinct from products because they are 'related to the discovery and estimation of future wants, to research, development and design, to the execution and co-ordination of processes of physical transformation, the marketing of goods and so on'; while capabilities are associated with knowledge, experience and skills because 'the capability of an organisation may depend upon command of some particular material technology, . . . , or may derive from skills in marketing or knowledge of and reputation in a particular market' (*ibid.*). However, Richardson explains only the principles of the organisation of industry as a system that consists of direction within firms, inter-firm co-operation and market transactions. He leaves unexplained the mechanisms by which the division of labour of industrial coordination between different possible institutional arrangements can be established.

In contrast, the Penrosean notion of diversification has to be understood in a dynamic sense, as a specialisation of productive activities because 'diversification and expansion based primarily on a high degree of competence and technical knowledge in specialized areas . . . together with the market position it ensures is the strongest and most enduring position a firm can develop' (Penrose, 1959, p. 119).

Then, this diversification movement leads not only to analysis of the diversity of the industrial structure but also to understanding of its evolutionary nature. Diversification is sustained for two reasons clearly brought to the fore by Penrose. The first, inscribed in the project of each firm, is its permanent intention of winning a dominant position, which pushes it to adopt new technological bases instead of reaching new market areas. The second is that 'even when a firm exploits to the fullest possible extent the opportunities for monopolistic gain available to it, the protection afforded, though often extensive, can neither be complete nor absolutely certain' (*ibid.*, p. 113). This paradox can be explained by the temporality of diversification. In other words, the diversification strategies developed by firms form the essence of an industrial dynamics which prevents steady state positions from being permanent, because of the new opportunities generated by the expansion process of firms. Nevertheless, the solution adopted by Penrose is closely linked to Richardson's approach to the competitive process, which conceives 'competition in terms of activity rather than structure' (Richardson, 1975, p. 359). As Richardson suggests, 'increasing returns may lead not to market concentration but to specialization and interdependence' (*ibid.*, p. 357). To the question: 'will we not find, at the end of the road, precisely that state of monopolistic competition described by Chamberlin, the only difference being that differentiation takes place in the vertical as well as the horizontal dimension?' Richardson answers that 'the end of the road may never be reached' (*ibid.*).

III. CONCLUSION

The originality of Penrose's approach lies in her reference to time. It is particularly in this field that her analysis differs from that of Coase. She considers that 'one of the primary assumptions of the theory of the growth of firms is that "history matters"' (Penrose, 1995, p. xiii). Her intention to develop a theory of the firm which is intrinsically dynamic, leads her to dissociate the issue of the expansion of the firm from the issue of the size of the firm. And she does not focus specifically on the boundaries of the firm, which are crucial in Coase (1937). Especially she emphasises in her book that 'growth is a process, size is a state' (Penrose, 1956, p. 88).

If the transaction cost theory answers exactly to the question of the size of the firm, it cannot deal with the growth of the firm and, more generally, with the problem of the organisation of the industry. From this point of view, the static theoretical framework proposed by Coase is irremediably closed. In contrast, in adopting an approach in which 'the influence of the "environment" was put on one side in the first instance in order to permit concentration on the internal resources of the firm' (Penrose, 1995, p. xiii), Penrose successfully develops a theory the purpose of which is to explain how and why firms 'move', questions that Coase's static framework cannot answer.

If this methodological choice requires an explanation of the behaviour of the firm in itself and not in reference to the environment, it also requires a different definition of the environment. Penrose solves this problem with the 'image' concept which allows her to consider that 'the relevant environment, that is the set of opportunities for investment and growth that its entrepreneurs and managers perceive, is different for every firm' (*ibid.*). Penrose's environment conception is radically different from Coase's. According to her, 'the environment is not something "out there", fixed and immutable, but can itself be manipulated by the firm to serve its own purposes' (*ibid.*). The result is that Penrose's analytical framework is an *open system* from which it is possible to tackle new subjects such as the organisation of industry.

REFERENCES

BOULDING, K. E. (1956). *The Image*, Ann Arbor, MI, University of Michigan Press.

COASE, R. H. (1937). The nature of the firm, *Economica*, Vol. IV, November, 386–405.

COASE, R. H. (1972). Industrial organization: a proposal for research, in *Policy Issues and Research Opportunities in Industrial Organization*, edited by V. R. FUCHS, NBER, New York, Columbia University Press.

COASE, R. H. (1988). *The Firm, the Market and the Law*, Chicago, University of Chicago Press.

COASE, R. H. (1992). The institutional structure of production, *American Economic Review*, Vol. LXXXII, No. 4.

FOSS, N. J. (1993). Theories of the firm: contractual and competence perspectives, *Journal of Evolutionary Economics*, Vol. 3, 127–44.

FRANSMAN, M. (1994). Information, knowledge, vision and theories of the firm, *Industrial and Corporate Change*, Vol. 3, No. 3, 713–57.

GEORGESCU-ROEGEN, N. (1971). *The Entropy Law and the Economic Process*, Cambridge, MA, Harvard University Press.

KNIGHT, F. H. (1921). *Risk, Uncertainty and Profit*, London, London School of Economics and Political Science [1933].

LANGLOIS, R. N. and ROBERTSON, P. L. (1995). *Firms, Markets and Economic Change: A Dynamic Theory of Business Institutions*, Routledge, London.

LOASBY, B. (1991). *Equilibrium and Evolution*, Manchester, Manchester University Press.

MARSHALL, A. (1920). *Principles of Economics*, 8th edn, London, Macmillan.

PENROSE, E. T. (1952). Biological analogies in the theory of the firm, *American Economic Review*, Vol. XLII, No. 5, December, 804–19.

PENROSE, E. T. (1959). *The Theory of the Growth of the Firm*, Oxford, Basil Blackwell.

PENROSE, E. T. (1995). *The Theory of the Growth of the Firm*, 3rd edn, Oxford, Oxford University Press.

RAVIX, J. L. (1998). Co-operation and competition: paradoxes in the theory of the organisation of industry, in *Economic Organization, Capabilities and Co-ordination, Essays in Honour of G.B. Richardson*, edited by N. J. FOSS and B. J. LOASBY, London, Routledge.

RICHARDSON, G. B. (1972). The organisation of industry, *Economic Journal*, Vol. 82, No. 327, 883–96.

RICHARDSON, G. B. (1975). Adam Smith on competition and increasing returns, in *Essays on Adam Smith*, edited by A. S. SKINNER and T. WILSON, Oxford, Clarendon Press.

SAMUELS, W. J. (1997). Kenneth Boulding's *The Image* and contemporary discourse analysis, in *The Economy as a Process of Valuation*, edited by W. J. SAMUELS, S. G. MEDEMA and A. A. SCHMID, Cheltenham, Eward Elgar.

11

REGIONAL GROWTH DYNAMICS: A CAPABILITIES PERSPECTIVE

MICHAEL H. BEST*

University of Massachusetts Lowell and Judge Institute of Management Studies, University of Cambridge

The purpose of the paper is to extend Edith Penrose's theory of the growth of the firm to account for inter-firm and regional growth dynamics. To explain the growth of the firm, Penrose developed a resource creation perspective based on a dynamic between productive capabilities and market opportunities. She was not the first to call attention to a resource creation process. As Brian Loasby (*Contributions to Political Economy* (1999) 18) has demonstrated, both Penrose's resource creation and the resource co-ordination perspective of conventional microeconomics can be traced to Adam Smith's *An Inquiry into the Nature and Causes of the Wealth of Nations* published in 1776. As the title of the classic text suggests, Smith was concerned with wealth creation. Loasby and George Richardson (this issue) make strong arguments that the two perspectives can be integrated. They may be right but it will require a substantial rethink of the relations between the firm and the market in ways which give integrity to industrial organisation and regional growth dynamics. In either case, Penrose's growth dynamics is a concept rich with implications for economic analysis that have yet to be drawn out.

I. ADAM SMITH AND THE WEALTH CREATION PROCESS

The division of labor is not a quaint practice of eighteenth-century pin factories; it is a fundamental principle of economic organization (Stigler, 1951, p. 193).

Wealth, for Smith, is created by a process of increasing specialisation and division of labour. Specialisation involves the decomposition of the commodity into an ever-greater number of constituent activities; each activity, in turn, is targeted for a refinement in skills and technique. Every increase in the extent of the market increases the number of activities that are subject to 'new improvements of art'.

* I wish to thank Jane Humphries, Aidan Gough and Christos Pitelis for major assistance in organising this paper. For many years I have discussed ideas on regional industrial processes while working on projects with Giovanna Ceglie, Robert Forrant, Cristian Gillen, Aidan Gough, Robin Murray, Rajah Rasiah, Frederic Richard and Sukant Tripathy. On Adam Smith and the resource creation perspective in the history of economic thought I am deeply indebted to Brian Loasby. Stanley Engerman made helpful suggestions on the industrial organisation argument. Elizabeth Garnsey, my co-author on a Penrose obituary, has left an imprint in the discussion of Penrose's growth dynamics. None, of course, are responsible for interpretations and errors.

In a resource coordination reading of Smith, changes in production methods, skills, and technology are outside the theory. But this violates Smith's view of production as an unfolding adjustment process. An increase in demand does not simply lead to an increase in production but to adjustments in production activities which are at the heart of the wealth creation process. An increase in demand, in Smith's words

. . . though in the beginning it may sometimes raise the price of goods, never fails to lower it in the long run. It encourages production, and thereby increases the competition of the producers, who, in order to undersell one another, *have recourse to new divisions of labour and new improvements of art*, which might never otherwise have been thought of (Smith, 1976, p. 748).

It is critical to note that lower prices are not a consequence of increasing returns to scale for an unchanging process of production, but due to the *adaptation of process* to meet the *opportunities* of an expanded market. Smith suggests, instead, an interactive dynamic between the emerging opportunities and evolving activities of production. With each increase in the extent of the market, the *subdivison of activities* proliferates and ever more activities become subject to specialisation and increasing returns.[1]

Mechanisation, technical change, and invention are part of the process of increased specialisation.[2] In the opening chapter of *The Wealth of Nations* Smith describes how increasing specialisation leads to simplification of production activities which, in turn, creates search opportunities for improvement and innovation in methods and machines:

A great part of the machines made use of in those manufactures in which labour is most sub-divided, were originally the inventions of common workmen, who, being each of them employed in some very simple operation, naturally turned their thoughts towards finding out easier and readier methods of performing it (Smith, 1976, p. 20).

And, specialist machine-making trades join in the division of labour and technical change process:[3]

What takes place among the labourers in a particular workhouse, takes place, for the same reason, among those of a great society. The greater their number, the more they naturally divide themselves into different classes and subdivisions of employment. More heads are occupied in inventing the most proper machinery for executing the work of each, and it is, therefore, more likely to be invented (Smith, 1976, p. 104).

Allyn Young, writing in 1928, adds a complementary aphorism to Smith's principle: 'the division of labour depends upon the extent of the market, but the extent of the market depends upon the division of labour' (1928, pp. 539–40). Young's aphorism is

---

[1] 'In consequence of better machinery, of greater dexterity, and of more division and distribution of work, all of which are the natural effects of improvement, a much smaller quantity of labour becomes requisite for executing any particular piece of work . . .' (Smith, 1976, p. 260).

[2] Brian Loasby uses the term 'discovery process' to encompass a series of innovation activities described by Smith (Loasby, 1997, p. 3). The term 'continuous improvement' is the English translation of *kaizen*, Japanese for a problem-solving model of work organisation also referred to as 'incremental innovation' (Best, 1990, ch. 5).

[3] 'Many improvements have been made by the ingenuity of the makers of the machines, when to make them became the business of a particular trade . . .' (Smith, 1976, p. 21).

a powerful insight into the endogenous wealth creation process. It means, quite simply, that Smith's wealth creating process has a potent feedback loop. An increase in the market leads to further division of labour and further division of labour leads to increased markets. The mediating variable is resource differentiation or increased specialisation of activities and skills. An increase in the market triggers further specialisation which is a process that simultaneously increases the size of the market for specialist skills and activities.

Edgebanding, a specialist activity in furniture making, illustrates the idea.[1] A specialist edgebander can supply the entire industry—no longer is edgebanding activity distributed across all firms in which the market for each edgebander is limited by the sales of the individual firm; instead, the size of the market for the edgebander is the entire furniture industry. The edgebander enjoys increasing returns and so does the district; organisational productivity has advanced by a reallocation of resources not by an accumulation of capital or labour.

II. PENROSE AND ENTERPRISE GROWTH DYNAMICS[2]

Young's extension of Smith's principle of increasing specialisation into a concept of self-sustaining growth has remarkable resonance with Penrose's theory of the growth of the firm. The enterprise growth dynamic derives from an extension of the principle of increasing specialisation from skills to 'productive services' (think capabilities) of enterprises.[3] Penrose draws a distinction between resources, which are homogeneous, and productive services which are heterogeneous: '. . . it is never *resources* themselves that are the "inputs" in the production process, but only the *services* that the resources can render' (Penrose, 1995, p. 25, emphasis in original).

The services of resources derive from the unique experience, teamwork and purposes of each enterprise. Consequently, every enterprise is unique:

The services yielded by resources are a function of the way in which they are used—exactly the same resource when used for different purposes or in different ways and in combination with different types or amounts of other resources provides a different service or set of services (Penrose, 1995, p. 25).

Productive services are potentially dynamic: '. . . the process by which experience is gained is properly treated as a process of creating new productive services available to the firm' (1995, p. 48). And the generation of new productive services is a knowledge-creating process: '[t]he very processes of operation and of expansion are intimately associated with the process by which knowledge is increased' (1995, p. 56). Pro-

---

[1] An edgeband is the strip of material that seals the edge of a table or desktop. An edgebanding capability requires knowledge about, for example, adhesives, substrate and banding materials, and edge-shaping tools. For a case study of two furniture districts, in which one firm specialises by capabilities and the other combines capabilities, unsuccessfully, within the same enterprise, see Best (1990, chs. 7 and 8).

[2] For an expanded description of Penrose's enterprise growth dynamic ideas see Best and Garnsey (1999).

[3] Penrose's term for capability was 'productive service'. G. B. Richardson first suggested the change in terminology (see below).

duction involves both the making of products or services and the creation of new production-related knowledge.

The new production-related knowledge is a form of unused productive services; it creates both an imbalance and an opportunity.[1] The source of the imbalance is inherent in the execution of business plans: '. . . the execution of any plan for expansion will tend to cause a firm to acquire resources which cannot be fully used . . . and such unused services will remain available to the firm after the expansion is completed'. However, the act of deploying unused resources will eventually set in motion the process whereby new knowledge is created and, with it, unused resources which, in turn, creates a new round of pressures to seek yet new activities. The resource creation process is endless.

Realisation of the growth process, however, is an entrepreneurial challenge. In an uncertain world, management must recognise and successfully pursue 'productive opportunities'. The pursuit of 'productive opportunities' links the firm to the customer in an interactive relationship in which new product concepts are developed. The advances in productive services can extend the firm's 'productive opportunities' by enlarging the members' capacity to recognise and respond to new product concept possibilities in the environment.

Experience . . . develops an increasing knowledge of the possibilities for action and the ways in which action can be taken by . . . the firm. This increase in knowledge . . . causes the productive opportunities of a firm to change . . . (Penrose, 1995, p. 53).

From the Penrosean perspective, the firm shapes the market as much as the reverse but within a moving, historically contingent environment. As firms develop and respond to productive opportunities they alter and further differentiate and, in the process, recharacterise the parameters (technological, product, organisational) of the 'market'.[2]

Penrose's case study of Hercules Powder (Penrose, 1960) illustrates the productive services and market dynamic with the company's technological base in 'extensive knowledge' of cellulose technology. Its deep knowledge of the technology leveraged a strategy of moving into and developing promising markets which, in turn, led the firm to invest further in the advance of cellulose technology.

But limits are part of the same process. Just as the scope for expansion on the basis of the firm's existing productive services comes up against market opportunities, so

---

[1] The process of creating new productive services, a by-product of goods production, engenders a balancing or coordination problem: 'only by chance [will] the firm . . . be able so to organize its resources that all of them will be fully used' (Penrose, 1995, p. 32). The coordination problem, however, is partly the outcome of planning limitations: 'In general there will always be services capable of being used in the same or in different lines of production which are not so used because the firm could not plan extensively enough to use them.'

[2] In contrast, technology is exogenous to the resource coordination treatment of the market and firm. The production challenge is to squeeze the maximum amount of output from a fixed pool of resources; the challenge can be met by satisfying a set of optimality rules involving inputs, outputs and prices for a *given* technology. 'Waste is economic sin'; optimality is the elimination of waste. Unused resources are not seen as part of a dynamic process of increasing a firm's productive services.

movement into new areas comes up against the problem of extending the firm's unique productive services to meet market opportunities, that is of '. . . building up an experienced managerial and technical team in new fields of activity'. Internal co-ordination needs to set a brake on the rate at which market opportunities can be pursued: '. . . the rate of growth is retarded by the need for developing new bases and by the difficulties of expanding as a co-ordinated unit'.

The reference to 'bases' is to the 'basic position' a firm must establish and protect (as distinct from merely achieving efficiency in production):

In the long run the profitability, survival, and growth of a firm does not depend so much on the efficiency with which it is able to organize the production of even a widely diversified range of products as it does on the ability of the firm to establish one or more wide and relatively impregnable 'bases' from which it can adapt and extend its operations in an uncertain, changing and competitive world. It is not the scale of production nor even, within limits, the size of the firm, that are the important considerations, but rather the nature of the basic position that it is able to establish for itself (Penrose, 1995, p. 137).

The drive to establish a basic position limits the productive opportunities that any single firm can pursue. But in an open system of firms (see below), such opportunities are not lost but instead are shifted into market 'interstices' and become opportunities for other firms, existing and new. In this way the growth dynamic is propagated to the larger population of firms. The interstices represent new opportunities for expansion that develop out of industrial change and innovation but which cannot be pursued by the originating enterprise: they are inconsistent with reinforcing the basic position, or unique productive services, of the firm in which they emerged.

III. INDUSTRIAL ORGANISATION: A CAPABILITIES PERSPECTIVE

George Richardson suggested replacing Penrose's terminology with that of activities and capabilities. In his words:

It is convenient to think of industry as carrying out an indefinitely large number of *activities*, activities related to the discovery and estimation of future wants, to research, development and design, to the execution and co-ordination of processes of physical transformation, the marketing of goods and so on. And we have to recognize that these activities have to be carried out by organizations with appropriate *capabilities*, or, in other words, with appropriate knowledge, experience and skills (Richardson, 1972, p. 888, italics in the original).

The concept of capability illuminates the process view of production and competition. It gives integrity to organisation as an economic concept. Capabilities can be neither reduced to individual skills nor purchased in the market. The concept of capability can be extended to address a range of issues in industrial organisation.

The growth process, from the capabilities perspective, is simultaneously a networking theory of industrial organisation.[1] Firms specialise in activities that utilise a similar capability and partner with other enterprises that specialise in complementary

[1] Instead of firms as islands in a sea, Richardson's image is of industry as a 'dense network of co-operation and affiliation by which firms are inter-related' (1972, p. 883).

activities. The boundary between the firm and the market becomes blurred as the firm takes on resource creation functions; the firm and the market are no longer simply substitute means of resource coordination. Networking emerges as a means of co-ordination that can enhance the resource creation activities of enterprises.

Open-system dynamics, or networking adjustment processes, in turn, foster capability specialisation, decentralisation and diffusion of design, and technological experimentation. In addition, the consequent regional diversity of enterprises and technologies increase the opportunity for local innovation, based on new combinations of existing technologies.

Production and inter-firm relations are no longer off stage. The production of a commodity involves hundreds if not thousands of activities. Increasing specialisation and further decomposition of the commodity reveal a dynamic population of firms themselves connected by 'markets'. The market in which the end-user receives a completed product is but one market in a long continuum of 'markets' in the form of relationships between producers and users of intermediate products and services between raw material and the end-user.

The principle of increasing specialisation is inconsistent with a static image of industry. The new image is one of a moving picture of increasing specialisation within networked systems of enterprises each specialising in complementary but distinctive activities. This has far-reaching implications for understanding the dynamics of industrial organisation including the mutual adjustment processes amongst firms, regional competitiveness, and industrial growth. For example, the concept of an industrial sector of replica firms producing replica products, and deploying replica production methods, a facilitating assumption of the economics of a stationary economy, denies a role for entrepreneurship or the identification of new product concepts from emerging market opportunities. But it also limits the understanding of the processes of enterprise specialisation and inter-firm integration which are central to understanding capability development and economic progress.

By introducing a structural relationship between the intra- and inter-firm organisation, the capabilities perspective converts the growth process from one of firms to one of regions. It implies that sustained regional industrial progress is predicated upon increasing the specialisation and diversity of capabilities. And just as the internal dynamic of the firm leads to unique capabilities which foster greater diversity of capabilities in the production of a commodity, a more open system of specialist enterprises fosters greater opportunity for increased regional capability specialisation.

Schumpeter came close to the formulation of a dynamic between intra- and inter-firm organisation.[1] This was a logical outcome of his rejection of the ideal of perfect

---

[1] Karl Marx spelled out a static structural relationship, not a dynamic, between the two: '. . . anarchy in the social division of labour and despotism in that of the workshop are mutual conditions the one of the other . . .' (Marx, 1961, p. 356). The powerful determining force of the market on the internal organisation of the 'workshop' is captured by Marx as follows: 'The *a priori* system, on which the division of labour, within the workshop, is regularly carried out, becomes in the division of labour with the society, an *a posteriori*, nature-imposed necessity, controlling the lawless caprice of the producers, and perceptible in the barometrical fluctuations of the market-prices' (Marx, 1961, p. 356). Anarchy in the social division of labour

competition: '. . . perfect competition is not only impossible but inferior, and has no title to being set up as a model of ideal efficiency' (Schumpeter, 1942, p. 106). His critique of perfect competition was that it could 'spread the bacilli of depression'. Fighting for survival, homogeneous firms would resort to cut-throat competition (selling at prices that only covered variable costs) and without investment, capitalism has no dynamism. Profit in the Schumpeterian perspective is not simply a trigger for the invisible hand to coordinate demand and supply but a return to innovation, a source of investment, R&D, growth and, in Schumpeter's term the source of 'future values'. This is a variation on the resource creation theme.

Penrose's dynamic capability perspective fills in the concept of an internal/external dynamic and anchors Schumpeter's innovation in business and industrial organisation. It creates space for the idea of product-led as distinct from price-led competition. Just as price-led competition and closed-system models of industrial organisation are conceptually linked, so too are product-led competition and open-system models of industrial organisation. The textbook ideal of perfect competition presumes the first structure and denies the second. Real-world economies offer both as possibilities. The fact that at least two possibilities exist offers a further possibility: competition across regional business systems each constituted by distinctive regional capabilities. A business system is constituted by the regional equivalent of the 'experience and teamwork' that constitute Penrose's enterprises.

IV. REGIONAL GROWTH DYNAMICS

Growth from the resource creation perspective is about processes of change in technology and the organisation of firms and industry. Several are identified above: Adam Smith's increasing specialisation fosters increasing returns including the 'improvement of art'; the technology/market dynamic of the entrepreneurial firm which simultaneously produces output, new ideas, and creates new imbalances between productive capabilities and market opportunities; and 'open-systems' networking which fosters capability specialisation, the diffusion of design, and, I will argue, greater opportunities for innovation.

The internal dynamics of the firm, complemented by internal/external dynamics of open-system networking models of industrial organisation, propagate a virtuous circle of regional economic growth dynamics particularly relevant to the 'knowledge-driven' economy. Growth involves some combination of new firms and growth of existing

footnote 1 continued from previous page

is Marx's expression for the market for commodities; perfect competition in the market imposes despotism inside the factory walls. Marx did not depart from the resource allocation perspective in presuming price competition as the only form of competition. Both Marx's 'law of value' and the 'discipline of the market' assume price competition which, in turn, imposes a single 'optimal' organisation on all firms. For Marx, the result was 'despotism' in the workplace, for conventional economics it is an equilibrium in the markets for labour, capital, and raw materials. In both cases the adjustment process is one of price movements. Management, strategy, and entrepreneurship are determined by the 'market'; the technology and organisation have no role in the theory. The internal/external dynamic is denied by assumption.

firms. New and rapidly growing firms, however, do not appear in isolation but emerge from and develop within, an industrial infrastructure constituted by a larger population of specialist and affiliated enterprises. A firm's capabilities, like an individual's skills, are shaped in a mutually interdependent process. These inter-firm processes involve simultaneously resource coordination and capability creation.

Figure 1 characterises the extension of the internal growth dynamics of the firm to the region and inter-firm networking processes. It is, at the same time, an extension of Adam Smith's principle of increasing specialisation from skills and occupations to capabilities as mediated by business and industrial organisation.

IV.a. *Entrepreneurial firms: propagating regional dynamics*

The entrepreneurial firm, represented by the box at the right of Figure 1, is driven by a *technology/market dynamic*.[1] The term dynamic connotes an ongoing historical process in which both technology and market are mutually redefined, a process that is built into the ongoing operations of the firm: firms pursue unique capabilities, often of a technological form, but the process of developing such capabilities creates new productive opportunities in terms of a refined match between product or service performance and customer demand, existing and inchoate.[2] In the process of redefining the product, the market, too, is recharacterised. These new 'market' opportunities feed back to motivate changes in productive capabilities setting in motion a new dynamic.

The internal dynamics of entrepreneurial firms enhance regional growth potential. Whether or not the potential is realised depends, in part, upon choices made within the entrepreneurial firm. Firms that experiment and develop unique and/or new capabilities simultaneously must choose which of the new possibilities to pursue as the basis of their competitive advantage. Bets must be placed on which technological possibilities are pursued and which are abandoned.

Any individual firm's choice of technology streams in which to pursue unique capabilities is constrained by cumulative and path-dependent character of capability

---

[1] The term technology/market dynamic is an abbreviation of Penrose's productive capability/market opportunity dynamic. The idea of the entrepreneurial firm, as an extension of the entrepreneurial function from an individual attribute to a collective or organisational capability, is developed in Best (1990).

[2] A variant of the technology/market is Intel's 'dynamic dialectic' as described by co-founder Andrew Grove (1996). Grove's 'dynamic dialectic' is built into a business model organised to combine recurrent phases of bottom-up experimentation and top-down direction. Phases of experimentation, which stimulate new ideas and innovation, are fostered by decentralisation of decision-making. The challenge of leadership is to allow enough time for free rein to stimulate the development of new ideas before managing a new phase during which the most promising ideas are pursued and the weaker ideas are abandoned. The challenge of leadership is to balance the phases of experimentation and direction so that the enterprise can benefit from the advantages of both bottom-up initiatives and top-down decision-making. Too much experimentation can result in chaos; too much direction can stultify innovation. Built into the challenge of leadership is the ability to manage organisational change; leaders must gain personal commitments to new directions, technologies, processes, and products. Without personal commitments from top to bottom, human energies will not be mobilised to drive the redirection of organisational resources. While experimentation demands turning everyone into a designer, direction demands that everyone enthusiastically accept the winning designs. This is no small organisational challenge.

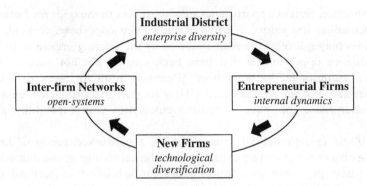

FIGURE 1. Model of regional dynamics.

development. The bets are constrained by the need to sustain unique capabilities based on the acquired technological skill base. The firm faces a dilemma: unique capabilities are both the source of competitive advantage and a constraint on future development. New opportunities which require activities which are not consistent with reinforcing the firm's basic position risk devaluing the firm's unique capabilities. The constraint may dominate in a region populated by vertically integrated enterprises with little opportunity for entry of new firms. On the other hand, the constraint at the level of the individual firm can be a growth opportunity for the region. It all depends upon the regional capability for firms, existing and new, to respond to new productive opportunities created in the 'interstices'.

Abandoned possibilities are simultaneously opportunities for new divisions within subsidiaries or spin-offs, or for new firm creation. The pursuit of new capabilities also pens new possibilities for partnering for complementary capabilities. Both of these processes contribute to *potential* regional techno-diversification which, if activated, can trigger industrial 'speciation' or the emergence of new industrial sub-sectors.

IV.b. Techno-diversification

The link between the box at the right and bottom in Figure 1 represents a 'techno-diversification' dynamic between entrepreneurial firms and 'new firms' or new activities within companies specialising in the requisite capabilities. As noted, in the process of pursuing its own goals the entrepreneurial firm simultaneously propagates new productive opportunities which can be either pursued internally or pushed outside the firm. Those not pursued internally become 'market' opportunities for other firms to advance their productive capabilities.

The originating firm can generate three types of new productive opportunities to other firms which can, in turn, foster secondary internal dynamics. First, because of the inherent uncertainty about future technological pathways, the entrepreneurial firm must place its R&D bets on some and forsake internal development of others. The choice to abandon certain opportunities may speak less to the odds for future success than to sustaining the value of firm's existing knowledge base. Second, the new opportunities may not offer the scale required by the existing organisation.[1] In both cases productive opportunities that have been created but not pursued represent market opportunities in the 'interstices' (Penrose's term for niche market opportunities that have not yet been pursued). They are potential productive opportunities for new enterprises, spin-offs, or existing enterprises with capabilities in similar activities.

Third, if the entrepreneurial firm is part of a networked group of firms each specialising on a complementary capability, a technical change at one link in the chain will create new pressures and opportunities for specialists in each of the complementary capabilities. In this way advances in design and technology are both diffused and interactive across production networks. In some cases the induced effect may be one of induced technical change which, in turn, may set off a secondary internal dynamic and consequent pressures for change across the network.

New firms are often the path to techno-diversification. Moreover, new firms and spin-offs can trigger a process of industrial 'speciation' or the emergence of new industry subsectors. Classic examples are the transistor, the telephone, the laser, the personal computer. None of the key technological innovations were developed by the companies that sponsored the original research but all became the basis for the emergence of a vast range of new enterprises based on unique capability development.

The process of techno-diversification can operate on a much smaller scale. In any case, new firm creation is often critical to the emergence of new industrial sectors and business models. For example, the resurgence of Massachusetts' Route 128, following the demise of the minicomputer industry, was driven by new firms pursuing new technologies with new business models. Moreover, the new open-system business model has itself fostered much greater technical diversity and industrial speciation.[2]

Finally, the exploitation of interstice opportunities by small and new firms limits the tendency to industrial concentration. This led Penrose to conclude '. . . in a steadily growing economy, or in an economy where expansion is more prevalent than stagnation, the process of concentration will come to an end and eventually reverse itself' (Penrose, 1995, p. 258). Ease of entry is critically important to regulating industrial districts given the structural link between ever greater specialisation and increasing returns.

In these ways the growth dynamic of entrepreneurial firms is propagated to the larger industrial system. But the story does not stop here. A growth impact is not automatic; it depends on choices made within the originating firm, inter-firm organis-

---

[1] Gordon Moore, a co-founder of Intel, makes the point that a single firm cannot pursue multiple technological capabilities simultaneously: 'integrated circuits, MOS transistors, and the like proved too rich a vein for a company the size of Fairchild to mine, resulting in what came to be known as the "Silicon Valley effect". At least one new company coalesced around and tried to exploit each new invention or discovery that came out of the lab' (Moore, 1996, p. 167).

[2] These themes are developed in Best (1999, 2000).

ation, and extra-firm infrastructure. Furthermore, the relations between the firm and regional growth dynamics are not one way; as we shall see they too are interactive.

IV.c. Open-systems networking

The box at the left of Figure 1 represents open-systems networking, commonly referred to as 'horizontal integration', multi-enterprise integration, cooperation, networking, or affiliated groups of specialist enterprises. Open-systems networking is an inter-firm counterpart to increasing specialisation of the entrepreneurial firm.

Inter-firm networking has evolved with the shift from price-led to product-led competition. This entails integration of manufacturing and new product development processes. But rapid new product development is not simply adding a product (multi-divisional diversification); increasingly it involves a whole group of specialist companies operating at different links along the product chain.

Inter-firm networking offers greater flexibility for new product development and innovation than does vertical integration.[1] Ironically, networking can foster the social relations necessary for effective co-location of specialist but complementary activities more easily than can vertical integration. While a vertically integrated company operates under a single hierarchy which can direct departments to co-locate, it does so within a bureaucracy and a set of technologies that were originally designed for different purposes. They become embedded in social systems and individual career paths within the firm which can offer resistance to organisational change. Open-systems networking offers a range of co-design possibilities without locking an enterprise into any one design possibility. The open-system organisational model is fostered by open systems in the form of standardised interfaces and shared design rules at the technological level.[2]

Furthermore, the open-system networking model represents a departure from both the vertical integration and the closed-system networking or *keiretsu* models of inter-firm co-ordination.[3] Open-system networking is a system of inter-firm relations particularly hospitable to rapid new product *concept* development. Whereas the Toyota production system and the *keiretsu* closed-system networks enhanced production flexibility, the open network business model fosters product concept flexibility within an organisational infrastructure that fosters easy entry of new firms, technological

---

[1] Horizontal integration, the term used by Andrew Grove of Intel to describe open-systems networking can be considered an inter-firm consequence of Intel's production concept of integrated manufacturing (Grove, 1996; Best, 1998).

[2] The idea of system integration suggests a common design principle which enables the integration of independently designed components. The term open system suggests that the system design rules are openly published. A closed system, in contrast, suggests that the challenge of integration is achieved by an overarching design principle which leaves no space for independently designed components. The IBM 360 computer, for example, was a closed system before an anti-trust ruling forced the publication of the system design principle and thereby began a process that led to an open system. The embedded and private operating system of Massachusetts mini-computer companies is another example.

[3] It also evokes and complements Marshall's industrial district with new dynamics of localisation (Marshall, 1920).

experimentation, and greater technological diversity which enhances opportunities for innovation via new technological combinations.

The idea of an internal/external dynamic captures the interactive process between firm and region which propagate growth. The capability-creating process of the entrepreneurial firm is the source of internal dynamics, but the firm exists within a large system of inter-firm relations and dynamics which condition its opportunities. Thus a firm's opportunities to specialise and develop its unique capabilities are shaped externally by the constellation of enterprises and capabilities with which it can partner. The virtue of the open-system networking model is that it institutionalises the internal/ external dynamic which, in turn, propels growth.

The *new firm creation process* is itself an aspect of mutual adjustment. Just as the dynamics associated with new product development involve a continuous redefinition of product concept, carrying out the process can foster a proliferation of firms' concepts. Diversity and the principle of variation, or increased speciation, means the creation of new firm concepts. This process is enhanced in open-system networks in which new specialist firms can readily plug into pre-existing product chains. This process suggests that strategies of firms are themselves shaped in the ongoing practice of refining a firm's concept or specific characteristic that distinguishes itself from other firms and thereby gives it market power.

IV.d. Regional capabilities and innovation

The upper box in Figure 1 signifies the extent of capability specialisation and diversity within a regional population of networked industrial enterprises. The figure highlights a range of intra- and inter-firm dynamic processes that underlie capability development at both the enterprise and regional levels. Regional specialisation results from path dependence, the unique combinations and patterns of intra- and inter-firm dynamics which underlie enterprise and regional specialisation.

The process of enterprise capability development is also a potential process of regional capability development. A firm's own capabilities are developed as part of a dynamic, inter-firm process of capability development. Regional capabilities are different from physical infrastructure, they are akin to regional 'social capital' in the form of shared production capabilities. But regional production capabilities are more than the networks and extra-firm institutions that support enterprises. They are constitutive of a region's industrial organisation like teamwork to a sports team; 'social capital' enables participants to advance specialist skills that could not be accomplished alone. They make the whole greater than the sum of the parts because the whole accounts for the effects of unique inter-relationships on skills and capabilities. Regional production capabilities underlie distinctive patterns of regional specialisation.[1]

Greater diversity is particularly relevant to innovation. An industrial district, unlike any single firm, offers the potential for new and unplanned technology combinations

---

[1] Elsewhere I have developed the idea of technology management capability to explain changing patterns of industrial leadership (Best, 1998).

which tap a variety and range of research and production-related activities. Open systems offer wider opportunities to foster creativity, fill gaps, replenish the knowledge pool and match needs to research.[1]

Regional production capabilities lie behind the competitive advantage of 'low-tech', high-income industrial districts common to the 'third Italy'. Such districts have developed a competitive advantage in design capabilities which have fostered industrial leadership in a range of design-led or 'fashion industries'. Recently, high-tech regions have developed similar capabilities for rapid design changes and industrial innovation. In fact, regions such as Silicon Valley and Route 128 have developed regional innovation capabilities embedded in virtual laboratories in the form of broad and deep networks of operational, technological and scientific researchers which cut across companies and universities. Silicon Valley project teams are continuously combining and recombining across a population of 6000 high-tech firms making it an unparalleled information and communication-technology industrial district.[2]

While the high-tech districts are unique in terms of specific technologies and research intensity they exhibit regional innovation characteristics in an exaggerated form that are common to the virtuous circle of regional growth. Examples follow.

First, the high-tech, open-system industrial district is, as well, a collective experimental laboratory. Networked groups of firms are, in effect, engaged in continuous experimentation as the networks form, disband and reform. Both the ease of entry of new firms and the infrastructure for networking facilitate the formation of technology integration teams in real time. However successful the industrial district has been as a mode of economic coordination in international competition, heretofore it has been considered appropriate only to 'light' industry such as the design-led, fashion industries of the 'third Italy' and the machine-tool and metal-working regions of Badem Wurtemburg in Germany.

Second, an open-system district expands the number of simultaneous experiments that are conducted. A vertically-integrated company may carry out several experiments at each stage in the production chain but a district can exploit dozens simultaneously. In this way a district counters the barriers to introducing new ideas in firms that already have well-developed capabilities around competing technologies.

Third, an open-system district fosters the decentralisation and diffusion of design capabilities. Design modularisation in the personal computer industry is an example. IBM got the process underway with the modularisation of the 360 computer which

---

[1] The regional model of innovation offers a decentralised, self-organising explanation of the success of high-tech regions but of industrial districts in general as an alternative to the linear, science-push model of innovation. In the latter, technology is thought of as applied science; in the regional model, technology is part of the industrial process. It is built into the process by which firms establish unique capabilities and network with other firms. The science-push model, in contrast, fails to capture the extent to which research is woven into the production, technology, and networking fabric of a region's industrial system as distinct from being an external, autonomous sphere of activity.

[2] Intel is not the only driver of new products. Approximately one in five of the Silicon Valley (and Route 128 in Massachusetts) publicly traded companies were gazelles in 1997 which means that they have grown at least 20 per cent in each of the last 4 years (the number for the US is one in 35). See Massachusetts Technology Collaborative (1998).

created an open system. This was greatly enhanced when the design modules for the operating system and the microprocessor were developed by Microsoft and Intel.[1] The resulting standards have created enormous market opportunities for specific applications software. But in addition the concept of design modularisation combines common interface design rules with decentralisation of component design. This diffusion of design capability increases collective innovation capacity. It can also strengthen the district model of industrial organisation, even enhance conversion from a closed to an open system.

V. CONCLUSIONS

Adam Smith focused attention on the process of increasing specialisation in the division of labour which fostered increasing returns from organisational improvements and innovations as well as increased dexterity and the introduction of machines. Penrose focused attention on the process of resource creation, in the form of productive capabilities, as a by-product of executing plans within the firm.

An individual firm's capabilities are formed in a process of mutual adjustment with other enterprises as described by Adam Smith's principle of increasing specialisation. The open-system 'industrial district' model of industrial organisation is populated by networked enterprises driven by an internal growth dynamics, an ongoing process in which new capabilities are created as a by-product of implementing production plans which, in turn, enable and foster a search for new 'market' opportunities; the refined characterisation of customer needs acts back on production plans and sets the process in motion again.

The integration of Smith's principle of increasing specialisation with Penrose's dynamic process of capability development is a major step in the extension of the resource creation perspective into a theory of industrial organisation. Firms specialise in activities that are consistent with the advance of a unique capability and network with partners for complementary capabilities. This is an extension of both Smith's principle of increasing specialisation to firms and of Penrose's capability-creation growth process from the firm to networked groups of firms.

Regional capability development is driven by a process of mutual adjustment across firms; the internal and external processes of the Penrosean firm are themselves interactive in an open-system model of industrial organisation. Increasing regional diversity of skills, capabilities, and technologies are expressions of Smith's fundamental principle of increasing specialisation but applied to regions.

The internal dynamics of the entrepreneurial firm do not depend upon industrial districts or regional dynamics. They are equally associated with vertically integrated enterprises or the *keiretsu* model of industrial organisation. However, the sustained vitality of open-system industrial districts, whether in the light industries of the 'third Italy' or the high-tech regions of Silicon Valley and Route 128, illustrate that the

---

[1] See Katz (1996, p. 15). Katz also describes network economies and increasing returns.

internal dynamics of the entrepreneurial enterprise combined with techno-diversification and open-system networking generate additional regional dynamics with powerful effects on growth.

These sub-processes are the forces behind the horizontal integration identified by both Saxenian (1994) and Grove (1996) as a major source of the organisational advantage of Silicon Valley. It is the model of industrial organisation most appropriate for product-led competition.

REFERENCES

BEST, M. (1990). *The New Competition*, Cambridge, MA, Harvard University Press.

BEST, M. (1998). Production principles, organizational capabilities and technology management in *Globalization, Growth, and Governance*, edited by J. MITCHIE and J. GRIEVE-SMITH, Oxford, Oxford University Press.

BEST, M. (1999). Silicon Valley and the resurgence of Route 128: systems integration and regional innovation, in *Regions, Globalization and the Knowledge-based Economy*, edited by J. H. DUNNING, Oxford, Oxford University Press, 459–84.

BEST, M. (2000). *The New Competitive Advantage*, Oxford, Oxford University Press.

BEST, M. and GARNSEY, E. (1999). Edith Penrose, 1914–1996, *The Economic Journal*, Vol. 109, No. 453, 187–201.

GROVE, A. (1996). *Only the Paranoid Survive*, New York, Doubleday.

KATZ, J. (1996). To market, to market: strategy in high-tech business, *Regional Review, Federal Reserve Bank of Boston*, Vol. 6, No. 4, 12–17.

LOASBY, B. (1997). Edith Penrose's place in the filiation of economic ideas, mimeo University of Stirling.

MARSHALL, A. (1920). *Principles of Economics*, London, MacMillan.

MARX, K. (1867). *Capital: A Critical Analysis of Capitalist Production*, Vol. 1, Moscow, Foreign Languages Publishing House [1961].

MASSACHUSETTS TECHNOLOGY COLLABORATIVE (1998). *Index of the Massachusetts Innovation Economy*, Westborough, MA.

MOORE, G. (1996). Some personal perspectives on research in the semiconductor industry, in *Engines of Innovation*, edited by R. ROSENBLOOM and W. SPENCER, Boston, Harvard Business School Press.

PENROSE, E. (1995). *The Theory of the Growth of the Firm*, Oxford, Oxford University Press.

PENROSE, E. (1960). The growth of the firm—a case study: the Hercules Powder Company, *Business History Review*, Vol. XXXIV, 1–23.

RICHARDSON, G. B. (1972). The organization of industry, *The Economic Journal*, Vol. 82, 883–96.

SAXENIAN, A. (1994). *Regional Advantage: Culture and Competition in Silicon Valley and Route 128*, Cambridge, MA, Harvard University Press.

SCHUMPETER, J. (1942). *Capitalism, Socialism and Democracy*, New York, Harper.

SMITH, A. (1976). *An Inquiry into the Nature and Causes of the Wealth of Nations*, edited by R. H. CAMPBELL, A. S. SKINNER and W. B. RODD, Oxford, Oxford University Press.

STIGLER, G. (1951). The division of labor is limited by the extent of the market, *The Journal of Political Economy*, Vol. LIX, 185–93.

YOUNG, A. (1928). Increasing returns and economic progress, *The Economic Journal*, Vol. 38, pp. 527–42.

12

MISMATCHING BY DESIGN: EXPLAINING THE DYNAMICS OF INNOVATIVE CAPABILITIES OF THE FIRM WITH A PENROSEAN MARK

MARGHERITA TURVANI

University of Venice, IUAV

I. INTRODUCTION

> In undertaking an analysis of the growth of the firm in the 1950s, the question I wanted to answer was whether there was something inherent in the very nature of any firm that both promoted its growth and necessarily limited its rate of growth. Clearly a definition of a firm with 'insides' was required (Penrose, 1995, p. xi) . . . It follows from the argument developed in *The Theory of the Growth of the Firm* that a firm's rate of growth is limited by the growth of knowledge within it (Penrose, 1995, p. xvi).

These few sentences, drawn from the introduction that Penrose wrote in 1995 for the reprint of her most well known book, set the agenda for this paper.

My attention will be addressed to the understanding of the features that are responsible both for promoting and limiting the capability of firms to innovate. Following Penrose's mark I will look at the process of generation and utilisation of knowledge within the firm to highlight the mechanisms at work that govern the day-to-day process of knowledge creation, mechanisms that may foster or hinder the development of innovative capabilities of the firm and its capacity to grow.

Penrose's book deals with a theory of the growth of the firm, where growth is attained by breaking any previously reached state of coherence among the activities taking place; firms grow by innovating and innovation occurs because firms create imbalances while pursuing their strategies. The very same mechanisms that generate imbalances explain both the weaknesses and the strengths of any firm. It is the attempt to adjust the mismatch existing at any time between the old available flow of productive services and those needed to implement new plans; it is the attempt to take advantage of the mismatch resulting in the realisation of plans, between the newly available flow of productive services and their ongoing utilisation. Imbalances, though, play the role of both constraints and opportunities. As Penrose puts it (1995, p. xiii): 'Growth is essentially an evolutionary process and based on the cumulative growth of collective knowledge, in the context of a purposive firm'. Many subscribe to this statement; I will attempt to explain why it is so, by referring first of all to some of the Penrosean ideas (Section II) and the subsequent developments in the modern theory of the firm, to understand how people's knowledge is managed in organisation (Section III); secondly by recalling and using some of the recent findings in the field of cognitive sciences to explain how knowledge is created in individual minds and how it is contextually developed (Section IV).

Penrose was aware of the importance of knowledge creation and utilisation and this is why she devoted much attention to the human resources organised within the firm. Again I will follow her and I will try to explain how knowledge that individuals hold undergoes a process of renewal when new plans and new strategies are implemented within the firm (Section V) and the importance of human resources policies in developing the innovative capabilities of the firm (Section VI). Conclusions follow (Section VII).

II. THE LIMITS TO GROWTH

Penrose's contribution to the understanding of the nature of the firm and of its evolution is now widely recognised in various streams of economic studies; it has offered a route to a better comprehension of the role of the firm in a knowledge-based economy. Yet, her contribution was for some time reduced to the discussion of the limit to growth, within the framework of the debate on the optimum size of the firm. In 1980 Slater wrote in the reprint of Penrose's book:

She assumes constant returns to the firm both on the supply side and on the demand side. In the long run therefore there is no single optimum size that the firm will tend to, since any size is as profitable as any other. What then does determine the actual size of the firm at any point in time? . . . The answer to this question is given by . . . increasing cost of growth. This is perhaps the innovation most characteristic of Mrs. Penrose, so much that it has come to be known by her name as the Penrose Effect. Although there are constant returns in the long run, this constancy will be achieved only when perfect adaptation of all inputs to a particular scale has been made. This will not be possible in the short run if certain inputs are difficult to vary, and the faster the firm attempts to grow, the less well adapted will the input structure tend to be. Thus a firm is prevented from growing as fast as it may like because there is a very distinct cost of rapid growth (Slater, 1980, p. xi).

Imbalances and mismatches in the process of growth are thus presented as a major constraint for the growth of the firm. In this chapter I wish to show that imbalances and mismatches are at the root of the process of growth itself;[1] this perspective appears especially fruitful if we look at the firm from a cognitive perspective, directing our attention to the process of generation of knowledge in the individual minds and its utilisation within the firm.

To grasp fully the potential of Penrose's approach we must shortly outline the mechanism of growth and the limits to it; we can then offer a reappraisal of the same

---

[1] The idea of disequilibrium generating growth is a Schumpeterian one. It is also the reading of Penrose proposed by Best (1990)

mechanisms in the light of recent advancement in the theory of the firm and in the field of cognitive studies, to obtain not only a more accurate description of what is going on in the firm but also to contribute to a better theory of the firm.

In Penrose's approach the firm is an administrative unit which exercises coordination and authoritative communication. It does so by bringing resources together in a certain cluster or structure, which changes over time. Intangible resources, and specifically managerial and human resources are put centre stage: they own the body of knowledge upon which the firm can draw.

Change is costly but also necessary if the resources available at a certain period are to be used to best advantage. In Penrose's model costs of growth are linked to the difficulties of increasing and adapting managerial resources to the changing state of affairs. In such a situation, the area of administrative coordination and authoritative communication cannot be expanded adequately with respect to the market. The reason for this must be sought in those features of growth that imply that only very rarely does a firm grow according to the logic of 'more of the same'. On the contrary, growth and innovation in the use of productive resources go hand in hand; innovation implies that new responsibilities and new decisions are required, often in less familiar directions. Penrose's emphasis on administrative coordination and authoritative communication as the distinctive features of the economic organisation we describe as a 'firm' defines both the nature and the limitations of such a body. A firm's growth is the result of certain mechanisms that govern expansion, and foremost among these are the managerial functions. They are the basis of the coordination and communication that hold together the resources the firm has at its disposal. There is an endogenous mechanism—the active search for better ways of using internal resources—which dynamically leads to the growth of the resources the firm has at its disposal, and thus further fuels growth.[1] Expansion, therefore, is a recurrent and unbalanced phenomenon.

III. KNOWLEDGE AND THE THEORY OF THE FIRM

III.a. Information and knowledge

Penrose summarises her argument by saying that the firm's rate of growth is limited by the growth of knowledge within it. In economics, though, there is an abuse of the term knowledge while the concept is insufficiently clarified. To begin with it is necessary to distinguish between information and knowledge. Here we adopt the definition—used by Nonaka and Takeuchi (1995)[2]—of information as a good capable of producing knowledge.

---

[1] Metcalfe (2000, p. 19) points to the positive role of imbalances when he says: 'imbalance can take the form not only of an under-utilized capability, but of a "missing" capability, need for the optimal exploitation of other capabilities which are present . . . The restless firm . . . creates capabilities where imbalances existed previously.'

[2] This definition has been developed by the philosopher Dretske (1981).

Economics tends to obscure the differences between the two concepts; more, there is a tendency to use the two terms indifferently or to see knowledge simply as piling up of information, highlighting a sequential line from information to knowledge. Information is described as a set of data about states of the world and attendant consequences due to the production of natural and/or social events. The data set is closed-ended and can potentially be acquired. Knowledge is then portrayed as a closed acquirable whole: this contrasts overtly with the open-ended nature of knowledge, the processes of discovery and creation of new forms of knowing (Loasby, 1991).

According to our definition, information may give rise to knowledge; we then introduce the possibility that this cannot or does not happen in a way that can be determined *a priori*. In the case of knowledge being effectively built, its connection with a specific closed-ended finite set of data cannot necessarily be reconstructed. The tight coupling between information and knowledge is broken, or partially inverted and there is scope for interpretative ambiguity. Fransman (1994, p. 716) remarks, 'knowledge (belief) will be used to interpret incomplete information', and the results of this process will always be subject to verification, revision and comparison. It is precisely for this reason that knowledge is an unfolding open-ended, incomplete and local process.

Ambiguity and looseness of ties in the relationship between information and knowledge introduces an interactive dimension in individual or social cognition: a shift of attention is needed from information and knowledge, understood as a freely combinable cumulable stock, albeit in a costly way, to the process that allows knowledge to be created always in new and different forms.

III.b. The theory of the firm

Modern theories of the firm have been developed on the idea that the most interesting things that happen within the firm regard information and knowledge issues. This is the reason why Penrose's theory has become so fashionable, being as it is and as she explicitly recognised 'a theory of the growth of the firm in relation to the role of knowledge' (Penrose, 1995, p. xiii). Yet, theories of the firm treat information and knowledge differently. On one hand, there are theories highlighting information problems connected with completing transactions—both internal and with the market. On the other hand, there are theories that see the firm as a collection of forms of knowledge, and stress the study of the elaboration and use of knowledge within and between firms. In the first case (information processing case) the economy is made up of individual agents who, with no uniformly distributed perfect knowledge, have to manage a variety of information problems, all implying costly solutions: the firm is called to solve this function. In the second case (knowledge generation case) agents are characterised by bounded rationality, which makes the optimising calculation unattainable and encourages the development of local scale problem-solving techniques. Specific pools of knowledge and know-how with strongly specific connotations develop at the technical, organisational and market level. It is specificity that makes the firm unique and different, in order to exist within a competitive environment. The first group of theories, those relating to information processing, includes all the variations connected with the idea of the firm as a nexus of contracts. In particular we may include in it both Alchian and Demsetz's work of team production (1972), and the principal-agent theory put forward by Jensen and Meckling (1976), developing out of the concept of information asymmetry. Elaborating the pioneering work of Coase (1937), Williamson (1985) develops the economics of transaction costs: in his view hierarchy solves information problems relying on complex incomplete contractual arrangements that are developed in time by cognitively bounded but far-sighted agents.

A different perspective is offered when firms are described as a set of local forms of know-how structured through 'doing' and the recurrent activity of problem solving. The firm takes on the features of a 'pool of forms of knowledge': the representation of these forms of knowledge and the way in which they are generated, selected and modified thus becomes the focus for analysis. This current in the modern theory of the firm, although alluding to Marshall and the role he gives to knowledge as a stategic factor of production, is linked to the studies on bounded rationality and organisational learning (Cyert and March, 1963) and to the pathbreaking analysis of firm by Penrose (1959, 1995).

Here the focus is on the learning activities taking place within the firm and the historical and local nature of the knowledge creation process is emphasised. Organisations learn and their experience is encoded in routine or develops in capabilities: we owe to the contribution of Nelson and Winter (1982) the elaboration of the concept of routine and its role in encoding knowledge by means of routine creation and modification; the resource-base analysis of the firm and the dynamic-capabilities view of strategy (Teece and Pisano, 1994)[1] also offers the conceptual tools to analyse the role of knowledge in the firm.

All these perspectives describe firms in term of routines or core capabilities: firms are built on the cumulative growth of knowledge, sometimes depicted as the repertoire of persistent patterns of behaviour, sometimes described in terms of capabilities that build on one another through time, by the different experiences of firms.

Theories that describe firms in terms of a pool of forms of knowledge insist that firms remember and know by doing; knowledge and its underlying structure preserve a character of tacitness for the people holding it (Nelson and Winter, 1982, p. 99). The knowledge that organisations handle and continuously create resides in the complex of routines, and in the pool of capabilities making up the firm. It is through this lens that organisations approach messages coming from within the organisation itself and also from its outside. Firm's routines and capabilities serve as a repertoire of abilities and skills that may be used within the firm i.e., they describe what the firm knows and how it knows it; but they also frame the way by which the outside is understood.

We can then interpret them as firm's cognitive apparatus and, at any time, their specific organisation as 'cognitive model or frame (or map)' (Langlois, 1995). It is

---

[1] There is a vast literature on the subject. For a survey, see Foss (1993) and Montgomery (1995). For an examination of the points of contact between the strategic approach and the approach based on resources, see Mahoney and Pandian (1992) and Prahalad and Hamel (1990).

through this lens that the firm, from time to time, filters what it 'sees' in the environment and it monitors and interprets its internal evolution. Firm's absorptive capacity
(Cohen and Levinthal, 1990) which is the ability to understand and take advantage of
new ideas (or knowledge) is moulded accordingly.

Describing firms as a pool of forms of knowledge we emphasise the role that organisations play in providing the frame in which individual knowledge is actually
developed; individual cognition takes place in a mind which is active in the context
defined by the firm's scope but the firm's scope is limited by the ability of the
organisation to evolve without threatening organisational coherence (Loasby, 1991,
p. 61).

IV. COGNITION IN CONTEXT

IV.a. The cognitive process

Cognitive processes reside in the mind of the individual, but the importance of the
interactive process between individual and society in cognitive development is not in
doubt.[1]

The interest in cognitive problems was first developed in economics by Simon,
whose work builds on a critique of the paradigm of perfect rationality. He proposes an
alternative model—procedural rationality—which is better able to take into account
the cognitive limits typical of individuals. In this model the alternatives are formu-
lated, compared and assessed, while learning plays a key role in improving the cognitive performance of individuals. According to Simon, cognitive activities are simplified
through the use of symbols, and symbolic structures are images of the setting they
belong to. Models of reality developed in the mind are more or less accurate trans-
criptions of the outside world rather than representations and interpretations. Following Simon's approach, we can proceed to study these complex structures through a
decomposition method: the typical mental activity of problem solving may thus be
analysed, since it is made up of a set of elementary information-handling processes.
With Simon's model of bounded rationality (Simon, 1987) the main assumptions of
cognitivism (Piaget, 1969) were imported into economic analysis; perception is
connecting individual and environment. The environment is the 'outside' of the individual who receives new perceptive stimuli that are assimilated in the existing individual schemas by means of a process of accommodation and feedback, making it
possible to combine the old schemas and create new ones.

A different perspective is offered by connectionism, which offers social sciences
hints for moving attention from individuals to an analysis of relations connecting
them. In the study of the mind the unit of analysis is not the neuron, but the connections between neurons and the plasticity of the system of connections becomes of
major importance. The consequent cognitive structures (frames, mental models) are

---

[1] This theme was developed by Rizzello and Turvani (2000), who discuss the importance of the cognitive
sciences for institutional economics.

perceptive filters and more importantly are the links between the individual and the environment, limiting and selecting mental activities that individuals are able to develop. Although an important tool for representing, when a particular cognitive frame is in action, it can no longer be the subject of cognitive activity; in such a way knowing develops by structuring, by creating new, more complex forms.

Social interaction becomes a clue in understanding the development of interpretative patterns and common frames used by individuals to relate to reality and develop their own cognition at individual level. In such a way shared forms of behaviour in a social environment and also in problem-solving activities may develop.

Despite the highly idiosyncratic subjective nature of cognitive processes, the development of intense communication, favoured by proximity (not necessarily only spatial but also cultural) leads to commonality in developing cognition, which takes on the character of dependency on the context in 'local and idiosyncratic' forms. As Witt (1998) has observed with reference to the studies carried out by Bandura (1986), interestingly these mechanisms produce shared behavioural models and this assumes considerable importance when we analyse the effects on the decision-making activities of individuals. 'The significance of such models lies in their vicarious nature, i.e. the fact that their rewarding or non rewarding consequences can be grasped by observation rather than own experimentation. Within intensely communicating groups, learning by observation then means that the group members tend to focus on much the same model. Consequently, their individual learning processes produce correlated results which, via the prevailing social models, can intuitively be grasped by newcomers' (Witt, 1998, p. 5). These mechanisms are important because they offer a vicarious (i.e., using another person's experience) method to appraise success and failures of deviant behaviour, i.e., behaviour diverging from that commonly established in specific contexts. In such a way, individuals may receive information to evaluate their own behaviour in alternative situations, previously not under consideration. A re-framing of choices may take place allowing a smooth transition from an older social model to more innovative ones (Witt, 1998, p. 6).

IV.b. Cognition and the institutional context

Economics has now largely accepted that institutions respond to the limited cognitive capacity of individuals: the construction of complex institutional and organisational systems (the outcome of more or less intentional efforts) helps individuals and groups in their economic activity, supplying an aid to decision-making and sustaining human action. This is the approach adopted by North (1990). North tries to connect economic process to individual cognitive activity; the action of competing agents is the driving force of change, but that action depends on the agents' perception of the possible outcome of their moves. Yet, perception depends critically on the information available and the way this information is processed. Given the cognitive abilities of the agents, information is often incomplete and is elaborated on the basis of past experience, creating the condition for a 'path-dependent' evolution. Institutional context and inherited rules affect individual cognition, precisely because individuals perceive the environment as a function of the mental constructs they use to interpret the world. These constructs reflect their past and culture, and, more importantly, the local context of problems. Thus acting and knowing in a given context contributes to limiting and modelling the possibilities of perception and choice for agents (North, 1990).[1]

This situation works as a mechanism that reinforces differentiation and heterogeneity between agents, allowing the formation of different mental constructs and different interpretations of similar information. Subjective interpretation may play a major role in the agents' capacity to understand a problem and to attribute meaning to a set of information and consequently make decisions and take actions. Just as institutions limit possible alternatives, so, too, mental constructs limit perception and the set of decisions (North, 1995, p. 25).

V. HUMAN RESOURCES AND THE CREATION OF KNOWLEDGE IN THE FIRM

V.a. The central role of human resources

Looking at the production processes of knowledge within the firm implies focusing on human resources as well as on the models and procedures used to mobilise such resources. We mentioned the open-ended nature of knowledge and we have shown that such open-endedness is the outcome of the process whereby knowledge is generated and expands. In this process individual skills and cognitive models establish relations with those of others, giving rise to forms of context-dependent cognition. The interactive contextual nature of the creation of cognitive models is thus typical of the process producing knowledge in the firm. This process generates knowledge that is renewed but is never purely self-propagation on a larger scale. The growth of the pool of knowledge (like the growth of an organism) available to the firm does not follow a logic of 'more of the same', since its form and contents are transformed (Winter, 1982).

In looking at human resources as strategic resources for the purposes of the production of knowledge, we must understand how the mobilisation of these resources depends on the capacity of the organisation to set in motion the production of the forms of knowledge and cognitive models required to achieve its objectives.

We know that knowledge resides in individual minds, and that human resources in the firm hold the knowledge available at any time. Firms may also acquire technological knowledge which is embedded in machinery; while this knowledge becomes

---

[1] North emphasises the role of institutions in reducing the costs connected with political and economic transactions: the performances of firms can be improved through a reduction of the transaction costs and this occurs when the appropriate institutional forms develop. The definition of a set of suitable rules (mainly for governing and protecting property rights) leads to the structure of incentives and the opportunities for action for agents and the possibility of creating organisations encouraging or hindering the reduction of transaction costs.

objective and is embodied in an object, knowledge that human resources hold cannot be separated from the physical person. The tacit nature of this form of knowledge, underlying individual competencies, implies that knowledge cannot be described completely and is very difficult to transfer (Polanji, 1962).

Human resources organised within a specific administrative unit, the firm, hold the knowledge which is put to use and developed within a 'context', the frame offered by each specific organisation. This reality diverges from the idea of human capital, entering as an input in the production process and continually re-allocated in order to optimise performance. Knowledge and its uses are inextricably linked and its value cannot be represented by the sum of individual 'human capital'. The value of the knowledge held by human resources within the firm is a function of the flow of different ways in which knowledge is actually applied. These ways are strictly dependent on cognitive models held by individuals (their capacity to interpret information and create knowledge) and on the cognitive frame offered by the organisation which models its capacity to develop alternative uses of human resources.

Given the open-ended nature of knowledge, the pool of forms of knowledge within the firm is developed by means of discretion and of judgement in carrying out individual cognitive activities. These spaces for discretion and judgement grow with the complexity of both the knowledge that individuals hold and the different possible uses within the firm. The process by which knowledge is developed then offers the firm a degree of 'liquidity' in the stock of knowledge, both at the individual level and at the organisational level.

The coherence of the organisation cannot be taken for granted but needs to be built by constructing suitable channels of communication. The organisation is the intelligent system operating round individuals, sustaining their cognitive abilities, safeguarding the mechanisms of accumulation, reproduction and plasticity of the pool of knowledge the firm is able to put to use.

The work of Penrose (1959) helps us to understand these dynamics through three aspects: the distinction between productive resources and the services provided by them; the idea of the firm as a nexus of directives, and the emphasis on design activities pursued by the firm (Richardson, 1972, 1998; Penrose, 1995).

V.b. Resources and resource services

Penrose advances the idea of the firm as a pool of resources agglomerated round a centre, whose function is to coordinate them. In so doing, the centre also models their form by identifying the interconnecting structure on any given occasion. This, then, is the managerial activity carried out in the firm, and it begins from the creation of a pool of managerial resources. Management is therefore a design activity for itself and for the whole firm. Management's skills and cognitive models at various levels are thus crucial in understanding how the firm works and how it is a pool of resources. The resultant picture is a firm with extremely plastic and modifiable boundaries. The plastic nature of the firm is linked to the fact that human resources are relatively malleable: a resource is found on the market, but only acquires distinctive specificity in the framework of the firm employing it, or in connection with a certain framework of directives: 'There are also human resources available in a firm . . . For some purposes these can be treated as more or less fixed or durable resources, like plants or equipment; even though they are not owned by the firm, the firm suffers a loss akin to a capital loss when such employees leave the firm . . . Strictly speaking, it is never resources that are the inputs in the production process, but only the services that the resources can render. The services yielded by resources are a function of the way in which they are used . . . The important distinction between resources and services is not in their relative durability; rather it lies in the fact that resources consist of a bundle of potential services and can, for the most part, be defined independently of their use, while services cannot . . . it is largely in this distinction that we find the source of the uniqueness of each individual firm' (Penrose, 1995, pp. 24–5).

The distinction between resources and their potential services gives rise both to the possibility that the firm can grow through an endogenous mechanism from given resources, and, as Penrose points out, the personality, culture or individuality of single firms. Resources can be defined independently of their use, but services cannot: the transformation of resources into services in the framework of a collective context like the firm implies the need for management activity to orient and coordinate, i.e., it implies their use in a specific context. The same resource used for different purposes, in different ways or in different combinations with other resources gives rise to different services and different sets of services.[1] There is thus a substantial difference between a set of resources representing the factors available to a firm at any time and the services that flow from those resources over time. Services extractable from human resources vary as a function of the capacities present in the firm to model them, to give them form and content—to give them a certain structure.

The structure of the organisation is thus linked to its endowment of resources and their performance, achieved in a dynamic and—above all—non-deterministic way. But, according to Penrose, 'there is every reason to assume that the problem of fully using all resources will never be solved . . . because new services will become available from existing resources—services which were not anticipated when the expansion was originally planned . . . The change in the service of managerial resources also changes the nature of the productive services available from other resources, as well as the significance to the firm's management of existing services' (Penrose, 1995, p. 74).

---

[1] The fact that a set of resources in the firm can be used with different degrees of efficiency was focused on in a study by Leibenstein (1966), who develops the idea of X-efficiency. Here the possible indeterminacy in the combination of productive services is seen in relation to the difficulty of pushing the worker to achieve an adequate productive effort. This differs from the notion of organisational slack, introduced by March and Simon (1958) to illustrate the organisational dynamics in conditions of changing objectives and expectations. The distinction put forward by Penrose casts light on the difficulty of reconstructing an unequivocal relationship between input and services. This difficulty, for example, opens up a series of problems in transaction cost economics because the very notion of the transaction is no longer independent of the ways transactions are organised.

Penrose therefore claims that no firm can envisage fully the variety of services that can be derived from a particular resource, because the type of service identified is in large part limited by the management's ideas concerning possible productive associations. This claim may be read in two different ways. First, firms live in an uncertain world, in Knight's sense of the term, since management is not only fallible (i.e., makes mistakes) but, most importantly, the management's world does not dispense with 'surprise' (Shackle, 1955). Secondly, firms exist as autonomous units, characterised by their integrity insofar as there is consistency between the firm (administrative unit) and the management's ideas. Penrose thus emphasises the connection between the 'soft side' of a firm and the 'hard side', i.e., the connection between the corporate culture of the firm (its vision) and its ability to perform activities that shape the continuous renewal of its capabilities.

We have discussed at length how individual cognitive abilities are expressed and modelled by relating to the context. The construction of shared mental cognitive models and their renewal is itself a cognitive activity in which all individuals participate, creating dense communications networks (Hutchins, 1995). The distinction between resources and the set of production services that arise from them helps us in understanding how knowledge that resides in individual minds may give rise to a flow of different forms of knowing, which are renewed and reframed by relating to the context they develop in.

V.c. The firm as a nexus of directives

This brings us to a discussion of the importance of defining the firm as a nexus of directives. Penrose provides the following definition of what constitutes a firm, connecting it to 'the essential function it serves in the economic system: that of using productive resources to supply goods and services on the basis of plans prepared and implemented within the firm itself' (Penrose, 1995, p. 15). The autonomy and integrity of the firm, in all its variety and complexity, is the discriminating element compared with spontaneous forms of coordination: 'The essential difference between economic activity inside the firm and the economic activity in the "market" is that the former is carried on within an administrative organization, while the latter is not' (Penrose, 1995, p. 15). The firm is therefore 'an autonomous administrative planning unit, the activities of which are interrelated and coordinated by policies which are framed in the light of their effect on the enterprise as a whole' (Penrose, 1995, p. 16).

The question of the boundaries of the firm is thus set in terms of the size of the area of coordination or of the transmission of directives, seen as detailed instructions and as definitions of policy criteria, procedures or aims creating a common area of administrative coordination (Penrose, 1959). Managerial functions and the expression of directives, seen as the outcome of the activities of orientation, coordination and communication involved in design activity, constitute the firm, or rather mould the firm according to a certain order. It can be said that Penrose sees the firm as producing images or—in the language we have adopted—shared cognitive models (Witt, 2000).[1]

Producing images sets in motion an endogenous mechanism that mobilizes the creation of new knowledge. Penrose (1995) describes this process as the activity of expressing directives performed by managerial resources; it is the activity that shapes and designs new productive services of the existing human resources, modifying and combining them over time. Expressing directives, as described above, is thus the specific skill of management. It is the way in which 'translation' of local, specific languages increases the possibility of communication between various activities and contexts in the firm. Management focuses simultaneously on various projects and the coherence of the executive group is a vital condition, but also a limit, because it restricts the set of opportunities the firm is able to perceive. The addition of new managerial resources, acquired on the market, undermines the teamwork created through shared activities. Managerial services are thus highly idiosyncratic and specialisations acquired outside are not readily integrated. This raises the problem of a mismatch in productive services, described as the specific and contextualised flows of knowledge, generated by managerial action. The mismatch will be more significant, the greater the need for new managerial resources and the greater the problems connected to their integration, and at any given time the pool of managerial services (knowledge) will limit the potential of the firm to develop its own innovative projects. The mismatch originating in managerial resources, however, is a dynamic temporary state connected to the phenomenon of renewal. New managerial resources are integrated into the executive group working on given projects, and so the very process of integration leading to the re-adaptation of the firm[2] produces learning and with it the possibility that new internal resources will be available for recreating the conditions for a renewal.

Implementing this process thus transfers a body of knowledge from one context to another, according to the mechanisms of vicarious experience—outlined in section IV above—and shows how objective knowledge and tacit personal knowledge combine in a unique way through group work, creating the unique nature of experience. 'The experience gained is not only of the kind . . . which enables a collection of individuals to become a working unit, but also of a kind which develops an increasing knowledge of the possibilities for action and the ways in which action can be taken by the group itself, that is the firm. This increase in knowledge not only causes the productive opportunity of a firm to change in ways unrelated to changes in the environment, but also contributes to the "uniqueness" of the opportunity of each individual firm' (Penrose, 1995, p. 53). Thus the presence of a given pool of resources does not preclude their being managed flexibly but to some extent the very rigidity (cohesion

---

[1] Witt (2000) stresses the importance of entrepreneurship and the cognitive underpinnings of firm organisation; the very existence of the firm is threatened when size increases because entrepreneurial conception is weakened by the necessary delegation of functions taking place within the firm as it grows.

[2] In this context reference may be made to the notion of learning as the adaptation of a complex system (Hutchins, 1995).

through a structure) of the organisation favours both the dynamic (flexible) use of resources and new forms of internal cohesion.[1]

VI. MISMATCH MANAGEMENT AND THE DEVELOPMENT OF NEW FORMS OF KNOWLEDGE

VI.a. New knowledge

We have outlined the working of two dynamic mechanisms underlying the process creating forms of knowledge in the firm: the design-implementation dynamics giving rise to directives and the consequent dynamics of productive-service resources. The firm works from the definition of projects (strategies) to their implementation and in doing so develops new services from the available resources. In this way the base of knowledge in the firm develops through practice, guided by design activity, giving rise to a way of knowing how that is specific to the firm and which in turn will influence future design guidelines.[2]

The distinction between information and knowledge, mentioned in Section III above, helps us to build a fuller understanding of the mechanisms whereby the introduction of human resources continues to generate new knowledge in the firm, precisely because it introduces elements of mismatch between the flows of existing production services and their new configuration. The introduction of a cognitive dimension allows us to reinterpret these flows as the appearance of new forms of knowledge. Individuals, previously endowed with a mere calculation ability, are now minded agents, in a world of bounded rationality and uncertainty, where innovation also exists: this is the world in which the firms compete and markets evolve.

Highlighting the loose connection between information and knowledge has shown important implications for the role organisations have in managing and producing the information and knowledge required for their functioning.

The decision-making process in the firm is characterised by a strategic assessment (aimed at achieving objectives) of the information available, and it does so by interpreting it in relation to the interdependencies that the firm is able to 'see' or 'perceive' between its own behaviour, the behaviour of other firms and of all the agents involved. The knowledge firms require is thus the outcome of a process of elaborating elementary

---

[1] In another work (Turvani, 1998) I discuss the economic significance of the relationship of long-term labour with the firm. Formal rigidity implies incompleteness of contractual terms: in that book rigidity–incompleteness was linked not only to the difficulty (in terms of costs and information available) of completing the contract in all its details, but also to the impossibility of using all the potential of human resources in a given moment through the detailed definition of the services to be extracted from them. To express this concept I use the notion of liquidity of resources, based on Simon (1951). In this sense I gave the labour contract a long-sighted character, absent in other contractual forms, in which the parties' effort is directed at predicting and incorporating into the contract the largest possible number of future circumstances.

[2] The distinctive skills of a firm influence not only the direction of expansion (according to the criterion of similarity) but also the modality of expansion, that is the choice between acquisitions and internal growth. The possibility of enjoying quasi rents, critically depends on the existing pool of knowledge and skills in the firm (Teece, 1986).

information, selected not only for its importance with regard to objectives, but above all organised in an interpretive model defined in relation to the objectives. There will never be that set of information, so popular in the economic approach, allowing the outcome of decision-making processes to be calculated, even though supplementary information may be acquired outside the firm. As Knight (1921) has already pointed out, decisions in a firm are based on judgement, and this judgement is a constituent element in knowledge-producing processes.

Judgement and experience acquired by individuals is mediated by cognitive models and they connect events and experiences to the production of knowledge. In this process the set of information is thus transformed into a cognitive model which is built upon not only the available information, but direct or vicarious experience and by conjecture. The strategy is built and with it also the behavioural hypotheses about the agents involved are produced, together with the set of objectives to be implemented. Strategy offers a frame that allows the articulation of partial objectives through an operation of 'translation', so that the various individuals (or functions) in the organisation may 'understand' the objectives in terms of their own cognitive apparatus, and thus 'identify' with the new strategies of which they can only know some partial elements.

What is going on within the firm is the activation of a process that utilises the dispersed knowledge and which gives individuals (and functions) the possibility to extract the value from their idiosyncratic knowledge (Hayek, 1945).

Thus, what we have described as a set of directives is a complex problem of translating new cognitive models into accessible terms for those working in agreement with behavioural and cognitive (organisational) models consolidated over time. This process requires major investments, since the greater specialisation of work and jobs tends to reduce communication between various functions and tends to specialise professional languages. Nonaka and Takeuchi (1995) describe this process as the creation of knowledge within the firm, providing an image of a spiral linking the transformation processes of knowledge from tacit to explicit and vice versa, beginning from the individual and collective processes of verification and validation of knowledge (the process of internalisation in individuals and externalisation between individuals). Such processes imply a fundamental role for communication.[1]

The new cognitive models underlying a given strategy undergo a process of translation into languages and cognitive models specific to various areas in the firm. The outcome is a set of directives (to use the image suggested by Penrose) or the expression and the formulation of a set of partial objectives.

A similar translation process also applies to the definition of the productive services flowing from human resources which must be modelled to implement strategies. Conceiving the set of potential services implies a refining of the knowledge held by

[1] Nonaka agrees on the importance in corporate communication of slogans unifying and summarising at times incongruent cognitive models, and gives various examples of this phenomenon (Nonaka and Takeuchi, 1995). On the importance of translating any rational statement into the language and schemas of the interlocutor in order to arrive at shared languages and shared forms of rationality, see Stich (1996).

individuals within the limits offered by the set of services available at any time, which, as Penrose points out, is never completely specified.

Moreover, the cognitive model giving rise to the potential demand for new productive services contains, as we have seen, elements of uncertainty and judgement. They will be all the greater, the more innovative the model is. In fact a cognitive model will be innovative in relation to the fact that it gives rise to a different set of resources services. Bearing in mind that productive services are an expression of the knowledge held by individuals and based on their cognitive activity, and that this is re-adapted according to new cognitive models which strategy attempts to transmit, then the adjustment of the ongoing cognitive frame for the organisation will be subject to an interactive-type process of verification and validity; hence the instability and margins of uncertainty in innovation strategies.

VI.b. The assortment of human resources

The possibility of relying on a malleable pool of knowledge within the firm depends on its policies with respect to the quantitative and qualitative turnover in human resources. The assortment and development of knowledge within the firm does not necessarily imply renewing human resources, even though labour turnover may be used as an important tool to remodel the knowledge that the pool of individuals hold. In particular labour turnover encourages the replacing of old skills with new in the demographic process affecting the human resources employed.[1] In addition to regulating employment levels, labour turnover plays a key role in guaranteeing assortment of the knowledge available according to the needs of the firm. While the malleability of the pool of knowledge required by the firm may be achieved through the acquisition of new human resources, other tools exist, such as internal training, the training and acquisition of skills on the market, either with or without the addition of new human resources to the firm (i.e., a new employee rather than a consultant). Firms discriminate between various tools not only in relation to the strategies being developed, but also with reference to the costs of the various options. The resultant complex decision-making picture features a variety of options. Yet, they cannot be freely combined since there exist restraints of an institutional type (the rules governing the labour market), market restraints (the conditions of the labour market outside the firm, but also the structure and conditions of competitiveness in the entrepreneurial system) and restraints imposed by the type of entrepreneurial strategy. Here we have concentrated on this last factor. By representing innovative activity as an activity for producing knowledge we have focused on the implicit transformation of cognitive models when an innovative strategy is introduced. The renewal of the knowledge held by individuals and put to use in the firm has been represented by the process of implementing an innovative strategy and it has been described as the establishment and spread of new

---

[1] In the literature of organisation there is talk of business process re-engineering, meaning the improvements in performance or the reorganisation of firms in difficulty. In the range of tools available to the firm management turnover is not mentioned, whereas there is a great emphasis on redesigning the organisational structure. See Migliarese *et al.* (1999).

cognitive models replacing the previous ones. Specific competencies connected to the process of converting or translating (i.e., communication activities), implicit in the adjustment of the cognitive models that individuals hold, are much needed. The vehicle for innovative strategy, they may be scarce in the firm, whereas they may be available in other firms; new human resources may be acquired on the market. This scarcity may be compelling when firms work for the implementation of highly innovative strategies, when in-house training, though enabling skill upgrading, may play a limited role in the knowledge renewal.[1]

It is important to note that enlarging the knowledge held by individuals by resorting to the acquisition of new entrants in the human resources incurs the well known 'knowledge paradox' (Shackle, 1955; Arrow, 1962). How do we ascertain the quality and value of knowledge we need, given that we are attempting to reduce our ignorance by buying it? In fact, if the resources were already present in the firm we could assess them, but would no longer need to acquire them.

The recurrence of this paradoxical situation does not suggest that firms play lotteries but rather that they must rely on those indispensable elements of judgement, implicit in any innovation strategy and in any process of knowledge production. Firms do not operate on an abstract entity called the 'market'; the 'market' is created by the operations and strategies of other firms. Then, the knowledge individuals hold is located in 'the market' or to be more explicit it has to be found in other firms. Taking it away from other firms means, therefore, having a strategy, or a model to use it that may be utilised to reshape the overall flow of productive services in a profitable way. It means that the organisation is able to absorb the knowledge which the new entrants hold in a way that sets the dynamics of re-encoding and re-framing the knowledge in the firm. In short, it implies 'seeing' opportunities that others do not, and therefore exercising the entrepreneurial action.

VII. CONCLUSIONS

In this chapter we approached the issue of the growth of the firm as the development of knowledge in the firm; we also described it as the renewal of the pool of individual

---

[1] Human resources policies in the firm are very important; different costs and different evaluation are implied. This is not the place to discuss them in detail and we may just refer to some of the issues. The assessment of training costs depends on the quality of skills to be modelled (technical, managerial) and the quality of human resources in play. These are the tools governing skill turnover within the internal labour market, subject to explicit and implicit rules, which vary from firm to firm. But training costs will vary mainly as a function of the features of human resources such as age, education and accumulated experience, not only because of the direct costs (acquisition of training) and indirect costs (reduced production), but especially because the contents of training will vary: from the straightforward transmission of new information and models to learning complex new languages. The assessment of the hiring costs again implies a comparative analysis of the make or buy options. Selection may be honed very finely in order to find those skills on the external market not present within the firm or a more approximate selection of skills may be made before integrating and specialising them through training according to the needs of the firm. In this case, too, the two paths are not equally practicable because the second implies that the strategy of the firm shapes a demand for clearly defined skills.

knowledge that human resources hold. The growth of the firm results in a cumulative collection of tacit or explicit, general or specific and technical or managerial forms of knowledge. Growth and the capacity for innovation that firms display are then strictly linked to the development of specific, local forms of knowledge. In other words, innovation is considered as an activity involving producing and using new forms of knowledge. Assuming the features of individual cognitive activity, the plasticity of individual knowledge needs to be connected with its development in 'context'. To grow by innovating may then be discussed in terms of the modification of knowledge held by human resources within the context of the firm.

How does knowledge in the firm change over time? Traditional action to improve the quality of human resources, such as training, learning by doing, further education and the acquisition of new skills on the market are mechanisms that give rise to changes in knowledge. Managing these processes in a firm becomes a strategic activity and the decisions in this field influence the set of forms of knowledge available to the firm as well as its capacity to innovate and to grow.

Human resources hold knowledge, i.e., meaningful information or information interpreted by cognitive models. They are malleable and they evolve, so does the knowledge held by individuals and used within the firm. To work for a better theory of the firm then implies to cross over into other disciplines and borrow concepts and theories able to take into account the cognitive activities occurring within the firm. The working of the mind and its structuring in relation to context are a necessary point of departure, albeit at a summary level of description, in order to highlight how agents' cognitive activities are supported by the economic activities conducted in the firm. In this chapter we suggested that agents possess knowledge and not just information. They have cognitive models which they use and renew through reciprocal interaction. They hold knowledge, which rather than simply being human capital is 'know-how' and 'ways of knowing how', guiding their understanding of problems and their capacity to solve them.

How can we apply processes typical of the mind to the economic activities of the firm? How can we apply economic analysis to the process of renewing knowledge that individuals hold?

In this chapter we tried to show that a bridge connecting economic and cognitive analysis may be built by re-reading Penrose (1959). The distinction between resources and services extractable from resources enables us to distinguish between human resources in the firm and the services of resources, which are specific flows of knowledge implemented over time by human resources. Governing these flows is the task of the management activity at various levels in the firm. Managerial skills themselves are a flow of services and are the way we can represent the cognitive models and abilities belonging to managerial resources, transmitted to the organisation through 'directives', to use one of Penrose's favourite terms. From this point of view, directives are communication media and also the translation of the management's cognitive model in terms accessible to the cognitive languages and models present in the organisation. Activating these media encourages communication, allowing the spread of knowledge and its renewal. In this picture, the renewal of knowledge within the firm is facilitated both by the entry of fresh managerial resources through the acquisition on the market, and—more often—by extending managerial services flowing from human resources already present in the firm. The capacity of firms to create new forms of knowledge is thus linked not only to the possession of technical knowledge, of an appropriate pool of know-how, but also to the development in the everyday activities of people of specific skills of orienting, translating and spreading 'ways of knowing how' (vision, cognitive models or the idiosyncratic reading of reality), which over time characterise the firm.

REFERENCES

ALCHIAN, A. and DEMSETZ, H. (1972). Production, information costs and economic organization, *American Economic Review*, No. LXII, 777–95.

ARROW, K. (1962). Economic welfare and the allocation of resources to invention, in *The Rate and Direction of Inventive Activity: Economic and Social Factor*, edited by R. NELSON, Princeton, Princeton University Press.

ARROW, K. J. (1974). *The Limits of Organizations*, New York, Norton.

BANDURA, A. (1986). *Social Foundation of Thought and Action. A Social Cognitive Theory*, Englewood Cliffs, NJ, Prentice Hall.

BEST, M. (1990). *The New Competition*, Oxford, Blackwell.

COASE, R. (1937). The nature of the firm, *Economica*, Vol. IV, 386–405.

COHEN, W. and LEVINTHAL, D. A. (1990). Innovation and learning: the two faces of R&D, *Economic Journal*, Vol. 99, September 1989, 569–96.

CYERT, R. and MARCH, J. (1963). *A Behavioral Theory of the Firm*, Englewood Cliffs, NJ, Prentice Hall.

DEZAU, A. and NORTH, D. (1994). Shared mental models: ideologies and institutions, *Kyklos*, Vol. 47, No. 1, 3–31.

DRETSKE, F. (1981). *Knowledge and the Flow of Information*, Cambridge, MA, MIT Press.

EGIDI, M. and TURVANI, M. (1994). *Le ragioni delle organizzazioni economiche*, Torino, Rosenberg & Sellier.

FOSS, N. (1993). Theories of the firm: contractual and competence perspectives, *Journal of Evolutionary Economics*, No. 3, 127–44.

FRANSMAN, M. (1994). Information, knowledge, vision and theories of the firm, *Industrial and Corporate Change*, Vol. 3, 713–57.

HAYEK, F. (1945). The use of knowledge in society, *American Economic Review*, Vol. XXXV, 519–30.

HAYEK, F. (1952). *The Sensory Order: An Inquiry into the Foundations of Theoretical Psychology*, London, Routledge & Kegan Paul.

HUTCHINS, E. (1995). *Cognition in the Wild*, Cambridge, MA, MIT Press.

JENSEN, M. and MECKLING, J. (1976). Theory of the firm: managerial behaviour, agency cost, and capital structure, *Journal of Financial Economics*, No. 3, 305–60.

KNIGHT, F. H. (1921). *Risk, Uncertainty and Profit*, Chicago, IL, University of Chicago Press.

LANGLOIS, R. (1995). Cognition and capabilities: opportunity seized and missed in the history of the computer industry, in *Technological Entrepreneurship: Oversights and Foresights*, edited by R. GARUD, P. NAYYAR and Z. SHAPIRA, New York, Cambridge University Press.

LEIBENSTEIN, H. (1966). Allocative efficiency vs. X-efficiency, *American Economic Review*, Vol. LVI, 392–415.

LOASBY, B. J. (1991). *Equilibrium and Evolution*, Manchester, Manchester University Press.

MAHONEY, J. and PANDIAN, R. (1992). The resource-based view within the conversation of strategic management, *Strategic Management Journal*, Vol. 13, 363–80.

MARCH, J. G. and SIMON, H. (1958). *Organizations*, New York, Wiley.

METCALFE, S. (2000). 'Towards an Epistemology of Innovating Firms and Economies', CRIC, University of Manchester, Mimeo.

MIGLIARESE P., FERIOLI, C. and IAZZOLINO, G. (1999). Dimensione organizzativa e business process reengineering: la necessità di un chiarimento, *Studi Organizzativi*, No. 2.

MONTGOMERY, C., ed. (1995). *Resource-based and Evolutionary Theories of the Firm: Towards a Synthesis*, London, Kluwer Academic Publishers.

NELSON, R. and WINTER, S. G. (1982). *An Evolutionary Theory of Economic Change*, Cambridge, MA, Belknap Press.

NONAKA, I. and TAKEUCHI, H. (1995). *The Knowledge Creating Company*, Oxford, Oxford University Press.

NORTH, D. (1990). *Institutions, Institutional Change and Economic Performance*, Cambridge, Cambridge University Press.

NORTH, D. (1995). Five propositions about institutional change, in *Explaining Social Institutions*, edited by J. KNIGHT and I. SENED, Ann Arbor, MI, University of Michigan Press.

PENROSE, E. (1959). *The Theory of the Growth of the Firm*, Oxford, Basil Blackwell.

PENROSE, E. (1995). *The Theory of the Growth of the Firm*, 3rd edn, Oxford, Oxford University Press.

PIAGET, J. (1969). *The Mechanism of Perception*, London, Routledge & Kegan Paul.

POLANYI, M. (1962). *Personal Knowledge*, New York, Harper & Row.

PRAHALAD, C. and HAMEL, G. (1990). The core competence of the corporation, *Harvard Business Review*, May–June, 78–91.

RICHARDSON, G. B. (1972). The organization of industry, *Economic Journal*, Vol. 72, No. 327, 883–96.

RICHARDSON, G. B. (1998). 'Production, Planning and Prices', Oxford, mimeo.

RIZZELLO, S. and TURVANI, M. (2000). Institutions meet mind: the way out of an impasse, *Constitutional Political Economy*, Vol. 11, No. 2.

SHACKLE, G. (1955). *Uncertainty in Economics*, Cambridge, Cambridge University Press.

SIMON, H. (1951). A formal theory of employment relations, *Econometrica*, No. 19, 293–305.

SIMON, H. A. (1987). Bounded rationality, in *The New Palgrave* Vol. I, edited by J. EATWELL, M. MILGATE and P. NEWMAN, 266–7.

SLATER, M. (1980). Foreword, in E. PENROSE, *The Theory of the Growth of the Firm*, Sharpe, New York.

STICH, S. (1996). *La frammentazione della ragione*, Bologna, Il Mulino.

TEECE, D. (1986). Firm boundaries, technological innovation, and strategic management, in *The Economics of Strategic Planning*, edited by L. THOMAS, Lexington, MA, Heath.

TEECE, D. and PISANO, G. (1994). Understanding corporate coherence: theory and evidence, *Journal of Economic Behavior and Organization*, Vol. 23, January, 1–30.

TURVANI, M. (1998). Black boxes, grey boxes: the scope of contracts in the theory of the firm, in *Evolution of Institutions, Organization and Technology*, edited by K. NIELSEN and B. JOHNSON, Cheltenham and Lyme, USA, Edward Elgar.

WILLIAMSON, O. E. (1985). *The Economic Institutions of Capitalism*, New York, Free Press.

WINTER, S. (1982). An essay on the theory of production, in *Economics and the World Around It*, edited by S. HYMANS, Ann Arbor, MI, Michigan University Press.

WITT, U. (1998). 'Cognition, Entreprenurial Conceptions and the Nature of the Firm Reconsidered', Jena, Max Planck Institute, mimeo.

WITT, U. (2000). 'Changing Cognitive Frames—Changing Organizational Forms. An Entrepreneurial Theory of Organizational Development', Jena, Max Planck Institute, working paper 0007.

13

INNOVATION, PROFITS AND GROWTH: PENROSE AND SCHUMPETER

JOHN CANTWELL*

Department of Economics, University of Reading

I. INTRODUCTION

In this chapter the Schumpeterian theory of profits and growth through innovation is revisited and recast, with explicit reference to the changing institutional form of innovation during the twentieth century. It is shown how many clues for the restatement and modernisation of Schumpeter's approach can be found in Edith Penrose's *The Theory of the Growth of the Firm,* her 1959 book having benefited from her reading of what for our purposes are the crucial aspects of Schumpeter (1943). The chapter has four parts following this introduction. Section II sets out an evolutionary and institutional account of how profits are created through innovation, which is contrasted with the standard interpretation of Schumpeter's theory found in the literature. It is argued that the standard interpretation does not do justice to Schumpeter's theory, but that the original theory requires adaptation in any case to better reflect the means by which capitalist institutions have promoted innovation during the twentieth century, and beyond, into the twenty-first. The third section reviews Penrose's work on the growth of the modern firm, demonstrating how she incorporated Schumpeter's insights into her thinking, and explaining how her approach provides a link between Schumpeter's theory and the modern institutional form of innovation in the large firm. Section IV illustrates and elaborates upon the argument through some evidence on the changing form of innovation in large firms in the major industrialised countries during a first phase roughly from 1900 to 1970, and a more recent phase from around 1970 onwards. A concluding Section V recaps on the commonality between Schumpeter and Penrose, and on how their analytical approach can be usefully developed in the light of the current evidence.

It is contended that innovation has relied on the creation of technological or social capability, through problem solving or learning activities principally within (and between) large firms. The development of new products and processes is the outcome of a path-dependent building upon established capabilities and achievements, by the critical revision of emergent new products or methods and the search for novelty of a

* I am grateful to Brian Loasby, Dick Nelson and Christos Pitelis for helpful comments on an earlier draft.

kind that is relevant to addressing in new ways producer problems or user needs. Hence, innovation must be understood as a continuous learning process in firms supported by other institutions, and not as a discrete event, whether an exogenous shock that gives rise to a monopoly or a flash of entrepreneurial alertness that requires no resources, nor as the implementation of a fully defined and foreseen strategy. Innovation is a problem-solving search that creates and continually renews technological or social capability within firms, and is not a search for positions of market power as such. This insight into the form of innovation is an amalgam of the conclusions of the work of Usher (1954) and Rosenberg (1976, 1982, 1994) on the history of technology, Nelson and Winter (1982) on the evolutionary theory of economic change, as well as Penrose's (1959) theory of the growth of the firm. Thus, innovation depends upon the generation of new capabilities made feasible as the outcome of problem solving and progressive experimentation, the operation of which capabilities adds new value to the existing circular stream of income, and thereby creates new profits and higher income.

By contrast, the standard interpretation of Schumpeter's theory of profits through innovation focuses upon the quasi-monopoly positions developed in markets by entrepreneurial firms that enjoy first mover advantages. This common approach to Schumpeter's theory renders it understandable within the conventional framework of market-based analysis, in which institutions are discussed only with regard to their role in the process of economic exchange, primarily through markets (or with reference to a hypothetical alternative market in the case of transactions within firms). Since the leading innovators establish a temporary monopoly within some output (product) or input (process) markets they obtain 'super' profits from innovation, typically associated with higher output prices and lower input prices or costs. But this brings us back to issues of the distribution of the circular flow of income, a flow that is sustained through markets, rather than the question of how that flow can be increased over time as new value-generating activities are added into the stream. The conventional interpretation also tends to disregard in this respect the difficulties raised by increasing returns in new fields of value creation for the analysis of the circular flow of income as an equilibrium system, as discussed by Young (1928), consistently with Schumpeter's view of innovation as a disequilibrium process. In other words the standard treatment reduces the means by which profits can be earned through innovation to a matter of the capacity for static appropriability through the exercise of market power, and hence analytically no different to any other kind of 'normal' profits. The relevant markets may be new, but their newness is significant only for its relationship to the scope for temporary monopolies. The distinctiveness of Schumpeter's notion of adding to the existing circular flow of income is lost.

So my interpretation of Schumpeter differs from the standard one, in holding that throughout his work Schumpeter distinguished the profits due to innovation from those due to market power, and that in this respect the distinction sometimes drawn between the 'early' Schumpeter and the 'late' Schumpeter has been greatly exaggerated. My interpretation of Schumpeter has antecedents in the understanding of the

Schumpeterian tradition provided by Nelson (1996) and Langlois (1998). Schumpeter's (1934) original theory of innovative profits emphasised the role of entrepreneurship (his term was entrepreneurial profits) and the seeking out of opportunities for novel value-generating activities that would expand (and transform) the circular flow of income, but it did so with reference to a distinction between invention or discovery on the one hand and innovation, commercialisation and entrepreneurship on the other. This separation of invention and innovation marked out the typical nineteenth century institutional model of innovation, in which independent inventors typically fed discoveries as potential inputs to entrepeneurial firms. Firms that innovated first would make profits from the new areas of value creation they established, which profits would generally be gradually eroded through the entry of competitors. After his early work on entrepreneurship, Schumpeter became only too aware of the rise of in-house corporate R&D in large firms in the twentieth century which meant a closer institutional association between invention and innovation, to the extent that the literature now distinguishes his 'Mark I' model of innovation from his 'Mark II' model in which innovation was envisaged as a more routinised process within large firms (Phillips, 1971). The Mark I model is associated with Schumpeter (1934) originally published in 1911, and the Mark II model with Schumpeter (1943). In the standard view this shift in Schumpeter's thinking towards the role of large oligopolistic firms as the key agents for innovation is supposed to reinforce the conventional interpretation of his earlier theory of innovative profits, since these large firms certainly exercise market power, and in what has become known as the 'Schumpeterian hypothesis' Schumpeter himself is now widely believed to have thought in terms of a link running from market power to the extent to which resources are devoted to innovation.

However, I argue in Section II that the so-called 'Schumpeterian hypothesis', which links profits based on market power with innovation, is a misunderstanding of Schumpeter (1943) which is due to an attempt to recast his insights from within what he labelled traditional theory; and that Schumpeter himself did not revise his earlier theory of profits when advancing his Mark II model, but in fact reiterated his view of the distinctiveness of profits from innovation as opposed to market power, although without working through the implications of endogenous innovation in large firms for his theory.[1] As argued by Nelson (1996), what changed between the Mark I and Mark II models was not as commonly thought any attempt to associate innovation with market power—Schumpeter continued to believe the reverse, namely that it was competition or rivalry in the classical sense of a process (not a competitive market structure) that leads to innovation, and innovation in its turn renders any supposed monopoly power ineffective—but rather that the sources of innovation were increasingly

---

[1] My attention was called to this by Richard Nelson's presentation in the plenary session 'Joseph Schumpeter 50 years on' at the International J. A. Schumpeter Society conference held in Manchester in June 2000, in which he argued that there was no evidence of what has subsequently become known as the 'Schumpeterian hypothesis' to be found in Schumpeter's own writings. See Nelson (1996), Chapter 3. What is new here is my explanation of how the 'Schumpeterian hypothesis' came about when Schumpeter is viewed misleadingly through traditional spectacles, and what Schumpeter actually meant in terms of his own theory by the arguments that have been misinterpreted as the 'Schumpeterian hypothesis'.

concentrated in large firms, which blurs the distinction between invention and innovation. The blurring of the distinction between invention and innovation implies a greater significance for inter-company knowledge flows given the more direct concern of firms with the development of the knowledge sources required for innovation, and for knowledge interchanges between firms and other institutions such as universities. This implication of a greater degree of interaction between firms in innovation was not properly identified or discussed by Schumpeter, and it affects the working of his theory, as it suggests the need for a more complex treatment of the distribution of the gains from innovation, since it is no longer clear that first movers will tend to capture the highest share of any profits from innovation. With endogenous innovation in large firms Schumpeter's retention of a sequential separation of innovative leaders from imitative followers appears questionable, so on this view Schumpeter's Mark II model should have revised the simple Mark I model more than it actually did. Another drawback of misleadingly interpreting Schumpeter's profits from innovation within the standard framework for the analysis of market structure is that, as also emphasised by Nelson (1996), one loses sight of Schumpeter's accompanying contention that innovation entails a disequilibrium process.

In any case, recent empirical research has cast doubt upon the alleged association between market power, firm size and innovation, and suggests that smaller firms may be highly innovative as well, especially through their interactions with large firms in the same industry (Pavitt et al., 1987; Audretsch, 1995). The consensus of empirical research is now that demand, technological opportunities and appropriability conditions play a greater role than market structure in influencing inter-industry differences in the rate of innovation (Cohen and Levin, 1989; Cohen, 1995); and that the relationship between innovation in large and small firms in each industry varies according to a sectoral taxonomy of corporate technological trajectories (Pavitt, 1984). In two typical cases in which small firm innovation in an industry is important, small technology-based firms may be spin-offs from larger firms, or small specialised suppliers may provide innovative machinery or equipment to large downstream user firms, which provide them with guidance and feedback. In other words, the critical contribution of large firms to modern innovation may lie not so much in their independent power but instead in their creation of novel technological capability run by skilled teams and developed through their continual problem-solving activity, which becomes a resource for other firms with which they cooperate as well as for themselves (Loasby, 1998). Large firms are repositories of competence and hence sources of innovation in the form of providing a reservoir of new opportunities for value creation for themselves and for those with which they are linked. This alternative perspective returns us to Schumpeter's original emphasis on creating new value-generating activities as a means of searching for higher profits from innovation, as opposed to statically maximising profits by appropriating higher rents from an existing income stream.

In Section III of the chapter it is argued that Penrose (1959) relied on an approach to profits and innovation in the firm that implicitly embodies the most important elements of Schumpeter's original theory (that is, the elements that are most important for our purposes), and she explicitly incorporated the role of in-house R&D and endogenous innovation in large firms. As such, she helps us to link together these two aspects and from that vantage point to expand upon Schumpeter's theory of innovation, profits and growth for a modern institutional setting. What is more, as is perhaps better known and is referred to in other chapters, she anticipated the recent approach to technological change and the firm with her resource-based perspective on corporate growth (of those such as Teece *et al.*, 1997). She also anticipated in this context the linkage between own-capabilities and what has become known as the absorptive capacity (Cohen and Levinthal, 1989) to receive and utilise the knowledge of others, and hence connected in-house R&D and learning with the scope for inter-firm exchange in innovation. Hence, we can trace to Penrose the foundations of the approach taken here, which aims to connect Schumpeter's theory of innovation, profits and growth to the changing institutional reality of innovation since the start of the twentieth century. It will be argued that given her incorporation of Schumpeter's perspective on innovation, profits and growth, Penrose's insights into corporate capabilities and absorptive capacity are just what are needed to revise his theory to accommodate the major institutional changes in innovation since the time he was writing, not least because she anticipated better than he did the nature of these changes.

In Section IV we relate our discussion to some evidence of recent studies on the changing institutional form of innovation over the last 100 years. By understanding how profits are created from innovation in an alternative evolutionary way through corporate learning and search processes, it can be appreciated how innovative profits are of steadily rising significance relative to the more traditional kind of profit derived from market power, given the way in which capitalist institutions have evolved during the twentieth century through to today. In the first phase or paradigm in about the first three-quarters of the twentieth century, science-based innovation in large-scale production facilities (as documented by Chandler, 1990) depended upon the capabilities that were associated with the rise of in-house corporate R&D in large industrial firms. Large firms became the key actors in combining the processes of invention and innovation, each individual firm being technologically specialised in a way that reflected the specific profile of corporate technological competence that it accumulated through cumulative path-dependent learning processes. An inter-company variety of capabilities gradually extended the reservoir of social capability for innovation, and hence broadened the foundations for the creation of profits through innovation.

In the most recent phase or paradigm from the latter part of the twentieth century onwards, science-based innovation has been combined with information and communication technologies in computerised and flexible production facilities. Large firms have remained the key actors in the accumulation of technological capabilities, but in an institutional context that now emphasises the economies of scope to be obtained from the fusion of interrelated capabilities (which is an extension of Schumpeter's 'new combinations'), and a new role for the internationalisation of economic activity.

Between firms this has taken the form of a growing number of inter-company alliances for the purposes of promoting innovation. Within large firms in-house R&D is increasingly directed to the emergent benefits of corporate technological diversification through novel and more complex combinations, and to the development of technological competence through internal international company networks. These latter international corporate networks for innovation mark a change in the institutional character of multinational firms in terms of how they innovate and organise their R&D. While in the past the internationalisation of firms was mainly a matter of the internationalisation of their markets and hence supporting through adaptation the wider exploitation of their established technological competences, it is now also becoming a matter of the internationalisation of their ability to create technological competences through the combination of geographically distinct lines of innovation. Of course, the greater scope for establishing such international research-based networks has depended upon organisational innovation and new types of managerial capabilities, as stressed by scholars of business strategy. The emergence of institutions that can accommodate successful international integration of corporate innovation implies a shift away from obtaining profits through the exploitation of established capabilities through new positions of market power abroad, towards the creation of innovative profits by building new capabilities through knowledge exchanges and co-operative learning, and thereby utilising cross-border networks for the establishment of new value-generating activities. These recent changes have further reinforced the growing relative importance of innovative profits of the original Schumpeterian or Penrosean kind, when the conceptualisation of innovative profits is suitably re-interpreted to fit the current institutional conditions for innovation.

In the concluding Section V, I return to Schumpeter and Penrose in the light of the evidence on historical and recent trends in corporate growth and innovation. It is argued that far from his Mark II model representing a fundamental revision of his Mark I approach the problem was that Schumpeter did not revise his model of innovation sufficiently to take account of the changed conditions, but Penrose has helped to provide us with the means to do so. It is shown that a more complex model of innovation can be constructed in the Penrose–Schumpeter tradition, and that this model is highly relevant to coming to terms with the new reality of corporate innovation and growth that has emerged during the twentieth century, a reality that has been reinforced by current trends. In particular, it allows us to better understand how the profits and growth from innovation relate to greater technological complexity and interrelatedness, amid a deeper sea of knowledge flows between firms and their external environment.

II. SCHUMPETER'S THEORY OF PROFITS THROUGH INNOVATION REVISITED

Schumpeter (1934) relied on a distinction between two realms of economic analysis, and corresponding to these realms are two different means of creating profits. The first realm is grounded on the circular flow of income, and this is the realm of traditional economic theory focused upon the determination of prices and quantities in the markets that link together the flows of inputs and products. In this realm the economy is most easily analysed as either stationary or as growing at a steady state in the form of a simple reproduction of at least some existing elements of the economy on an expanded scale. Profits derive from positions of market power (some might say from market imperfections), since in perfectly competitive conditions profits would be driven to zero. However, Schumpeter (1943) argued that perfectly competitive markets had never existed and would never exist, so comparisons with this hypothetical state as a welfare ideal are unhelpful, but it can still usefully serve as a theoretical benchmark to indicate the minimum conditions for positive profits in markets in the absence of innovation. So let us suppose that there is some irreducible degree of market power inherent in every market in practice, which is then associated with positive 'normal' profits (in equilibrium, leaving aside issues of recession or declining demand and the like). In this traditional context of price and quantity setting in the realm of circulation or markets, an increase in profits to values above the 'normal' must be attributed to a rise in the degree of market power. Now when profits are achieved through the market-based adjustment of prices and quantities by firms that possess the market power to do so, then it is appropriate to use the conventional apparatus for the analysis of profit maximisation and optimisation by rational economic agents. This is the sphere of most mainstream or orthodox economic analysis.[1]

The second realm is that of novelty-creating economic activity which generates new sources of value-adding productive endeavour, and which disturbs the circular flow of income. In this realm growth must be understood as an inherently disruptive rather than a smooth process, which the later Schumpeter (1943) termed 'creative destruction' (although this term is also often misunderstood, as the disruption referred to relates to the circular flow and established market structures, but the creative process itself is likely to be cumulative and incremental, as argued below and by Cantwell and Fai, 1999). Profits derive from creating new fields of productive activity, given that there is an inertia in the wages of the firms responsible, such that their wage costs only rise with a lag. A traditional theorist might reply here that such profits are still conditioned on the fact that markets do not adjust instantaneously (wages being bid up immediately in the labour market to match higher productivity), but the source of the profits is the creative process that added new value to the income stream—and this type of departure from a hypothetical absence of any kind of market power is highly socially beneficial, since everyone enjoys higher income in the long run as a result. Following successful innovation workers do earn higher wages on average, but only high enough to leave room as well for a return to the creation of a collective social capability, which is jointly exercised, especially in large firms, and does not initially accrue to the individuals that make up an innovating firm's team. The value created

---

[1] It might be noted in passing that it is also the sphere of a lot of non-orthodox economic analysis, such as that of those Marxists which emphasise the market power of multinational firms, or those neo-Ricardians which emphasise the bargaining strengths of capitalists as against workers.

by the collective operation is greater than the sum of the parts, and the individuals concerned could not appropriate a higher return on their own particular knowledge or contribution were they to set up independently, so in this sense there is no market failure or departure from rational behaviour. Since it is relevant to Penrose's subsequent linkage of profits from innovation to the growth of the firm as an institutional or administrative entity, it can be noted here in passing that the orthodox theory of the firm typically excludes any such notion of collective operation as itself adding value or raising productivity (which depends upon the organisational innovation that usually supports technological innovation). While Williamson (1975) allows for the superiority of collective operation through hierarchies this is to overcome externalities in established activities rather than to facilitate innovation and coordinated learning or problem solving, and otherwise everything is settled by contracts between individuals. Even more clearly for Hart (1989) the firm supports a nexus of contracts simply by vesting the ownership of critical assets in appropriate hands; all decisions remain strictly individual. In the Schumpeter–Penrose case in which the new organisational form that accompanies innovation is part of the creative process, we have a synergy or 'superadditivity' in which the additional productivity or income cannot be attributed to individual elements, but is instead due to the connections between them (Loasby, 2000B; Potts, 2000).

In this case of raising income per head through the collective operation of novel areas of production, the increase in profits over and above the 'normal' is not due to an increase in the degree of market power at a given point in time, but rather is due to a continual process of creating new value-adding activities. It is a process of incessant change and improvement despite a tendency within markets for such profits to be subsequently whittled away (were there no further change) through technological competition. The elements of market power in the position of an innovating firm are not the source of the new value, but are rather a coincidental by-product (when viewed from the perspective of the realm of circulation) of the uneven character of the creative process (which stems principally from the realm of production). Given the uncertain and the experimental nature of the process of innovation which follows a course of trial and error including an inevitability of some mistakes (Nelson and Winter, 1982), this second category of profits is best characterised analytically in a framework of a search for higher profits, and not within the standard framework of profit maximisation strictly interpreted (even if stylised as 'long run' profit maximisation, since this is not achievable as a behavioural strategy in the context of experimentation, and as pointed out by Alchian (1950), long-run profit maximisation is not a meaningful concept in conditions of uncertainty in the sense of Knight (1921)).

It might be further noted in passing that while the standard neoclassical analysis of profit maximisation supposes that agents are rational in the sense that the outcomes of actions are capable of being foreseen and the context is sufficiently simple or familiar that behaviour can be treated as mechanical and the outcomes can be solved as equilibrium conditions, in decision-making that breaks new ground as in the experimental search for innovative profits, rationality is necessarily bounded (Nelson and Winter,

1982; Nelson, 1996). However, rationality may be bounded either in the sense of Simon (1982) that human information-processing capacity is limited although in principle a more pure rationality might be conceived as feasible in a higher state of knowledge (computerised solutions of chess games get better), or instead rationality may be bounded in the sense that empirical knowledge is open-ended and never complete. Langlois (1987, 1998) maintains that Schumpeter mixed up these two notions of the nature of knowledge as rationalist or empirical and hence implicitly confused the two alternative forms of bounded rationality in his explanation of why innovation is not amenable to analysis in terms of the framework of equilibrium and maximisation. Langlois claims that this is why Schumpeter mistakenly believed that innovation would tend to become a more routinised process within large firms, which holds only if knowledge can be treated as rationalist rather than as empirical and open-ended. Since knowledge is open-ended innovation has not and cannot become more routine, but innovation has become increasingly embedded in and reliant upon a complex network of external knowledge connections entailing transfers between firms, and between firms and other institutions, as alluded to earlier.

It can further be observed that although creative innovative activity generates distinctive profits from new value added in the second realm of analysis, it may also impact upon profits in the first realm, since the process of creative destruction may over time shift the degree of market power prevailing in an industry. Schumpeter seems to have regarded this as a second order effect, but it is of course this aspect that is emphasised in the conventional 'Schumpeterian hypothesis'. Yet even in this respect the 'Schumpeterian hypothesis' reverses Schumpeter's own principal line of causation from innovation to market power rather than the other way round (although this criticism of the 'Schumpeterian hypothesis' is better recognised in the literature—see e.g., Pitelis, 1991), and there may be cases in which innovation erodes rather than increases the overall extent of market power at the industry level, as suggested by the very idea of creative destruction. This connection between the two realms is also critical to discussions of the 'appropriability conditions' for innovation. In the standard analysis of the first realm patents are regarded essentially as devices for temporary monopoly that confer market power to innovators (the effectiveness of which in general has been questioned by empirical survey evidence). Instead, in the second realm patents may function as a means of regulating knowledge exchanges between those that contribute to knowledge creation under conditions of technological complexity and interdependence, the returns to which creativity come mainly from the capabilities to identify and utilise suitable combinations of knowledge in new activities (for a further discussion of these alternative approaches to appropriability see Cantwell, 1999).

Taking appropriability conditions in the sense of the degree of market power exercised by the firms in an industry as given (disregarding for a moment the disruptive effect of creative destruction), by extending the scope of value creation innovation can generate profits for firms without any change in their ability to exercise downstream market power. Indeed, it will be argued below that to more accurately reflect Schumpeter's line of argument, insofar as innovation does also impact on the observed degree of market power the main objective of innovating firms may be not so much to earn additional profits from increased market power, as to help lay the stable institutional foundations for inter-company relationships (or, in Schumpeter's case, relationships between innovative firms and consumers of new products) within which the learning processes of each can function more effectively, thus raising the sustainable rate of innovation and the stream of future profits expected from it. A parallel could perhaps further be drawn between Schumpeter's conceptual distinction of two types of profits, and the distinction sometimes made between rent-seeking and profit-seeking behaviour. At first glance this alternative terminology might seem tempting, since profits from greater market power are by definition an exclusive extraction by the privileged owners of scarce resources from some existing stream of income-generating activity, whereas innovation is a creative process that has the social benefit of providing the potential to raise the income of others as well as generating new profits for innovating firms. However, if rents are defined as attributable to the ownership of scarce but specific assets they may be captured from both types of profits. So it is better to distinguish between the realms of profit maximisation through the exercise of market power, and profit-seeking through innovation.

These two different realms of profits, which correspond to two different realms of economic analysis with alternative focuses of attention and corresponding methodologies, can be linked as well with two different functions of the firm, as discussed by Penrose (1959) in the early sections of her book. The firm is in part a price and output decision-maker, in which guise it can earn higher profits through increasing its degree of market power, but the firm is also a device for innovation, problem solving and cumulative learning in production, the incentive for which is to generate higher profits through creating new areas of social or productive capability. While Schumpeter suggested that the first realm of the market-based allocation of resources and co-ordination could be left to the closed system of Walras and his own contribution was focused instead mainly on the second realm of innovation, for her part similarly Penrose suggested that the first realm could be left to the conventional theory of the firm which was thus set apart from her separate theory of the *growth* of the firm ('so long as it cultivates its own garden and we cultivate ours', Penrose, 1959, p.10) (see Loasby, 2000A). Given what has been said above about the construction of the orthodox theory of the firm on the framework of individual decisions and contracts, which is an extension of standard price or market theory, rather than building upon ideas of collective value creation and social capability or competence, it might be argued that Schumpeter and Penrose were wise to erect a barrier against the standard theory. In any event, they both emphasised that the theoretical issues raised by value creation, innovation and firm growth are distinct from those addressed by the standard analysis of markets and the co-ordination of individual decisions in established activities. An inappropriate fudging of the two realms may cloud both analytical and policy judgement (Richardson, 1990).

It turns out that both the common misunderstandings of Schumpeter's two primary arguments to which we have referred earlier can be reduced to attempts to recreate his contentions about the source of innovative profits and their relationship to market power within the framework and constraints of the first realm of analysis, when they can only be properly and fully comprehended when placed in their original and more appropriate context of the second realm. His theory of innovative profits depends upon creating new fields of productive endeavour to add to and restructure the established circular flow of income and cannot be understood by means of a simple reference to the building of temporary positions of market power within that circular flow. Likewise, Schumpeter's view that when areas of market power are the occasional effects of a continuous stream of innovation from within large firms such monopolistic positions in the market are an incidental by-product and not the source of innovative profits, became inverted in the so-called 'Schumpeterian hypothesis', which held that market power is the cause of innovation by providing resources and safeguarding against the potential downside of risk-taking activity. The attribution of the latter idea to Schumpeter (1943) seems to come from a misreading of his Chapter 8 on Monopolistic Practices, in which his criticisms of the notion of perfect competition and its inconsistency with a regime of innovation and creative destruction were taken as an advocacy of the benefits of market power and imperfect markets for innovation, when viewed through the lens of conventional market-based analysis. Instead, Schumpeter's real point was that the very intellectual framework which gave rise to the hypothetical concept of perfect competition as an extreme point on a spectrum of market structures was misplaced when it comes to the analysis of innovation, but this point could hardly be absorbed by those whose objective it was to try and introduce reference to the scope for innovation within the confines of the traditional analysis of market structure and resource allocation. His idea was simply too revolutionary to be taken on board, and so it could only be accommodated partially and hence in the process in a misleading fashion.

So in order to better understand and appreciate the significance of Schumpeter's own view let us allow him to speak for himself. First of all, it is important to understand the context that was set for his chapter on Monopolistic Practices by the preceding Chapter 7 on The Process of Creative Destruction. In a quotation that is now well known amongst evolutionary economists, as part of an effort to show the need for the analysis of a process in its own right and not through the device of comparative statics, Schumpeter says:

The essential point to grasp is that in dealing with capitalism we are dealing with an evolutionary process. It may seem strange that anyone can fail to see so obvious a fact which moreover was long ago emphasised by Karl Marx[1] (Schumpeter, 1943, p. 82).

---

[1] In turn, in order to understand the context for this remark it should be recalled that the first four chapters of Schumpeter's book were devoted to an appraisal of Marx, from which there are three points worth noting for our purposes. First, Schumpeter draws a contrast between the evolutionary Marx as an economist and the revolutionary Marx as a sociologist, and believed that it was possible to isolate the evolutionary economic aspects of his work: 'To say that Marx, stripped of phrases, admits of interpretation in a conservative sense is only saying that he can be taken seriously' (Schumpeter, 1943, p. 58). Although Schumpeter's view runs against the whole tenor of twentieth century Marxism, perhaps it might be more

footnote continued overleaf

The profits that derive from evolutionary processes over time can be best characterised as the outcome of profit-seeking activities, in contrast to the strict notion of profit
maximisation which better describes the profits that result from coordinating activities
at a given point of time and with given technology; and over the longer term innovative
profits are probably more important:

A system—any system, economic or other—that at *every* given point in time fully utilises its
possibilities to the best advantage may yet in the long run be inferior to a system that does so at
no given point in time, because the latter's failure to do so may be a condition for the level or
speed of long-run performance (Schumpeter, 1943, p. 83).

Crucially, in order to understand how to interpret Schumpeter's subsequent discussion of monopolistic practices (which follows immediately in his book), he goes on
to argue that when monopolistic positions are created through innovation they are not
intentionally profit maximising despite any appearance of being so since they are
merely one aspect of a wider process of transformation, and so innovation requires
competition to be analysed in a fundamentally different framework:

In other words the problem that is usually being visualised is how capitalism administers existing
structures, whereas the relevant problem is how it creates and destroys them . . . However, it is
still competition within a rigid pattern of invariant conditions, methods of production and forms
of industrial organization in particular, that practically monopolizes attention. But in capitalist
reality as distinguished from the textbook picture, it is not that kind of competition which counts
but the competition from the new commodity, the new technology, the new source of supply,
the new type of organization . . . (Schumpeter, 1943, p. 84).

Schumpeter claims that with innovation output expands despite restrictive practices,
unlike in the conventional account of the effect of monopoly, but this is not because
these restrictive practices are themselves the source of the incentive to innovate. In this
context the motive for restrictive practices is to facilitate further learning and inter-
company knowledge transfer in an environment of rapid change, within which such
practices provide some temporary stability so as to be able to introduce new products
or processes more gradually (in an evolutionary and path-dependent fashion, learning
in the process) and hence more effectively, so as to raise the longer term growth of

---

footnote 1 continued from previous page

readily accepted today. Second, Schumpeter criticises Marx's supposed solution to the question of the
origins of profit (surplus value) through the concept of a market for labour power in place of the conventional labour market. But in doing so Schumpeter focuses his attack on Marx's use of this device to explain
the creation of absolute surplus value through lengthening the working day or increasing the intensity of
work; while arguably the more significant application was how Marx explained the creation of relative
surplus value through productivity-raising innovation, and hence introduced the distinction between the two
types of profit emphasised by Schumpeter himself. Third, indeed, just as Marx learned much through an
intense criticism of Ricardo (as noted by Schumpeter), so Schumpeter appears to have learned much from
his detailed critique of Marx, and this is reflected in his own theory of innovation, which incorporates many
of Marx's insights (on which see Rosenberg, 1976, 1982, 1994). Just as Marx took the parts of Ricardo that
he needed and critically demolished the redundant parts that didn't fit with his argument, so Schumpeter
did the same with Marx.

output. Restrictive practices may provide the stable background within which the generation and selection of variants can work reasonably well, and it is this process of variety generation and learning in the development of each variety that adds new value, rather than the restrictive practices themselves, which are in this context merely one supporting prop. There is an analogy here with the cycles of innovation within large successful firms, in which knowledge becomes codified and associated with the creation of new institutions in the form of organisational routines (Nelson and Winter, 1982), in which codes are shared, and which then constitute the stable framework needed for a further round of evolution and learning that involves the renewed creation of tacit knowledge. While institutions themselves evolve only slowly, they thereby contribute the stable framework that is a prerequisite of evolution (Loasby, 1999). Thus, any monopolistic price and output decisions in a changing market are a coincidental (and transitory) source of profits, and should be distinguished from the profits due to innovation as such:

> Thus it is true that there is or may be an element of genuine monopoly gain in those entrepreneurial profits which are the prizes offered by capitalist society to the successful innovator. But the quantitative importance of that element, its volatile nature and its function in the process in which it emerges put it in a class by itself. The main value to a concern of a single seller position that is secured by patent or monopolistic strategy does not consist so much in the opportunity to behave temporarily according to the monopolistic schema, as in the protection it affords against temporary disorganization of the market and the space it secures for long-range planning (Schumpeter, 1943, pp. 102–3).

This is a quotation which we can be sure does not appear in orthodox references to the 'Schumpeterian hypothesis', since it disputes the so-called 'Schumpeterian hypothesis' and challenges the very notion of analysing innovation as the outcome of the degree of market power associated with one particular kind of market structure as opposed to another. Schumpeter does then go on to show how the hypothetical regime of perfect competition would be even less desirable with respect to technological dynamism, but he did so in the context of a thorough critique of the conventional market structure framework as a whole, rather than (and indeed explicitly not) to associate innovative profits with market power. From here, in what would become a link with Penrose, Schumpeter argued that the large firm was more an innovator and an organisational device for learning beyond being a price and quantity decision-maker, and within innovation (much more clearly than in his first book) he stresses the centrality of technological change in production:

> What we have got to accept is that it [the large-scale establishment or unit of control] has come to be the most powerful engine of that [economic] progress and in particular of the long-run expansion of total output not only in spite of, but to a considerable extent through, this strategy which looks so restrictive when viewed in the individual case and from the individual point of time . . . Was not the observed performance due to that stream of inventions that revolutionized the technique of production rather than the businessman's hunt for profits? The answer is in the negative. The carrying into effect of those technological novelties was of the essence of that hunt (Schumpeter, 1943, pp. 106, 110).

So we have seen that far from abandoning his earlier theory of innovative profits in his analysis of the later phase of trustified capitalism, Schumpeter reasserted that theory and continued to draw a clear conceptual distinction between the profits that are due to the market power of monopolistic or oligopolistic firms and the profits they earn from their capacity to innovate, and indeed to insist on the priority of the latter over the former. However, in this event the original theory was in need of extension in order to accommodate the significance of inter-company technological cooperation as well as competition, and the blurring of the distinction between innovation and imitation to which Schumpeter continued to adhere, but which is by now more obviously unsustainable (since with greater technological complexity imitation requires the related absorptive capacity that comes from innovation, and innovation always incorporates some elements of imitation). In this respect the problem with the original formulation of the theory is that it stresses the need to identify the original sources of innovation as opposed to subsequent imitation in order to determine the distribution of innovative profits, with the initial leaders (innovators) earning the higher share. It is true that a useful recent literature has continued in this tradition to distinguish between 'Schumpeter Mark I' and 'Schumpeter Mark II' technological regimes, according to whether innovations are introduced mainly by new entrants or by established firms (Malerba and Orsenigo, 1995; Breschi *et al.*, 2000). Yet moving beyond this to link innovation with the distribution of profits and growth across firms, a drawback of Schumpeter's approach is that the first mover in a successful innovation does not always perform best.

Empirical evidence indicates that among large firms technological leaders tend to retain leadership positions from one phase of development to another (they are the companies specialised in the fields of technological opportunity), but at the level of the industry innovative profits and technology-based growth is highest the more quickly that other firms catch up (Cantwell and Andersen, 1996). This suggests that innovative profits are created by 'followers' and not just by 'leaders'. What is more, the technological leaders are not in general the firms that earn the highest profits within the industry or experience the most rapid growth (Teece, 1992; Andersen and Cantwell, 1999). Now none of this need be a problem once we accept that although social capability is created through internal learning processes within firms, such learning is interactive and involves continuous exchanges of knowledge, whether through deliberate cooperation in learning or independent exchanges through licensing, imitation or the like (Cantwell and Barrera, 1998). Defining innovation to be what is new to a firm with its own differentiated area of expertise or what is new to a particular local context rather than as something new to the world as a whole (Nelson, 1993), the most effective corporate innovators are not necessarily the technological leaders whose expertise is focused on the leading edge fields as such. They may be other firms that have found the most productive industrial applications of the leading edge technologies, which applications themselves require further innovation and other supporting capabilities—linked in part to the process of critical revision of new technologies which enhances their workability and effectiveness, as emphasised by Usher (1954) and Rosenberg (1982).

This line of argument is now entirely intelligible in terms of the most recent litera-ture on the evolutionary approach to technological change which has stemmed from the work of Nelson and Winter (1982) and Rosenberg (1982), and in the process rediscovered the contribution of Penrose (1959). In the evolutionary theory of techno-logical change innovation is always context specific and localised, since learning builds incrementally through guided experimentation on what is already known, and so capabilities—including the capabilities to envisage new productive opportunities—are always context specific. The division of labour is associated with a spatial and firm-specific differentiation of detailed knowledge and skills, as can now be supported by the emphasis in evolutionary psychology on the development of distributed skills and competences as represented by different neural networks (Loasby, 1999). Hence, a technology developed in one context requires the cost of further innovation to be transferred into some other, but the cost or difficulty of subsequent innovation depends upon the initial degree of technological relatedness or complementarity between the activities (Cantwell and Barrera, 1998), and upon the degree of absorptive capacity in the recipient or imitating firm (Cohen and Levinthal, 1989). When firms have a higher degree of technological complementarity between their profiles of specialisation they will each have a greater absorptive capacity with respect to taking advantage of the knowledge being created by the other, and so they will be better able to mutually make use of technology-based alliances and the external capabilities that can be accessed through inter-firm cooperation (Cantwell and Colombo, 2000). In the network of inter-company interaction in innovation the greatest profits are likely to accrue to the firm with the best fit between initial capabilities and the new field of opportunity, as opposed to the firm that first initiates a new line of innovation. The greatest benefits go not to the 'first to discover' or the 'first to commercialise' a core technology with important implications, but rather to firms whose social capabilities are best adapted to absorb and to further develop and entrepreneurially to apply the new lines of inno-vation that emerge from the areas of greatest technological opportunity to novel con-texts and in new combinations with other branches of (and perhaps more traditional) technology.

According to this view Schumpeter's theory of innovative profits should be retained, but his underlying theory of innovation needs to be strengthened in the modern institutional context. It is not necessarily technological leaders that become the best innovators, let alone the only innovators, but rather the firms that succeed in making the most effective combinations between new and old technologies and uncovering the most conducive new fields of application. Actually, abolishing the hard distinction between innovation and imitation only goes to reinforce Schumpeter's point about the distinctiveness of innovative profits which has been stressed above. Innovative profits should not be understood as the returns to the temporary positions of monopolistic market power enjoyed by first movers, but rather represent a return to the creation of the social capability that enables firms to experiment with new technological combina-tions and solve the problems that arise in doing so, and hence to learn and to innovate in production successfully.

III. PENROSE ON PROFITS, GROWTH AND PATH-DEPENDENT
LEARNING IN LARGE FIRMS

Penrose identified herself clearly with those such as Schumpeter who were interested
mainly in the second realm above of innovation, productive experimentation and
novel creativity, rather than in the first realm of coordination, exchange and market
power. She focused on innovation as the source of profits, which would be achieved
through learning to develop new applications of the current resource base of the firm,
as opposed to profits due to the market positioning of the firm or the rents achieved
through market power. So like Schumpeter before her and Nelson and Winter subsequently, Penrose stylised the firm as a profit seeker rather than as a profit maximiser.
She argued that in the most successful and longest standing firms (on which her
attention was concentrated) profits were typically desired for the sake of the firm itself,
to facilitate a stream of continued longer term profit creation through the expansion of
the firm, by developing and taking advantage of the opportunities provided by the
firm's capabilities or resources. It is (sustaining a flow of creative) opportunities, rather
than (safeguarding against) opportunism, that drives the growth of a Penrosean firm
(Loasby, 2000b). Thus, she claimed that the goals of profit seeking and raising through
appropriate investment the long-term rate of growth of the firm became equivalent,
since each was derived from the innovative adaptation and extension of the firm's
resource base. Echoing Schumpeter's view that innovation is the only reliable basis for
longer term corporate growth as distinct from the shorter term gains that might be
made from monopolistic practices or market power she states:

Examples of growth over long periods which can be attributed *exclusively* to such protection
[market power] are rare, although elements of such protection are to be found in the position of
nearly every large firm (Penrose, 1959, p. 113).

While acknowledging that the firm could create new opportunities through marketing
and advertising by better exploiting its established competence as well as through the
development of new technological competence, Penrose suggested that such a strategy
is only feasible within its existing market areas. To diversify into new areas of specialisation requires the appropriate technological base to do so. In other words she asserted
the ultimate primacy of the realm of productive and technological competence over
that of exchange and selling relationships, since purely market exploiting activity
works only within the confines of an established market area, and so must sooner or
later run up against limits. Like Schumpeter she contrasted the attainment of a monopolistic market position and technological progressiveness as conceptually alternative
(although in practice quite possibly correlated) routes to profitability and corporate
survival. She argued that the seeking of the innovative profits needed for longer term
survival led to a wider range of diversification based on the underlying technological
complementarity or relatedness of activities:

Firms . . . 'specialize' . . . in a much wider sense than the logic of industial efficiency [cost
minimisation and price competition] would suggest, for the kind of 'specialization' they seek is
the development of a particular ability and strength in widely defined areas which will give them

a special position *vis-à-vis* existing and potential competitors. In the long run the profitability, survival and growth of the firm does not depend so much on the efficiency with which it is able to organize the production of even a widely diversified range of products as it does on the ability of the firm to establish one or more wide and relatively 'impregnable' bases from which it can adapt and extend its operations in an uncertain, changing and competitive world (Penrose, 1959, p. 137).

Hence, Penrose had little use for her purposes for the standard model of the firm as a price and output decision-maker (the coordination-based theory of the firm), which was not designed for the analysis of a firm that is free to internally vary the kind of products it produces as it grows and to innovate by creating from within new products and processes (as in the capabilities-based approach to the firm, which she effectively initiated). She remarked:

we will be dealing with the firm as a growing organization, not as a 'price-and-output decision maker' for given products (Penrose, 1959, p. 14).

She goes on to explain how the very nature of the social capabilities of the firm as embodied in its organisational structure is transformed as the technological base of the firm is expanded and increased in complexity. She was especially interested in large successful firms not because she believed them to be representative of the population of firms as a whole (indeed she was careful to distinguish between the two), but rather because it is these large firms that encapsulate a repository of competence for the economy of which they are part, which today we might refer to as social capability but which she termed administrative organisation. Penrose spent some time discussing the relationship between the productive competence of firms and market demand, acknowledging of course that external changes in the structure of demand may be responsible for growth opportunities. However, as noted above, she argued that a firm that lacked an adequate technological base would lack the capability to diversify, while conversely a firm with a strong degree of technological competence would find its opportunities for expansion likely to be so prevalent that it would have to choose carefully between many different possibilities of action. Indeed, in a view that anticipates quite well the later Cohen and Levinthal (1989) argument about the role of absorptive capacity, she emphasised that the very ability to perceive opportunities in the firm's external environment (including new market opportunities) depended upon the initial capabilities and resources of the firm, given that to appreciate the potential relevance of external achievements or events connections must be made with the firm's existing differentiated knowledge base:

I have placed the emphasis on the significance of the resources with which a firm works and on the development of the experience and knowledge of a firm's personnel because these are the factors that will to a large extent determine the response of the firm to changes in the external world and also determine what it 'sees' in the external world (Penrose, 1959, pp. 79–80).

The different fields and distinctive scope of their established capabilities regulates differences between firms and individuals in what they are able to envisage as 'the imagined, deemed possible' (Shackle, 1979), or perhaps more in line with Penrose's view of how the productive purpose to which resources are put may be shifted over time, 'the imagined deemed capable of being made possible' (Loasby, 2000B). In this context Penrose referred as well to Schumpeter's contention that new products may be forced on consumers by the initiative of entrepreneurs where the latest fashion or model comes to be desired in its own right, but this is not necessary to understand why she wished to focus on the firm as a repository of capabilities or resources as opposed to the coordination functions of the firm. Rather she believed that innovation-based profitability and growth is essential to the firm's longer term survival, that it is 'built into' or inherent in the characteristics of every successful firm, and hence that the firm can be most usefully depicted as a device for learning and the posing and solving of new problems in its field of expertise and production. New products and processes are in her view created through learning from the established resources and technological base of the firm, by extending and adapting it for novel purposes:

> Consequently if we can assume that businessmen believe there is more to know about the resources they are working with than they do know at any given point in time, and that more knowledge would be likely to improve the efficiency and profitability of their firm, then unknown and unused productive services [from existing resources] immediately become of considerable importance, not only because the belief that they exist acts as an incentive to acquire new knowledge, but also because they shape the scope and direction of the search for knowledge . . . both an automatic increase in knowledge and an incentive to search for new knowledge are, as it were, 'built into' the very nature of firms possessing entrepreneurial resources of even average initiative. Physically describable resources are purchased in the market for their known services; but as soon as they become part of a firm the range of services they are capable of yielding starts to change. The services that resources will yield depend on the capacities of the men using them, but the development of the capacities of men is partly shaped by the resources men deal with. The two together create the special productive opportunity of a particular firm (Penrose, 1959, pp. 77–9).

Indeed, Penrose began her book by remarking that there are forces inherent in the nature of firms that induce expansion even if all external conditions remain unchanged, and that the internal interaction between inherited resources and managerial perception is a dynamic process that encourages continuous corporate growth, but which constrains the achievable rate and direction of growth. In other words, excess or available resources are themselves a sufficient incentive for endogenous innovations as a continual source of newly created profits and growth in large firms. From all this the centrality to her perspective of capabilities, corporate learning and innovation is quite clear, and as with Schumpeter, conceptually separate from issues of market power or other aspects of the exchange and coordination of established products or activities. What is interesting here is not just her linking of Schumpeter's second realm of innovative profits to the resource-based theory of growth of the firm, but also the way in which as a result she depicted the direction of corporate learning and growth as a path-dependent resource-constrained process. In this respect she anticipated current ideas on the evolutionary approach to technological change, and in particular the notions of corporate technological trajectories (Dosi, 1982; Teece et al., 1997), corporate technological diversification (Pavitt et al., 1989; Granstrand and Sjölander, 1990; Granstrand *et al.*, 1997) and corporate coherence in diversification (Teece *et al.*, 1994). Anticipating as well the argument of Cantwell and Fai (1999) that since innovation is rooted principally in internal learning within the firm, technological competence evolves gradually and changes much less dramatically than the composition of down-stream products or markets, Penrose claimed that each successful firm had a continuity which was provided by its capabilities or resources:

In practice the name of a firm may change, its managing personnel and its owners may change, the products it produces may change, its geographical location may change, its legal form may change . . . [yet] the identity of the firm can be maintained through many kinds of changes, but it cannot survive the dispersal of its assets and personnel nor complete absorption in an entirely different administrative framework. . . . The general direction of innovation in the firm (including innovation in production) is not haphazard but is closely related to the nature of existing resources . . . and to the type and range of productive services they can render. . . . The Schumpeterian process of 'creative destruction' [of established products] has not destroyed the large firm; on the contrary, it has forced it to become more and more 'creative' [in the adaptation and application of its capabilities] . . . when it [a firm] develops a specialized knowledge of a technology which is not in itself very specific to any particular kind of product . . . it [research] enables at least the large firms to turn aside the process of 'creative destruction' and to thrive on the novelty which might otherwise have destroyed them (Penrose, 1959, pp. 22–3, 84, 106, 115).

From here Penrose was led to forecast that there would be an increasing impetus to innovation as a means of creating profits (which implies a greater reliance on innovative profits, compared to profits due to market power) as the role of technological competition rises, and this is what has actually occurred, as argued at the start of this chapter. However, at the same time she noted that greater technological competition would compel firms to specialise in a narrower range of basic areas of production, since resources would be increasingly tied up in continual innovation which would restrict the rate at which they can diversify their fundamental activities (Penrose, 1959, pp. 106–7). This was a remarkable anticipation of the modern trend towards corporate technological diversification or more properly a restructuring of diverse technological capabilities around the clusters of greatest interrelatedness (Cantwell and Santangelo, 2000), accompanied by greater product concentration (less diversification across products or lines of business activity). However, as was more appropriate to the historical period in which she herself was writing, and to the earlier stages of large firm growth, Penrose tended to stress how technological diversification would in general facilitate and support greater product diversification:

There is no reason to assume that the new knowledge and services [from corporate exploration and research] will be useful only in the production of a firm's existing products; on the contrary, they may well be useless for that purpose but still provide a foundation which will give the firm an advantage in some entirely new area (Penrose, 1959, pp. 114–15).

As has now been described in a detailed historical survey by Chandler (1990), and as Penrose had observed from her own collection of case study evidence, for most of the twentieth century large firms grew through a combination of technological diversification from their initial resource base linked to related product diversification, or what

Chandler later depicted as the interlinkage between the economies of scale and scope. Penrose indicated as well that the greater market spread that accompanied technological diversification through innovation from within the resource base of the large firm may entail either product diversification or geographical diversification. Hence, industrial diversification or internationalisation could be considered substitutes for a particular firm at a given point in time given the resource constraint upon its overall rate of growth (see Cantwell and Piscitello, 2000, for a discussion of how this relationship later shifted from one of substitutability to one of complementarity). However, it should be underlined that for Penrose the substitutability between different types of new market entry was not a result of her trying to focus on the realm of market exploitation rather than competence creation (selecting between alternative ways of exploiting a given competence), but was on the contrary the outcome of her focus upon the nature of competence creation in the specific historical context in which she was writing, at which time any diversification of the basic market area(s) of the firm tended to require a supporting diversification of the firm's technological base. This emphasis upon capability formation from the resource base of the firm rather than market exploitation should already be clear from what has been said of Penrose's central focus upon the firm as a device for innovation and knowledge creation rather than as a means of coordinating established activities. However, she believed that the scope of feasible technological diversification would not only regulate the degree of market spread, but a move into markets that require a complementary technological base may result in positive feedback to further innovation. If there were no such feedback the long-term rationale for common ownership in the firm would be weak and the character of the investment association between the parts of the enterprise would be qualitatively different from that in a coherent corporate group:

expansion by acquisition does not necessarily, or perhaps even usually, mean that a firm is entering a field for which it would otherwise have no qualifications. Acquisition is often a profitable process precisely *because* the firm has peculiar qualifications in the new field. . . . In some cases [of exceptions in which foreign subsidiaries operate independently of their parents] . . . the acquisition of foreign subsidiaries should be treated . . . simply as an investment akin to investments in financial assets [as a portfolio rather than as a direct investment] (Penrose, 1959, pp. 129, 193).

Thus, Penrose asserted the need for coherence in the technological and productive activities of the firm from the perspective of the capacity to continue to innovate and grow as a combined organisation, although not necessarily from the perpective of the pure realm of coordination, which may be instead essentially a financial perspective. By thinking in terms of feedback to subsequent growth perhaps Penrose also to some extent anticipated the notion of intra-firm networks of competence creation, but at least at an international level for cross-border innovative feedback to become fully effective was to wait for the new historical phase of integrated multinational firms only from around 1980 onwards (Cantwell, 1989; Cantwell and Piscitello, 2000).

IV. THE CHANGING INSTITUTIONAL FORM OF INNOVATION

Something has already been said of the way in which the institutional conditions for innovation in large firms have shifted since Schumpeter's day, and to some extent since Penrose's early contribution as well, although she said much that anticipated these changes. The import of this shift in institutional regime has been to reinforce the significance of innovative profits as against the profit and growth strategies associated with market power. Hence, the arguments of Penrose and Schumpeter are still more relevant today than at the time they were writing, once their essential themes are related to the modern context. The two major reasons why innovative profits are now even more relatively important are first that innovation has been increased in what modern Schumpeterians such as Freeman (1987) have termed a new techno-economic paradigm, and with a greater intensity of international competition positions of pro- tected market power are increasingly under threat; and second, the firm must now rely on more complex combinations of related technologies to serve even more narrowly defined product markets, so relative to some given level of cost of the innovative development of resources the opportunities for establishing downstream monopolistic positions are reduced in this way as well. A by-product of these changes is that inter- company cooperation between large firms is increasingly motivated by the need for mutual technology-based exchanges and coordinated learning relative to the more traditional collusion to secure jointly exercised positions of market power.

Taking a step back for a moment, in the first phase of the growth of large firms that ran up until about 1970, as described by Chandler (1990) there was an interleaving between the economies of scale and scope. Large firms grew by diversifying their tech- nological base, and in the process diversified their product markets in similar propor- tion, and together this combined diversification supported a rise in the scale of output. An essential plank behind this process was the rapid growth of in-house corporate R&D in the largest firms, as stressed by Schumpeter, which improved their innovative search activities, and their capacity for problem solving and learning in an age in which science and technology began to become more interdependent with one another. Using the patents that large firms are granted from these problem-solving activities, we can trace their individual profiles of technological specialisation (Cantwell, 1993; Cant- well and Fai, 1999; Cantwell, 2000). What emerges from these studies is that these patterns of corporate technological specialisation are differentiated and firm specific, that groups from common countries of origin have certain country-specific features in the form of their expertise, and that the profiles of specialisation persist over time, reflecting a path-dependent technological accumulation or corporate technological trajectories.

One other aspect of this empirical evidence on the history of corporate innovation patterns is worth emphasising in particular. This is the relationship between the degree of diversity of technological activity in the firm and the overall scale of technological effort, the latter also serving as a proxy for the scale of output of the firm since the various measures of size tend to be correlated across firms. It is well established that the degree of diversification rises with size, and we can plot a size–diversification relationship across large firms (size is measured by the total number of patents granted, and diversification by the reciprocal of the coefficient of variation across sectors in the index of corporate technological specialisation—see e.g., Cantwell and Santangelo, 2000). Now the interesting point is that whether we are working with firms (Cantwell and Fai, 1999) or at the level of countries (Cantwell and Vertova, 1999) for most of the twentieth century the size–diversification frontier didn't shift very much. That is, most growth took the form of the joint achievement of increased scale and greater scope through a movement along the frontier, as diversification and growth went hand in hand, a statistical confirmation of the case study conclusions of Penrose and Chandler. However, in more recent times the size–diversification frontier has shifted quite markedly (Cantwell and Santangelo, 2000), which shows how the relationship between technological diversification and growth has become less simple than in the past. For firms the size–diversification frontier has tended to shift upwards (so the average extent of technological diversification controlling for size has risen), but it has also shifted rotationally so that the very largest firms have tended to reduce the diversity of their technological profiles. There is now an impetus for firms to achieve a minimum threshold degree of technological diversification to take advantage of greater interrelatedness, but the combinations constructed by the firm must also be coherent enough to focus upon potential linkages in which interrelatedness or technological complementarity is at its highest. At the level of countries instead the size–diversification frontier has shifted downwards, which shows that on average countries have increased the extent of their technological specialisation for any given size (Cantwell and Vertova, 1999). This is likely to be due to the effects of internationalisation of activity, leading locations to become more focused in their efforts while the largest firms span more technological fields and more geographical areas.

The increasing significance for firms of technological interrelatedness and fusion is one aspect of the historical shift mentioned earlier as having been termed a new techno-economic paradigm (Freeman and Perez, 1988). In this context a techno-economic paradigm is a system of scientific and productive activity based on a widespread cluster of innovations that represent a response to a related set of technological problems, relying on a common set of scientific principles and on similar organisational methods. The old paradigm until around 1970 was based on energy and oil-related technologies, and on mass production with its economies of scale and specialised corporate R&D. In recent years this has gradually been displaced by a new paradigm grounded on the economies of scope as distinct from scale, and derived from the interaction between flexible but linked production facilities, and a greater diversity of search in R&D. Individual plant flexibility and intra-company network linkages both depend upon the new information and communication technologies (ICT). An analogy might be drawn here between the increasing tendency in the most recent paradigm for innovation to take the form of novel combinations of newly related technologies, and Lachmann's (1986) conception of capital as a set of complementary elements within a given structure, which elements would instead be substitutes for one another if used for some other purpose with a different structure. As argued earlier, in general innovation embodies the creation of this kind of 'superadditivity' of new combinations being more than the sum of their parts, and so the modern increase in technological interrelatedness provides a new potential which has increased the scope and rate of innovation.

Part of the reason for the increased extent of technological interaction within and between firms lies in the more sophisticated modern system of production as well as in the more intensive linkages between science and technology in the current techno-economic paradigm, which relies on flexibility through computerisation and diversity through new combinations drawing upon a wider range of disciplines. Firms increase the returns on their own R&D through suitably adapting their underlying tacit capability so that they can absorb and apply the complementary knowledge acquired from other locations or from other firms more intensively in their own internal learning process. Technological diversification and internationalisation have become positively related in more internationally integrated multinational corporations (MNCs) since around 1980 (Cantwell and Piscitello, 2000). Apart from the rise in techno-logical interrelatedness, the potential opportunities for cross-border learning within MNCs have been enhanced by an increased take-up of ICT technologies (Santangelo, 1999). ICT specialisation seems to amplify the firm's technological flexibility by enabling it to fuse together a wider range of formerly separate technologies. In this sense, in the current ICT-based paradigm government intervention is better geared towards the promotion of cross-firm and cross-border knowledge flows (presuming that firms follow the model of a continually interactive search for better methods and improved products, and hence a search for higher profits through experimental inno-vation in the fashion of Schumpeter and Penrose); rather than to provisions to protect the monopolistic and separate exploitation of knowledge by those that have independ-ently invested in its creation (which could be more easily represented through an underlying model of static profit maximisation by firms through the exercise of market power) (Cantwell, 1999). Yet as stressed above, the theory of innovative profits needs revising in this era of greater technological interrelatedness in which firms must not only sustain an adequate spread of coherent in-house diversification but must be able to access other related capabilities through partnerships. It is not leadership in ICT as such that is likely to count for most, but rather the capacity to blend ICT with other technologies as a means of fusing them together and creating new combinations.

However, the creation of technology may be locationally concentrated or dispersed, according to the degree of complexity embedded in it. Some kinds of technologies are geographically easily dispersed, whilst the uncodified character of others makes cross-border learning within and across organisations much more difficult. Thus, although multinationals have shown a greater internationalisation of their R&D facilities recently, it depends upon the type of technological activity involved. The development of science-based fields of activity (e.g., ICT, biotechnology and new materials) and an industry's core technologies appear to require a greater intensity of face-to-face inter-action (Cantwell and Santangelo, 2000). Nevertheless, it may sometimes still be the case that science-based and firm- and industry-specific core technologies are dispersed internationally. The main factors driving the occasional geographical dispersion of the creation of these kinds of otherwise highly localised technologies are either locally embedded specialisation that cannot be accessed elsewhere, or company-specific global strategies that utilise the development of an organisationally complex international network for technological learning (Cantwell and Santangelo, 1999).

The more typical pattern of international specialisation in innovative activity within the MNC is for the development of technologies that are core to the firm's industry to be concentrated at home, while other fields of technological activity may be located abroad, and in this sense the internationalisation of research tends to be complementary to the home base. Thus, when science-based technology creation is internationally dispersed it is most often attributable to foreign technology acquisition by the firms of 'other' industries—for example, chemical industry MNCs developing electrical technologies abroad, or electrical equipment MNCs developing specialised chemical processes outside their home countries (Cantwell and Santangelo, 1999, 2000; Cantwell and Kosmopulou, 2001).

Evidence has now emerged that the choice of foreign location for technological development in support of what is done in the home base of the MNC depends upon whether host regions within countries are either major centres for innovation or not (termed 'higher order' or 'lower order' regions by Cantwell and Iammarino, 1998, 2000). Whereas most regions are not major centres and tend to be highly specialised in their profile of technological development, and hence attract foreign-owned activity in the same narrow range of fields, in the major centres much of the locally sited innovation of foreign-owned MNCs does not match very well the specific fields of local specialisation, but is rather geared towards the development of technologies that are core to the current techno-economic paradigm (notably ICT) or earlier paradigms (notably mechanical technologies) (Cantwell et al., 2000). The need to develop these latter technologies is shared by the firms of all industries, and the knowledge spillovers between MNCs and local firms in this case may be inter-industry in character. Thus, ICT development in centres of excellence is not the prerequisite of firms of the ICT industries, but instead involves the efforts of the MNCs of other industries in these common locations.

It may also be the case that the development of the capability to manage a geographically complex international network lies in a firm's specialisation in ICT. The opportunities created for the fusion of formerly unrelated types of technology through ICT has made feasible new combinations of activities, the best centres of expertise for which may be geographically distant from one another. The enhanced expertise in ICT seems to provide a company with greater flexibility in the management of its geographically dispersed network, and an enhanced ability to combine distant learning processes in formerly separate activities. If this is the case for manufacturing companies in general, it is all the more true for electrical equipment and ICT specialist companies. Affiliate networks are increasingly used to source new technology. Accordingly, global learning has become an important mechanism for corporate technological renewal within MNCs.

The key importance of ICT to the now more complex management of innovation in MNCs is that it enables firms to better exploit their corporate technological diversification across national boundaries (Cantwell and Piscitello, 2000), owing to the role of ICT as a means of combining fields of knowledge creation that were previously kept largely apart (or what Kodama, 1992, terms technology fusion). However, while this use of ICT has led many smaller firms to extend the breadth of their technological diversification to create new combinations, in some of the very largest MNCs the extent of technological diversification has been reduced, so as to focus better on the most promising possible combinations from amongst the broader initial dispersion of innovative activity that such companies have inherited from the past (Cantwell and Santangelo, 2000). Thus, we find some convergence in the average degree of technological diversification across large firms, including amongst others, in the pharmaceutical industry (Cantwell and Bachmann, 1998).

Freeman and Perez (1998) had argued that in the latest techno-economic paradigm ICT has become a 'carrier branch' or a 'transmission belt' for the transferral of innovation across sectors, analogous to the role played by the capital goods sector in the mechanisation paradigm in the nineteenth century (Rosenberg, 1976). Company evidence now suggests more than this that ICT has become also a core connector of potential fields of technological development within firms (or between firms in technology-based alliances) that facilitates the technological fusion of a formerly disparate spread of innovative activity. Thus, while in the past the machine-building industry simply passed knowledge of methods from one field of mechanical application to another, ICT potentially combines the variety of technological fields themselves and so increases the scope for wider innovation. Hence, innovation has become a still more central part of corporate development in the ICT age. Internationalisation through the MNC to connect together in a network related streams of locationally specialised innovation, in-house technological diversification and inter-company technology-based alliances, and the corporate development and application of ICT have become intertwined in a new era of innovative capitalism.

V. FROM SCHUMPETER TO PENROSE TO THE SOCIALISATION OF INNOVATION BETTER UNDERSTOOD

Nelson (1996) has remarked on how Schumpeter provides a good analytical starting point for the development of the theory of innovation as an evolutionary process, but that his own theory was too little developed. In particular, Schumpeter did not take sufficient account of the relationship between science and technology, of the diversity of institutions that contribute to technological change, and of course he could not have been expected to have adapted his theory to the changes in the nature of technologies and in the institutional landscape that have occurred since the time he was writing. Before concluding on how Penrose's contribution has helped us to develop the evolutionary theory of innovation to make it more complex and better able to handle the modern institutional reality, I would like to recap on my own version of the respects in which Schumpeter's own approach is most in need of further development. Schumpeter (1943) was right on the role of large firms in the twentieth century coming to lead systems of innovation through their increasing responsibility for the sources of innovation and the lesser degree of relevance of their market power (contrary to the standard interpretation, as discussed at length above), but he did not work through the consequences of this for his theory of innovation, which was left largely intact from his earlier work. Schumpeter was also right on the greater degree of socialisation of innovation, but as mentioned above he worked through the implications of this wrongly (to suggest that innovation becomes a more routine process), rather than to treat its implications for the closer relationship between science and technology, and between the institutions involved in technological change. I will argue that this heritage left Schumpeter's theory of innovation most in need of further development or alteration in four areas.

First, Schumpeter's model of innovation was too simplistically sequential, from innovative leaders to imitative followers, and in the process involved an overly rigid separation of innovation and imitation. No doubt this had something to do with his association between innovation and entrepreneurship and his conception of the entrepreneur as an innovative leader, and his continued use of the term entrepreneurial profits (for what I have here called the profits from innovation) reflected his reluctance to break this association despite his arguments about the gradual diminution of the role of the entrepreneur in innovation within large firms. The modern evolutionary literature on technological change has instead followed the Penrosean tradition in supposing that there is a direct relationship between a firm's own capabilities for innovation and its absorptive capacity to be able to imitate others. So across firms innovation and imitation tend to be complements rather than substitutes, and at the level of the individual firm any strategic choice between innovation and imitation is a matter of degree rather than a matter of kind. Since we now better understand Penrose's point that corporate learning processes are cumulative and incremental and so technological development is path dependent (Nelson and Winter, 1982; Rosenberg, 1982), it is the experience gained in creating and using current technology that gives established players an advantage in the innovation and imitation of the next generation of technologies, rather than their control over any particular isolated and discrete invention. This is also why there are multiple interactive innovators rather than a single first mover, each of whom has joined the club at some stage with their own specific differentiated contribution. The ability of each firm to derive profits from the continuous process of innovation depends upon how well their own specific capabilities 'fit' with the main avenues of innovative potential in the wider system of which they are part. Thus, the analysis of the cross-company distribution of the profits from innovation is more complex than in Schumpeter's first mover case, and depends upon the strength of individual firm capabilities and on the characteristics of the combination of capabilities and the nature of the availability of technological opportunities.

In fact, Schumpeter's sequential model of evolution from innovation to diffusion is so simplistic that there is a case for arguing that Schumpeter should not have gained the exclusive kind of pride of place that he occupies in modern evolutionary economics (Hodgson, 1993). Indeed, Schumpeter shared with Penrose (1952) a scepticism about the use of biological analogies in economics, although the lack of an association with Darwin does not render Schumpeter (or Penrose) any less an evolutionary economist (Hodgson, 1997; Potts, 2000). However, the reluctance to accept Schumpeter as the primary founder of contemporary evolutionary models of technological innovation seems overly harsh given Schumpeter's clear drawing of the distinction of the evolutionary realm of profits from innovation from the standard realm of profits from market power, which anticipated Penrose's similar distinction between the evolutionary realm of the growth of the firm and the standard realm of the theory of the firm, as emphasised above. Hodgson (1993) refers to Schumpeter's admiration for Walras and his desire for an ultimate synthesis between the two realms of analysis, but in what he had to say about innovation and growth Schumpeter insisted instead on the need to separate these two realms to avoid confusion, as argued above. This having been said, it is clear that Penrose had a much better appreciation than did Schumpeter of the institutional (and hence modelling) complexity that is involved in social or cultural as opposed to biological evolution.

Second, Schumpeter's model of innovation was too product oriented, and dealt insufficiently with process innovation, change of organisational form, and technological complexity and interrelatedness. It is surely no accident that Schumpeter's model of innovative leaders, followers and diffusion is remarkably similar to the subsequent product cycle model (Vernon, 1966), in which firms and technologies are assimilated to products, and there is a single locus of innovation (for a further critique of the product cycle model see Cantwell, 1989, 1995). It is also no coincidence that subsequent survey evidence has found that first mover advantages are more important for product than for process innovations, but then only for industries such as chemical compositions and some devices in which the product is reasonably well defined and innovation is less systemic (Nelson, 1996). It is because of this difficulty that Schumpeter's (1939) analysis of long waves is most in need of amendment, although to develop this point fully lies beyond the scope of this paper. Suffice to say that while Schumpeter had thought of a long wave as being initiated by a bunching of related new product innovations, modern Schumpeterians have emphasised the role of organisational restructuring at the outset of a new techno-economic paradigm (Freeman and Perez, 1988), and so it is rather new processes and the associated inventions that lead a new wave, while new product families follow during the course of the wave (von Tunzelmann, 1995).

Third, once more accurately understood, the socialisation of innovation has meant that firms are less independent than they were, and that they now all float in a much deeper sea of background knowledge, which Nelson (1992) refers to as the 'public' element of technology. There are at least four aspects to this: inter-company knowledge flows have increased, there is a growing role for governments and other non-corporate institutions in knowledge development and transfer, the importance of science for technology has risen and diversified in its impact, and there has been a tendency towards more rapid codification and the formation and spreading of professional and scientific communities. We can now think of firms and the individuals aboard them like ships floating in a sea of public knowledge which connects them, or more accurately potentially public knowledge since the extent that they can draw upon it depends upon their own absorptive capacity and on their membership of the appropriate clubs (whether inter-company alliances or professional associations and the like). Over time, especially since Schumpeter's day, firms have been designed to float deeper down in the water, but they still always leave a critical part comprising their own tacit capabilities above the surface, which does not sink down or fall into the general mass. Indeed, it has been argued under the first point above that holding stronger capabilities above the surface is positively related to the depth into which one can reach below the surface, both for the absorptive capacity to extract complementary knowledge and for the extent to which one contributes oneself to the public knowledge pool. Universities and governments have increasingly contributed to the sea of public knowledge as well. Additionally, among firms that deliberately co-operate through technology-based alliances, personnel can be exchanged so as to coordinate learning efforts.

There is a connection between the first and the third points, since the sharing of knowledge between firms implies not just that technology must be developed through an interactive social and cultural evolution rather than through a biological evolutionary process involving competition between genetically independent entities (consistently with the use of the concept of evolution in economics favoured by both Schumpeter and Penrose), but also that the distribution of the gains from innovation in a complex system requires a more complex model than Schumpeter's. For example, knowledge developed in one context may ultimately prove to have a bigger impact in another, which was not foreseen by the originator or even perhaps initially by the most innovatively successful recipients. Firms now also devote much greater effort to attempts to understand their own technological practice and that of others. Codification of knowledge is the outcome of a conscious effort, shifting back the dividing line between what is potentially public and what is tacit (Cohendet and Steinmueller, 2000; Cowan et al., 2000). So firms that become especially adept at codification may find that this is a source of competitive advantage since they can then more readily draw on the public pool.

Fourth, the socialisation of innovation has not meant its routinisation in large firms, but rather a greater degree of interaction between firms in a continued experimental search for new diversity and variety generation. To engage in this inter-company interaction fruitfully, firms must maintain an adequate diversification of their in-house technological efforts, since the closer that knowledge is to the proprietary interests of a firm the more likely that it will only be shared in return for something else that is probably technologically complementary, which is what each firm needs to join the relevant corporate club (Cantwell and Barrera, 1998). The entrepreneurial function is not eliminated but it is more institutionally embedded in an ability to network and make new connections (for a discussion of the role of imagination and leadership in the modern context see Witt, 1998). Yet just as I have been arguing that Schumpeter (1943) retained his theory of entrepreneurial profits (indeed, I am claiming that he did not revise it sufficiently in the light of the other points just made), Langlois (1987) has provided the complementary observation that Schumpeter (1934) had already espoused the likelihood of an erosion of the entrepreneurial function under the anticipated future socialisation of innovation. As Langlois also suggests, the problem here seems to have been Schumpeter's desire to invert Marx (socialism would evolve from the success of capitalism) just as Marx had stood Hegel on his head, and the linking of his attempt to do so to the vitality of independent entrepeneurship within capitalism. This may have blinded him to the nature of the effects of the increasing impact of science on technology, the codification of knowledge and the greater sharing of knowledge.

On all the four points we have listed, Penrose provides ideas in the same evolutionary tradition as Schumpeter that help us to develop the theory of innovation, and to better apply it in the modern institutional context. Her themes of corporate capabilities and the related absorptive capacity provide us with the foundation needed to construct a more complex and meaningful model of the innovation process, and of the distribution of profits from innovation. The greatest gains will be realised not by first movers but by firms whose capabilities are best suited to the prevailing distribution of technological opportunities, and best placed to construct novel technological combinations. They will be the firms most able to dip into the public knowledge pool and integrate what they find with their own capabilities. So Schumpeter's simplistic sequential model of isolated innovations and diffusion can be replaced, with a greater focus on the characteristics of technologies and the organisational processes that accompany them, rather than on the resultant products as such. This framework calls attention as well to the greater complexity of the relationship between corporate capabilities and public knowledge. Although Penrose (1959) did not discuss to any great extent the role of inter-company networks outside the specific context of mergers and acquisitions, she became more interested in the extension of her work in this respect later. In what remains I elaborate briefly how Penrose provided the threads needed to help tie up each of the four major loose ends left by Schumpeter, and how in each case she did so in a way that establishes connections with the most recent institutional characteristics of the innovation process.

First, the need to replace Schumpeter's simple sequential model of leaders and followers, and his accompanying separation of innovation and imitation or diffusion. By linking Schumpeter's theory of profits through innovation to the resource-based theory of the growth of the firm, Penrose introduced the much more appropriate scheme of multiple competent innovators engaged in cumulative evolutionary learning and related absorption from their environments. This quite rightly leads us to focus on the capacity of innovative firms to blend technologies together in new combinations (a variant of Schumpeter's original idea of innovation as new combinations), rather than on being first in any given new product market, and in doing so underlines still more sharply Schumpeter's distinction between the profits from innovation and those due to market power. In a world of greater technological interrelatedness and the role of ICT in technology fusion this Penrosean kind of redrawing of the basic Schumpeterian model is entirely appropriate.

Second, the need to bring to centre stage the role in innovation of process restructuring, organisational form and the technological complexity of interdependent systems, in place of Schumpeter's concentration on product innovation. For Penrose, the identity of the large firm is described essentially by the composition of its technological competence or resources, which she saw as changing much less rapidly than its products or markets, a proposition for which we now have statistical support (Cantwell and Fai, 1999). Thus, the competence or capabilities of the firm constitute the basis for its systematic evolution, as opposed to the periodic revolutionary transformations of product markets in the other realm of analysis, that of market coordination, by the process of creative destruction. Again, it turns out that Penrose was remarkably perceptive in this respect as well, in that new organisational forms for the better management of technological complexity have co-evolved with the ability of firms to renew their competence base and to develop the connections between a greater diversity of related technologies. So corporate technological evolution as envisaged by Penrose has been elevated to a higher level, and now carries still greater significance. The organisational form of MNCs has recently shifted from the loosely connected network of affiliates needed for the exploitation of a given competence in international markets, to a more closely integrated network of affiliates designed to facilitate complementary paths of innovation and new competence creation (Cantwell and Piscitello, 2000).

Third, the recognition of the deeper sea of knowledge and of institutional interaction within which firms now evolve than at the time Schumpeter was writing. Penrose rightly foresaw the increasing extent to which firms would come to rely on the profits from innovation as technological competition increased relative to price competition, so making more urgent the need for an evolutionary mode of analysis. The availability of innovative opportunities has been increased by the extension of public knowledge and knowledge exchange, thus reinforcing Penrose's point. The relatively greater significance of innovation has more recently been further enhanced by the erosion of traditional positions of market power by increased internationalisation, the arrival of a new paradigm in which science has a closer relationship with technology and the role of ICT in taking advantage of wider technological relatedness, as mentioned already.

Fourth, the need to incorporate the greater institutional embeddedness of the entrepreneurial function, in place of Schumpeter's dead end of the supposed routinisation of innovation. Penrose's perspective emphasises how entrepreneurship needs resources to explore new profit-seeking opportunities as opposed to maximising the efficiency of current operations, and these resources are generally provided from within the firm or (especially in the case of smaller firms) by drawing on the resources of others through inter-company agreements. It follows from this that the direction of entrepreneurship depends on the existing composition of resources of the firm and its established learning programme and problem-solving agenda, rather than being some random act of inspiration in a vacuum. Penrose's rationale for corporate diversification (and by extension, for technology-based inter-firm cooperation) is that innovation in one area at the same time creates entrepreneurial opportunities in related fields, and the scope to find new combinations. In terms of modern trends it seems that the dual face of R&D is increasingly the outward-looking absorption of new opportunities into an existing programme, as well as internal inward-looking creativity.

Overall, the perspective of Penrose and Schumpeter on innovation, profits and growth in the large firm has not only stood the test of time, but provides a crucial theoretical backdrop to the analysis of the latest form of innovative and international capitalism. While Schumpeter laid the foundations he left much still to be done, and Penrose got much further in extending the evolutionary approach to the reality of large firms and the modern institutional context. Her work can be seen as a decisive link between Schumpeter and the current evolutionary analysis of innovation. No wonder that many of her insights are only now being rediscovered today.

REFERENCES

ALCHIAN, A. A. (1950). Uncertainty, evolution and economic theory, *Journal of Political Economy*, Vol. 58, 211–21.

ANDERSEN, H. B. and CANTWELL, J. A. (1999). 'How Firms Differ in their Types of Technological Competence, and Why it Matters', University of Manchester ESRC Centre for Research on Innovation and Competition Discussion Papers, No. 25.

AUDRETSCH, D. B. (1995). *Innovation and Industry Evolution*, Cambridge, MA, MIT Press.

BRESCHI, S., MALERBA, F. and ORSENIGO, L. (2000). Technological regimes and Schumpeterian patterns of innovation, *Economic Journal*, Vol. 110, 388–410.

CANTWELL, J. A. (1989). *Technological Innovation and Multinational Corporations*, Oxford, Basil Blackwell.

CANTWELL, J. A. (1993). Corporate technological specialization in international industries, in *Industrial Concentration and Economic Inequality: Essays in Honour of Peter Hart*, edited by M. C. CASSON and J. CREEDY, Aldershot, Edward Elgar.

CANTWELL, J. A. (1995). The globalisation of technology: what remains of the product cycle model?, *Cambridge Journal of Economics*, Vol. 19, 155–74.

CANTWELL, J. A. (1999). Innovation as the principal source of growth in the global economy, in *Innovation Policy in a Global Economy*, edited by D. ARCHIBUGI, J. HOWELLS and J. MICHIE, Cambridge, Cambridge University Press.

CANTWELL, J. A. (2000). Technological lock-in of large firms since the interwar period, *European Review of Economic History*, Vol. 4, 147–74.

CANTWELL, J. A. and ANDERSEN, H. B. (1996). A statistical analysis of corporate technological leadership historically, *Economics of Innovation and New Technology*, Vol. 4, 211–34.

CANTWELL, J. A. and BACHMANN, A. (1998). Changing patterns in technological leadership—evidence from the pharmaceutical industry, *International Journal of Technology Management*, Vol. 21, 45–77.

CANTWELL, J. A. and BARRERA, M. P. (1998). The localisation of corporate technological trajectories in the interwar cartels: cooperative learning versus an exchange of knowledge, *Economics of Innovation and New Technology*, Vol. 6, 257–90.

CANTWELL, J. A. and COLOMBO, M. G. (2000). Technological and output complementarities, and inter-firm cooperation in information technology ventures, *Journal of Management and Governance*, Vol. 4, 117–47.

CANTWELL, J. A. and FAI, F. M. (1999). Firms as the source of innovation and growth: the evolution of technological competence, *Journal of Evolutionary Economics*, Vol. 9, 331–66.

CANTWELL, J. A. and IAMMARINO, S. (1998). MNCs, technological innovation and regional systems in the EU: some evidence in the Italian case, *International Journal of the Economics of Business*, Vol. 5, 383–408.

CANTWELL, J. A. and IAMMARINO, S. (2000). Multinational corporations and the location of technological innovation in the UK regions, *Regional Studies*, Vol. 34, 317–32.

CANTWELL, J. A., IAMMARINO, S. and NOONAN, C. A. (2000). Sticky places in slippery space— the location of innovation by MNCs in the European regions, in *Inward Investment, Technological Change and Growth*, edited by N. PAIN, London, Macmillan.

CANTWELL, J. A. and KOSMOPOULOU, E. (2001). What determines the internationalisation of corporate technology?, in *Critical Perspectives on Internationalisation*, edited by M. FORSGREN, H. HÅKANSON and V. HAVILA, Amsterdam, Elsevier.

CANTWELL, J. A. and PISCITELLO, L. (2000). Accumulating technological competence—its changing impact on corporate diversification and internationalisation, *Industrial and Corporate Change*, Vol. 9, 21–51.

CANTWELL, J. A. and SANTANGELO, G. D. (1999). The frontier of international technology networks: sourcing abroad the most highly tacit capabilities, *Information Economics and Policy*, Vol. 11, 101–23.

CANTWELL, J. A. and SANTANGELO, G. D. (2000). Capitalism, profits and innovation in the new techno-economic paradigm, *Journal of Evolutionary Economics*, Vol. 10, 131–57.

CANTWELL, J. A. and VERTOVA, G. (1999). 'Historical Evolution of Technological Diversification', paper presented at the Annual Conference of the European International Business Academy, Manchester, December.

CHANDLER, A. D. JR (1990). *Scale and Scope: the Dynamics of Industrial Capitalism*, Cambridge, MA, Harvard University Press.

COHEN, W. M. (1995). Empirical studies of innovative activity, in *Handbook of the Economics of Innovation and Technological Change*, edited by P. STONEMAN, Oxford, Basil Blackwell.

COHEN, W. M. and LEVIN, R. C. (1989). Empirical studies of innovation and market structure, in *Handbook of Industrial Organization*, Vol. II, edited by R. SCHMALENSEE and R. D. WILLIG, Amsterdam, North Holland.

COHEN, W.M. and LEVINTHAL, D. A. (1989). Innovation and learning: the two faces of R&D, *Economic Journal*, Vol. 99, 569–96.

COHENDET, P. and STEINMUELLER, W. M. (2000). The codification of knowledge: a conceptual and empirical exploration, *Industrial and Corporate Change*, Vol. 9, 195–209.

COWAN, R., DAVID, P. A. and FORAY, D. (2000). The explicit economics of knowledge codification and tacitness, *Industrial and Corporate Change*, Vol. 9, 211–53.

DOSI, G. (1982). Technological paradigms and technological trajectories: a suggested interpretation of the determinants and directions of technical change, *Research Policy*, Vol. 11, 147–62.

FREEMAN, C. (1987). *Technology Policy and Economic Performance: Lessons from Japan*, London, Frances Pinter.

FREEMAN, C. and PEREZ, C. (1988). Structural crises of adjustment, business cycles and investment behaviour, in *Technical Change and Economic Theory*, edited by G. DOSI, C. FREEMAN, R. R. NELSON, G. SILVERBERG and L. L. G. SOETE, London, Frances Pinter.

GRANSTRAND, O. and SJÖLANDER, S. (1990). Managing innovation in multi-technology corporations, *Research Policy*, Vol. 19, 35–60.

GRANSTRAND, O., PATEL, P. and PAVITT, K. L. R. (1997). Multi-technology corporations: why they have 'distributed' rather than 'distinctive core' competencies, *California Management Review*, Vol. 39, 8–25.

HART, O. D. (1989). An economist's perspective on the theory of the firm, *Columbia Law Review*, Vol. 89, 1757–74.

HODGSON, G. M. (1993). *Economics and Evolution: Bringing Life Back into Economics*, Oxford, Polity Press.

HODGSON, G. M. (1997). The evolutionary and non-Darwinian economics of Joseph Schumpeter, *Journal of Evolutionary Economics*, Vol. 7, 131–45.

KNIGHT, F. H. (1921). *Risk, Uncertainty and Profit*, Boston, Houghton Mifflin.

KODAMA, F. (1992). Technology fusion and the new R&D, *Harvard Business Review*, July–August, 70–8.

LACHMANN, L. M. (1986). *The Market as an Economic Process*, Oxford, Basil Blackwell.

LANGLOIS, R. N. (1987). 'Schumpeter and the Obsolescence of the Entrepreneur', paper presented at the History of Economics Society Annual Meeting, Boston, June, University of Connecticut Department of Economics Working Paper No. 91-1503, November 1991.

LANGLOIS, R. N. (1998). Personal capitalism as charismatic authority: the organizational economics of a Weberian concept, *Industrial and Corporate Change*, Vol. 7, 195–214.

LOASBY, B. J. (1998). The organisation of capabilities, *Journal of Economic Behavior and Organization*, Vol. 35, 139–60.

LOASBY, B. J. (1999). *Knowledge, Institutions and Evolution in Economics: The Graz Schumpeter Lectures*, London, Routledge.

LOASBY, B. J. (2000A). 'Connecting Principles, New Combinations and Routines: Reflections Inspired by Schumpeter and Smith', paper presented at the Biannual Conference of the International J. A. Schumpeter Society, Manchester, June–July.

LOASBY, B. J. (2000B). 'Industrial Dynamics: Economics in Time', paper presented at the UK Network of Industrial Economists Meeting on Industrial Dynamics, Reading, December.

MALERBA, F. and ORSENIGO, L. (1995). Schumpeterian patterns of innovation, *Cambridge Journal of Economics*, Vol. 19, 47–65.

NELSON, R. R. (1992). What is 'commercial' and what is 'public' about technology, and what should be?, in *Technology and the Wealth of Nations*, edited by N. ROSENBERG, R. LANDAU and D. C. MOWERY, Stanford, CA, Stanford University Press.

NELSON, R. R., ed. (1993). *National Innovation Systems: A Comparative Analysis*, Oxford and New York, Oxford University Press.

NELSON, R. R. (1996). *The Sources of Economic Growth*, Cambridge, MA, Harvard University Press.

NELSON, R. R. and WINTER, S. G. (1982). *An Evolutionary Theory of Economic Change*, Cambridge, MA, Harvard University Press.

PAVITT, K. L. R. (1984). Sectoral patterns of technical change: towards a taxonomy and a theory, *Research Policy*, Vol. 13, 343–73.

PAVITT, K. L. R., ROBSON, M. and TOWNSEND, J. (1987). The size distribution of innovating firms in the UK: 1945–1983, *Journal of Industrial Economics*, Vol. 35, 291–316.

PAVITT, K. L. R., ROBSON, M. and TOWNSEND, J. (1989). Technological accumulation, diversification and organisation in UK companies, 1945–1983, *Management Science*, Vol. 35, 81–99.

PENROSE, E. T. (1952). Biological analogies in the theory of the firm, *American Economic Review*, Vol. XLII, No. 5, December, 804–19.

PENROSE, E. T. (1959). *The Theory of the Growth of the Firm*, Oxford, Basil Blackwell (1968).

PHILLIPS, A. (1971). *Technology and Market Structure*, Lexington, MA, Lexington Books.

PITELIS, C. N. (1991). *Market and Non-Market Hierarchies*, Oxford, Basil Blackwell.

POTTS, J. (2000). *The New Evolutionary Microeconomics: Complexity, Competence and Adaptive Behaviour*, Cheltenham, Edward Elgar.

RICHARDSON, G. B. (1990). *Information and Investment*, Oxford and New York, Oxford University Press. [Originally published in 1960.]

ROSENBERG, N. (1976). *Perspectives on Technology*, Cambridge and New York, Cambridge University Press.

ROSENBERG, N. (1982). *Inside the Black Box: Technology and Economics*, Cambridge and New York, Cambridge University Press.

ROSENBERG, N. (1994). *Exploring the Black Box: Technology, Economics and History*, Cambridge and New York, Cambridge University Press.

SANTANGELO, G. D. (1999). 'Multi-technology, Multinational Corporations in a New Socioeconomic Paradigm based on Information and Communications Technology (ICT): the European ICT Industry', PhD Thesis, University of Reading.

SCHUMPETER, J. A. (1934). *The Theory of Economic Development*, Cambridge, MA, Harvard University Press. [Originally published in German in 1911; reprinted by Transaction Publishers, New Brunswick, NJ, 1997.]

SCHUMPETER, J. A. (1939). *Business Cycles: A Theoretical, Historical and Statistical Analysis of the Capitalist Process*, New York and London, McGraw-Hill. [Reprinted by Porcupine Press, Philadelphia, PA, 1989.]

SCHUMPETER, J. A. (1943). *Capitalism, Socialism and Democracy*, London, Allen and Unwin. [Originally published in the USA in 1942; reprinted by Routledge, London, 1994.]

SHACKLE, G. L. S. (1979). *Imagination and the Nature of Choice*, Edinburgh, Edinburgh University Press.

SIMON, H. A. (1982). *Models of Bounded Rationality*, Vol. II of *Behavioural Economics and Business Organization*, Cambridge, MA, MIT Press.

TEECE, D. J. (1992). Strategies for capturing the financial benefits from technological innovation, in *Technology and the Wealth of Nations*, edited by N. ROSENBERG, R. LANDAU and D. C. MOWERY, Stanford, CA, Stanford University Press.

TEECE, D. J., DOSI, G., RUMELT, R. and WINTER, S. G. (1994). Understanding corporate coherence: theory and evidence, *Journal of Economic Behavior and Organization*, Vol. 23, 1–30.

TEECE, D. J., PISANO, G. and SHUEN, A. (1997). Dynamic capabilities and strategic management, *Strategic Management Journal*, Vol. 18, 537–56.

USHER, A. P. (1954). *A History of Mechanical Inventions*, 2nd edn, Cambridge, MA, Harvard University Press.

VERNON, R. (1966). International investment and international trade in the product cycle, *Quarterly Journal of Economics*, Vol. 80, 190–207.

VON TUNZELMANN, G. N. (1995). *Technology and Industrial Progress: The Foundations of Economic Growth*, Aldershot, Edward Elgar.

WILLIAMSON, O. E. (1975). *Markets and Hierarchies: Analysis and Antitrust Implications*, New York, Free Press.

WITT, U. (1998). Imagination and leadership—the neglected dimension of an evolutionary theory of the firm, *Journal of Economic Behavior and Organization*, Vol. 35, 161–78.

YOUNG, A. A. (1928). Increasing returns and economic progress, *Economic Journal*, Vol. 38, 527–42.

14

THE US INDUSTRIAL CORPORATION AND *THE THEORY OF THE GROWTH OF THE FIRM*

WILLIAM LAZONICK*

University of Massachusetts Lowell and The European Institute of Business Administration (INSEAD)

I. INTRODUCTION

Today, more than four decades after its initial publication, Edith Penrose's *The Theory of the Growth of the Firm* fits the definition of a 'classic' work: a book that everyone cites but few have read. Written in an abstract theoretical style that reflects the author's training as an economist, the book is nevertheless rich in 'real-world' insights into a complex social phenomenon. The prime purpose of this chapter is to argue that to make full use of *The Theory of the Growth of the Firm* requires an understanding of both the substance of her theory of the modern industrial corporation and the actual social conditions of the industrial corporation that, when the book was written in the late 1950s, her theory sought to capture.

Penrose wrote the book in the United States in an era when the US economy, with its powerful industrial corporations, was unchallenged in its dominance of the world economy. At the same time, especially in the United States, the dominant theory that guided the thinking of economists on the role of the large firm in the economy was the 'monopoly model'. Derived from the neoclassical theory of the firm, the monopoly model had much to say about the impact of 'imperfect competition' on prices and output in the economy but virtually nothing to say about the *growth* of the firm—that is, how a particular company, by accumulating productive resources, came to be dominant in its industry (see Lazonick, 2000). The empirical reality of the long-lived and dominant corporate enterprise in the United States in the 1950s and the obvious gap in microeconomic theory that it implied combined to set the stage for Penrose's classic book.

* This chapter reflects ongoing research on corporate governance, innovation and economic performance carried out in collaboration with Mary O'Sullivan at INSEAD (see http://www.insead.fr/projects/CGEP) with funding from The European Commission DG-XII (Contract No.: SOE1-CT98-1114; Project No.: 053), Commissariat Général du Plan of the French Government and INSEAD Association for Research (sponsoring a conference on 'The Penrosian Legacy: The Work of Edith Penrose, Her Contributions to INSEAD, and Research on Business Enterprise and Economic Development', INSEAD, Fontainebleau, France, May 10–11, 2001).

I shall argue that *The Theory of the Growth of the Firm* represents a seminal work on 'the theory of innovative enterprise' (see Lazonick, 2000; O'Sullivan, 2000A). As such, Penrose's theory of the growth of the firm can provide the basis for a devastating attack on the relevance of neoclassical price theory to an economy in which such growth on the basis of innovative enterprise is a characteristic feature. For if a firm grows into a position of dominance in an industry through technological and market transformations that give it the ability and incentive to produce more output at lower prices than 'perfectly competitive' firms, then the orthodox performance comparison of a restrictive, price-making monopolist with a more expansive, price-taking perfect competitor no longer holds (see Lazonick, 2000; also Schumpeter, 1950, pp. 90, 109). Yet the purported prescriptive power of neoclassical price theory, which in turn has justified the overwhelming preoccupation of economists with so-called 'market imperfections', assumes that an economy characterised by more rather than less 'perfect competition' yields superior economic performance.

The monopoly model can and has been criticised without invoking a theory of innovative enterprise. In particular, over the past three decades the monopoly model and the structure–performance–conduct approach to industrial organisation to which it gave rise have been the subject of a major critique by proponents of transaction-cost economics, foremost among them Oliver Williamson. Transaction-cost models advance beyond the monopoly model by making behavioural and cognitive assumptions concerning the binding constraints that firm decision-makers face in choosing the particular activities that their organisations should undertake. In transaction-cost theory, there is an absence of perfect competition, not because a theory of innovative enterprise makes it irrelevant, but because the existence of 'opportunism' (the behavioural assumption) and 'bounded rationality' (the cognitive assumption) render it impossible. From this perspective, it is 'human nature as we know it' (Williamson, 1985, p. 80), not monopoly, that accounts for the absence of perfect markets. Nevertheless, in sharp contrast to the innovative-enterprise critique of the monopoly model, the transaction-cost critique admits that perfect competition would yield superior outcomes if 'human nature as we know it' did not make it a Utopian fantasy (see Lazonick, 1991, chs 6–7, 2001A).

Given these behavioural and cognitive conditions, the transaction-cost approach (like the monopoly model that it critiques) ultimately relies on exogenously determined 'sunk costs'—Williamson's 'asset specificity'—to explain the scale and scope of the modern industrial enterprise. In contrast to the transaction-cost model, the theory of innovative enterprise shows how by choosing to engage in innovative investment strategies that rely on the *transformation* of behavioural conditions ('opportunism') and cognitive conditions ('bounded rationality') for their success, an enterprise can potentially generate superior economic outcomes—specifically more output at lower unit costs—that enable it to emerge as dominant in the industries in which it competes. Instead of viewing the firm's assets as exogenously determined, a theory of innovative enterprise analyses them as strategic investments. Unlike constrained optimisation models, be they of the monopoly or transaction-cost varieties, a theory of innovative enterprise seeks to analyse how, in certain times and places, a business enterprise strategically seeks to transform organisational (cognitive and behavioural) and industrial (technological and market) conditions that might otherwise impose constraints on its ability to generate higher quality, lower cost products.

As such, a theory of innovative enterprise does not provide a set of decision rules for choosing the 'optimal' course of action; by definition, the innovation process generates a superior course of action that has yet to be learned—otherwise it would not be innovation—and hence what might be viewed as 'optimal' at the outset of the innovation process will likely be viewed as 'sub-optimal' at the end. Rather than be trapped by adherence to existing 'optimality' conditions, the theory of innovative enterprise provides a set of principles concerning the social organisation of the innovation process that constitutes a basic framework for analysing the relation between business organisation and economic performance in particular historical—i.e., institutional and industrial—contexts (Lazonick, 1991, 2000).

In this chapter I shall evaluate *The Theory of the Growth of the Firm* as a contribution to a theory of innovative enterprise.[1] First I shall identify three essential social conditions—organisational integration, financial commitment and strategic control—that are at the core of a theory of innovative enterprise. Next, I shall summarise Penrose's theory of the growth of the firm as a contribution to the theory of innovative enterprise. Then I shall provide an historical overview of how the social conditions of innovative enterprise have changed in the US corporate economy over the past 40 years. On this basis, I shall evaluate Penrosean theory in historical perspective by focusing on its ability to explain (i) the overextension of the scale and scope of corporate activity, especially through conglomeration; (ii) the emergence and impacts of new international competitors; (iii) corporate responses to a new financial environment oriented toward 'shareholder value'; and (iv) the changes in the dynamics of the growth of the US corporate enterprise in the shift from the 'old economy' to the 'new economy'. Finally, I shall show that, writing later in her career, Edith Penrose herself understood that the continued relevance of her perspective depended on the employment of a research methodology that could integrate the elaboration of the theory of the growth of the firm with the ongoing study of the historical evolution of the modern corporate economy.

II. SOCIAL CONDITIONS OF INNOVATIVE ENTERPRISE

The productive process that generates innovation requires three social conditions: *organisational integration, financial commitment* and *strategic control* (Lazonick, 2000; O'Sullivan, 2000A). These three social conditions of innovative enterprise provide the

---

[1] The perspective that I present in this chapter is broadly consistent with arguments that the foundation of Penrose's theory of the growth of the firm is a dynamic and path-dependent organisational learning process. See Pitelis and Wahl (1998). A theory of innovative enterprise seeks to identify the social conditions related to finance, employment and regulation that not only permit dynamic and path-dependent organisational learning to occur but also transform it into sustained competitive success. See Lazonick (2000).

institutional support for business organisations (i) to integrate human and physical resources into an organisational process that develops and utilises technology; (ii) to commit resources to irreversible investments with uncertain returns; and (iii) to vest control over strategic decision-making within corporations with those who have the incentives and abilities to allocate resources to innovative investments.

Organisational integration is the social condition that creates incentives for participants in the hierarchical and functional division of labour to apply their skills and efforts to engage in interactive learning in pursuit of organisational goals. As a social condition for innovative enterprise, the need for organisational integration derives directly from the collective character of the innovation process. Hence, a theory of innovative enterprise must show how, given the collective character of the transformation of technology and markets in particular industrial activities, social institutions support business organisations in creating incentives for participants in a hierarchical and functional division of labour to engage in interactive learning.

Financial commitment is the social condition that allocates financial resources to sustain the process that develops and utilises productive resources until the sale of the resultant products can generate financial returns. As a social condition for innovative enterprise, the need for financial commitment derives directly from the cumulative character of the innovation process—that is, from the need for learning. For an enterprise that has accumulated capabilities, financial claims can take on an existence that, for a time at least, is independent of the need to reproduce or augment those capabilities. But, for innovation to occur within an enterprise, a basic social condition is financial commitment from some source for a sufficient period of time for developmental investments to generate returns. A theory of innovative enterprise must show how, given the financial requirements of the transformation of technology and markets in particular industrial activities, institutions and organisations combine to provide the requisite financial commitment.

Strategic control is the social condition that enables people within an enterprise who have access to financial commitment and who determine the types of capabilities that must be integrated into the organisation to allocate resources and returns in ways that can transform technologies and markets to generate innovation. As a social condition for innovative enterprise, the need for strategic control derives directly from the uncertain character of the innovation process (O'Sullivan, 2000A). Hence, a theory of innovative enterprise must show how, given the uncertain character of the transformation of technology and markets in particular industrial activities, control over financial commitment and organisational integration rests with those people within the enterprise who, as strategic decision-makers, have the incentives and abilities to use that control to invest in processes that can result in the innovative transformation of technologies and markets.

The social conditions of innovative enterprise—organisational integration, financial commitment and strategic control—vary across institutional environments and industrial sectors as well as over time with the economic development process. Entrepreneurial individuals, innovative enterprises and developmental states all play roles in the economic development process. But a theory of economic development that lacks a theory of innovative enterprise will fail to comprehend how, why and to what effect individual activity is transformed into social activity in a modern economy. Penrose's *The Theory of the Growth of the Firm* provides an important foundation for a theory of innovative enterprise, not least because of when and where it was written and the reality that it captured.[1] By the same token, however, the value of the work can only be realised if we use it as a foundation to explore (rather than ignore) the changes in the social conditions of innovative enterprise that have taken place since the time Penrose wrote.

III. THE PENROSEAN THEORY OF INNOVATIVE ENTERPRISE

In *The Theory of the Growth of the Firm*, Penrose argued that the modern corporate enterprise has to be viewed as an organisation that administers a collection of human and physical resources. By far the most critical resources reside in humans because they render services that can enable the firm to make unique contributions as a generator of higher quality, lower cost products. People contribute these unique labour services to the firm, not merely as individuals, but as members of teams who engage in learning about how to make best use of the firm's productive resources—including their own. At any point in time, this learning endows the firm with experience that gives it productive opportunities that are unavailable to other firms, even in the same industry, that have not accumulated the same experience.

The Penrosean firm is an innovative enterprise. As Michael Best (1990, p. 125) has put it, 'Penrose's theory of the growth of the firm is based upon two assumptions: everything cannot happen at once, and a person cannot do everything alone.'[2] For Penrose (1995, chs 5, 7 and 8), the accumulation of developmental experience enables the firm to overcome the 'managerial limit' that in the theory of the optimising firm causes the onset of increasing costs and constrains the growth of the firm. This developmental experience enables the firm to transfer its existing productive resources to new productive opportunities and even to shape the market for its products to generate new market opportunities. As Penrose (1995, p. xii) states in the Foreword to the 1995 edition of *The Theory of the Growth of the Firm*:

[T]he growing experience of management, its knowledge of the other resources of the firm and the potential for using them in different ways, create incentives for further expansion as the firm searches for ways of using the services of its own resources more profitably. The firm's existing human resources provide both an inducement to expand and a limit to the rate of

---

[1] See the discussion in Kor and Mahoney (2000). Penrose herself stated that her theory of the growth of the firm 'is concerned only with the incorporated industrial firm operated for private profit and unregulated by the state . . . and is applicable only to an economy where the corporation is the dominant form of industrial organisation; historically, therefore, only to the period since the last quarter of the nineteenth century. To be sure, the corporation was widely used in certain areas much earlier, but it did not dominate the field of manufacturing as it has since, at least in the western world' (Penrose, 1995, p. 6, also p. 218).

[2] See Best (1990), ch. 4 more generally. For a complementary analysis of the cumulative and collective character of the innovation process and the implications for economic theory, see O'Sullivan (2000A).

expansion. Even growth by acquisition and merger does not escape the constraints imposed by the necessity of using inputs from existing managerial resources to maintain the coherence of the organization.

Each move into a new product market enables the firm to utilise unused productive services that have been accumulated through the process of organisational learning. These unused productive services can enable a firm to grow not only through diversification of its products but also through the acquisition and absorption of other firms that have developed complementary productive services. But 'the growth of the firm' is not based simply on the *utilisation* of excess resources available to the firm. New product *development* also requires investments in the creation of new productive services that are the basis for the continuing growth of the firm. Key to the determination of which productive opportunities the firm pursues, and, in the case of merger and acquisition, which firm absorbs which, is the possession of what Penrose (1995, pp. 182–9) calls 'entrepreneurial services' on the part of the growing firm.

But, in a growing enterprise, entrepreneurial services will not yield success if the productive resources under the firm's control are not administered in an integrated way. Hence the critical importance to the growth of the firm of 'managerial services' that are devoted to 'administrative integration' (Penrose, 1995, pp. 189–94). Through this process of the growth of the firm, innovative capability becomes embedded in its very operations (Penrose, 1995, pp. 112–16).[1] What enables the firm to grow over time is this continual ability of the enterprise to utilise its unique pool of resources to generate new products, developed in the past, while building on this existing pool of resources to generate *new* unique capabilities.

'The proposition', says Penrose (1995, p. 88), 'that enterprising firms have a continuous incentive to expand and that there is no limit to their absolute size (other than that imposed by our conception of the nature of an industrial firm) stands in sharp contrast to the notion of an "optimum" size of firm'. How then can one characterise the economies that result when the 'enterprising firm' is able to utilise the productive resources that it has developed? Penrose (1995, p. 89) makes a distinction between 'economies of size' and 'economies of growth', arguing that 'growth is a process; size is a state'. 'Economies of size', she argues,

are present when a larger firm, because of its size alone, can not only produce and sell goods and services more efficiently than smaller firms but also can introduce larger quantities of new products more efficiently. In discussion of the economies of size, so-called 'technological economies', derived from producing large amounts of given products in large plants, are commonly distinguished from 'managerial' and 'financial' economies, derived from improved managerial division of labour and from reductions in unit costs made possible when purchases, sales, and financial transactions can be made on a large scale.

[1] In the identification of the routinisation of industrial research, Penrose's views are close to those of Joseph Schumpeter in his later work. Although Penrose (1995, n. 112) does not cite Schumpeter directly on this point, she does refer to the well-known article by Carolyn Solo (1951), influenced by the Schumpeterian perspective, as 'a perceptive discussion of research and innovation as an "ordinary" business activity'.

After a discussion of these different types of 'economies of size' and their interaction, Penrose makes it clear that, once one recognises that firms can differ in their managerial capabilities, it is futile to attempt to specify the 'optimum' size of the firm, without reference to these particular capabilities. 'Only for firms incapable of adapting their managerial structure to the requirements of larger operations', Penrose (1995, p. 98) argues, 'can one postulate *an* optimum size.' In other words, a unique 'optimum' only applies to those firms that, because of inferior managerial capabilities, must take as constraints the technological and market conditions that other enterprises, with superior managerial capabilities, are able to transform.

For such an enterprise with superior managerial capabilities, and hence the potential to continue to grow, 'the economies of growth' that it can reap depend on a unique process that evolves over time. Hence the outcome of this process—an eventual 'optimal' state that derives from the process of growth—cannot be known before the productive resources that can generate this growth have been developed and utilised. As Penrose (1995, p. 99) puts it:

Economies of growth are internal economies available to an individual firm which make expansion profitable in particular directions. They are derived from the unique collection of productive services available to it, and create for that firm a differential advantage over other firms in putting on the market new products or increased quantities of old products. At any time, the availability of such economies is the result of the process [discussed earlier in the book] by which unused productive services are continually created within the firm.

Penrose (1995, pp. 99, 100) adds that economies of growth 'may or may not be also economies of size' because 'under given circumstances, a particular firm may be able to put additional output on the market at a lower average cost than any other firm, whether larger or smaller'. 'For one of the significant characteristics of the economies of growth', she goes on to explain, 'is that they depend on a particular collection of productive resources possessed by the particular firm, and the exploitation of the opportunities provided by these resources may be quite unrelated to the size of the firm.'

The analytical strength of Penrose's theory of the growth of the firm derives from her explicit recognition of the theoretical difference between the innovative enterprise and the optimising firm. The 'growth of the firm' relies on the *process* of innovation, whereas the size of the firm is an *outcome* (which orthodox economic theorists often, tautologically, call 'optimal') that has been achieved through this process of growth. Underlying this distinction between the growth of the firm and the size of the firm is Penrose's understanding that a firm is a unique social entity that can engage in learning that is both collective and cumulative (see O'Sullivan, 2000B, pp. 15–17, 63–5). She also emphasised the dynamic relationship between the development of productive resources and their utilisation, and hence the potential conflict between the achievement of high quality and low cost. She understood, that is, that innovative strategies can place the firm at a competitive disadvantage if the productive resources that it develops are not sufficiently utilised.

In the last two chapters of *The Theory of the Growth of the Firm*, Penrose drew out the implications of her perspective for understanding the relationship between the growth of the firm and the growth of the economy. Elaborating on her earlier proposition that 'enterprising firms have a continuous incentive to expand' (Penrose, 1995, p. 88), she approved of arguments that lauded 'the technological achievements of "big business"', viewing market power as 'necessary to induce them to engage in extensive and expensive research, larger-scale capital investment, and long-range programmes of industrial development' (Penrose, 1995, p. 232).[1] As she continued:

'Big Business' competition, through different from the classical competition among small firms, is held to be even more beneficial to the economy and more effective in meeting consumers' wants. The competition between the large firms involves large amounts of investment and undoubtedly results in an increasing quantity and variety of goods and services which become cheaper and cheaper under the pressure of competition.

Those who see in this kind of competition the chief source of the enormous progress in real income in the more advanced capitalist countries urge, and quite correctly, that the large firms could not afford the competitive innovating race if small firms were immediately able to take the fruits of their research, copy their innovation and put products on the market at prices which did not reflect the costs of developing them. It follows that the control of output, the control of markets and the control of price must remain in the hands of those who bear between them the 'development cost' required for constantly increasing output and continually improving products (Penrose, 1995, p. 233).

Penrose (1995, pp. 233, 233–4n) concluded that '[these] results so widely deemed beneficial flow from competition, from competition among the few to be sure, but competition so effective and intense that no large firm can afford to act the role of a contented monopolist greedily exploiting the economy'.

Penrose's endorsement of the benefits of 'big business' competition were embedded in an analysis that argued that the dominant industrial positions of firms that had already grown large lay in their superior access to finance (in large part through internal funds), research (as a critical element of the learning process) and consumer markets (mainly through the maintenance of brand-name reputation) rather than in the 'artificial' barriers to entry postulated by proponents of the monopoly model (Penrose, 1995, pp. 218, 236). She contended, moreover, that a growing economy presented opportunities for the emergence of new innovative firms that, although small in size at the outset, could potentially grow large over time according to the principles of 'the theory of the growth of the firm'.[2] According to Penrose (1995, pp. 222, 253, 260), in a changing and growing economy the extent of new profitable opportunities that open up outstrip the ability of existing large firms, limited by their managerial

---

[1] In articulating what was to become known in the field of industrial organisation as 'the Schumpeterian hypothesis', Penrose (1995, pp. 232–3n) cited and quoted John Kenneth Galbraith and Almarin Phillips, but made no reference to Schumpeter himself.

[2] In referring to differences in firm size as differences in stages in the growth of the innovative enterprise, rather than as permanent distinctions among different types of firms, Penrose uses the terms 'large firms' and 'small firms'. In my view, the evolutionary character of her argument would have been clearer if she had rather used the terms 'going concerns' and 'new ventures' to capture the fact that her 'small firms' will not necessarily remain perpetually small, but are small only because they have as yet to grow large through successful innovation.

resources, to take advantage of all of them, thus creating opportunities for small inno-vative firms to emerge in what she called the 'interstices' of the economy.

Indeed, to drive home the argument, Penrose (1995, pp. 225ff) invoked the logic of the 'principle of comparative advantage', arguing that even if large firms were to have absolute advantage in all innovative activities, a rapidly expanding economy would compel them to focus on those lines of business in which their comparative advantage was greatest. The door to engage in those innovative activities in which the absolute advantage of the incumbent firms was least would thus be left open to new ventures, some of which, through innovation, might succeed over time in becoming going con-cerns that would themselves then have the absolute advantages of incumbents. As Penrose (1995, p. 253) put it:

> The interstices in the economy which provide the opportunities for the growth of smaller firms appear as opportunities to expand the production of specific products or to enter specific industries. Since under our assumption these are, by definition, the opportunities the large firms ignore, their significance for smaller firms depends upon the type of difficulty which must be overcome in taking advantage of them—upon barriers to entry. The competitive advantages of the older and larger firms in an industry may or may not be of a kind which make it difficult for newcomers to enter the interstices, but the most common type of advantage seems, as we have seen, to be associated either with the amount of capital required or with the possibilities of attaching consumers' loyalty through branding and advertising. Although there may not be sig-nificant technological advantages associated with large multiplant operations, the mere fact that individual plants are fairly large militates against the easy entry of new firms and facilitates the rapid expansion of existing large firms. And branding and advertising give competitive advan-tages very largely because they impede easy entry of newcomers. Thus we should expect con-centration to develop most rapidly in industries where this type of advantage is greatest or where the size and growth of the market are such that large firms can expand sufficiently rapidly to leave no interstices.

IV. THE EVOLUTION OF THE US CORPORATE ECONOMY IN THE LATE TWENTIETH CENTURY

IV.a. The hey-day of the 'managerial revolution'

When *The Theory of the Growth of the Firm* was published, the US industrial corpora-tion, and the US economy with it, dominated the global economy. In 1959, 44 of the world's 50 largest corporations in terms of revenues were US-based, with the remain-ing six headquartered in Europe (Kaysen, 1996, p. 25), and the United States was still far ahead of fast-growing nations such as Germany and Japan in terms of per capita income (Maddison, 1994, p. 22). In 1959, corporations represented 9·6% of all US business enterprises, but 79·3% of all business revenues. US corporations with assets of $100 million or more accounted for one-tenth of 1% of all corporations, but 55·4% of all corporate assets, 54·5% of before tax corporate profits, and 67·9% of all cor-porate dividends (US Bureau of the Census, 1976, Series V, 182–96).

During the generally prosperous quarter century after World War II, the key social characteristics of the US industrial corporation were:

(i) organisational integration of salaried administrative, technical and professional—that is, managerial—personnel to ensure the development and utilisation of technologies and markets;

(ii) financial commitment through the reinvestment of earnings to build organisations that could develop and utilise mass-production technologies in a growing diversity of products and for a growing number of national markets; and

(iii) strategic control by salaried managers (as distinct from owner-entrepreneurs or public shareholders) over the allocation of corporate resources, with an increasing tendency within multidivisional organisational structures to locate strategic control at corporate headquarters rather than in product divisions.

IV.b. Organisational integration

The outstanding social characteristic of the successful US industrial corporation in the post-World War II decades was its managerial organisation, made up of professional, technical and administrative personnel, which could develop new products and processes and which could seek to ensure that what the company mass produced could be sold on mass markets. In-house corporate investments in both management development and R&D were linked closely with the US system of higher education. A distinctive feature of the US system of business organisation was the extent to which production workers tended to be excluded from organisational learning processes, with the integration of personnel to engage in learning being generally confined to the managerial organisation. Indeed, investments in managerial learning often had as a prime goal the development of processes that would obviate the need to rely on the skills and efforts of production workers (Lazonick, 1990, chs 7–10). The co-operation of production workers, however, remained important to ensure the high throughput *utilisation* (as distinct from development) of mass-production facilities.

During the post-war decades, these corporations not only relied primarily on retained earnings to reinvest in productive resources (see 'financial commitment' below), but also allocated corporate returns to provide both managerial employees and unionised workers with employment security and rising real incomes. For both managers and workers, the major corporations provided both substantial employment security (in effect 'lifetime employment' for senior production workers in the major industrial companies) as well as pay increases that were connected with movements up internal job ladders. Yet the distinction between 'salaried' and 'hourly' employees captured the sharp social segmentation between 'managers' and 'workers' within US corporate enterprises, a distinction that was based on the assumption that production workers were interchangeable commodities in the allocation of labour to work on 'hourly-rated' tasks. 'Hourly' employees were typically members of unions that bargained the scale of hourly rates on the internal job ladders, and for these employees the climb up these ladders was generally based on seniority rather than achievement. Salaried employees by contrast were members of management (and hence did not according to US labour law have the right to union representation) and their movements up and around the organisation were based much more on their capabilities, ambitions and achievements than on seniority.[1]

Top managers tended to be integrated into the managerial organisations of their companies by the practices of promotion from within the company and the structure of managerial salaries. Into the 1970s the basic salaries of top managers were linked to the scale of salaries within the managerial structure, in contrast to what was to come in the 1980s and 1990s when the remuneration of top corporate managers exploded relative to the rest of the employees in the organisations over which they exercised control. Through the 1970s remuneration in the form of stock options went to only a small group of top corporate managers. The granting of stock options was a practice that began with changes in the tax law in 1950, and that became an increasingly important form of top management remuneration in the 1950s and 1960s. During this period, income from the exercise of stock options came to be seen, along with salaries, as part of the basic remuneration of top managers (whereas bonuses were used as compensation for superior performance). When the stock market turned down sharply at the beginning of the 1970s, basic salaries were increased to make up for the loss of stock option income, thus contributing to a process that was increasingly to segment the interests of top managers of major US corporations from the rest of the managerial organisation (Lewellen, 1968, 1971; see also Lazonick and O'Sullivan, 2000A).

IV.c. Financial commitment

The prime source of financial commitment in the US industrial corporation in the quarter century after World War II was retentions (undistributed earnings and capital consumption allowances). By one estimate (based on samples of 50 large companies), over the period 1948–71 retentions in the US corporate manufacturing sector accounted for 77% of the sources of funds, while net debt issues accounted for 14%, gross stock issues 13% and net stock issues 8%. Indeed, during this 24 year period, it was in 1954–9 (when Edith Penrose was researching and writing *The Theory of the Growth of the Firm*) that corporate retentions, at 81%, were at their peak as a source of funds (Ciccolo and Baum, 1985, p. 86).

Another major form of financial commitment during the post-war decades was the allocation of corporate returns to provide employment security and higher real incomes to the growing numbers of corporate personnel who came to see themselves as 'lifetime' employees. In general, large US industrial corporations expanded employment during the 1950s and 1960s. There was some contraction of employment among the large corporations in the 1970s, but the major contractions that dramatically changed the expectations of employees concerning the possibilities for career-long attachment to a single corporation took place in the 1980s and early 1990s. For example, the 50 largest US industrial corporations by sales employed 4·1 million people in 1959 (an employment increase of almost 10% from five years earlier), a figure that increased to

[1] For references, and comparisons with corporate organisation in other advanced economies, see Lazonick and O'Sullivan (1997).

6·4 million people in 1969 and was at 6·2 million in 1979, but which declined to
5·8 million in 1990 and 5·1 million in 1995. The total sales of these 50 companies
accounted for 19·9% of US GNP in 1959, 29·8% in 1979 and 22·0% in 1995
(*Fortune*, various years).

IV.d. Strategic control

The primary role of the highly liquid stock market that had developed in the United
States during the first three decades of the century and which, with increased regu-
latory oversight, survived the Great Depression was to enable the separation of asset
ownership from managerial control by transferring share ownership to the investing
public from owner-entrepreneurs. In thus transforming their corporate shareholdings
into liquid assets, these owner-entrepreneurs typically gave up control of their companies to the salaried managers whom they had employed in building their enterprises
into dominant going concerns. The prime interest of public shareholders was in the
liquidity of their stock holdings; they would not have held corporate stock if shareholding had required that they exert time and effort to generate returns, or, for that
matter, if the law did not provide them with limited liability. This transfer of ownership from owner-entrepreneurs to shareholders whose prime interest was in the liquidity of their shareholdings resulted in insider control by salaried managers of existing
corporations (going concerns) and enabled the separation of ownership from control
when younger companies (new ventures) went public. Rather than function as a means
of raising capital for corporate expansion, therefore, the stock market served to vest
control over the allocation of corporate revenues in the hands of salaried managers.

   Public shareholding, moreover, was widely diffused across households. Into the
1970s institutional investors were still relatively unimportant holders of publicly
issued corporate stock. The proportion of total corporate stock outstanding held by
pension and mutual funds combined was 8·7% in 1960 and 14·4% in 1970, rising to
20·8% in 1980 and 31·3% in 1990. Meanwhile, the proportion held by households
was 85·6% in 1960 and 68·0% in 1970, declining to 59·6% in 1980 and 51·2% in
1990. The widespread distribution of stock ownership among US households was
abetted by a set of institutions that promoted confidence in, and hence helped to
ensure the liquidity of, the stock market. The listing and disclosure requirements of
the New York Stock Exchange as well as Securities and Exchange Commission regu-
latory oversight of trading practices, including the ban on insider trading, gave the
small shareholder confidence in the fairness of the market. At the same time, the rights
of shareholders to intervene in the affairs of the corporation were minimal. Shareholders accepted this lack of control because the strength of the corporations whose
stock they held meant that they could count on a steady stream of dividends and could
expect capital gains over the long term. Meanwhile the widespread distribution of
shareholding, itself a result of the confidence that came from the regulation of the
stock market and the stability of the underlying corporations, enhanced the liquidity
of publicly traded shares (Lazonick, 1992; O'Sullivan, 2000B, Ch. 3).

In the 1940s and 1950s multidivisional and multinational corporations grew by investing in new lines of business that bore some technological relation to existing lines of business in which the corporation had become dominant. Although the multi-divisional structure helped to decentralise responsibility for corporate performance and relied on divisional managers for information in the process of strategy formulation, this organisational structure nevertheless tended to centralise authority over corporate resource allocation in corporate headquarters (Chandler, 1962). The conglomerate movement of the 1960s, in which unprecedented merger and acquisition activity brought large numbers (in some cases hundreds) of different, and often unrelated, lines of business under one corporate roof, greatly increased this centralisation of control (O'Sullivan, 2000B, ch. 5).

IV.e. 1970s: the US corporate economy under pressure[1]

(i) The problem of centralised control
Despite the decentralisation rhetoric of the 'multidivisional structure' the continuous growth of the US industrial corporation in the post-war decades was often, and probably typically, accompanied by a centralisation of strategic control. In the 1960s and 1970s, top executives who controlled the allocation of corporate resources and returns often had little knowledge of, or the ability to evaluate, the investment require-ments of the organisations whose allocation decisions they controlled. Centralised control thus tended to create a segmentation of strategic decision-making from the learning organisation, which then made it difficult for those who exercised strategic control to make, or approve, decisions to allocate resources to innovative investments that could enable the company to generate higher quality, lower cost products. Such 'strategic segmentation' was especially true of conglomerates that had engaged in unrelated diversification, but it also prevailed in many companies that had expanded through a combination of internal growth and acquisition to engage in (what appeared to be) related diversification.

(ii) The problem of new international competition
The old modes of organisational learning that had enabled US industrial corporations to become the most powerful mass producers in the world did not result in com-petitive outcomes when confronted by more innovative modes of organisational learning—that is, learning that could generate higher quality, lower cost products even while raising the wages of the employees involved. The strongest innovative challenge to US dominance in mass production came from the Japanese in the very industries in which US mass-production corporations had seemed unassailable as world leaders. At first in the 1970s, Americans attributed the Japanese challenge in industries such as automobiles, consumer electronics, machine tools, semiconductors and steel to

---

[1] The following material on the transformation of the US corporate economy in the 1970s, 1980s, and 1990s draws on Lazonick (1992), Lazonick and O'Sullivan (2000A), O'Sullivan (2000B), chs 4–6 and the references to the literature therein.

low-wage competition. But as Japanese wage levels rose during the 1970s and into the 1980s, it became apparent that the Japanese were able to develop and utilise productive resources more effectively than their American competitors. The Japanese beat out the Americans by investing in broader and deeper skill bases; specifically, skill bases that integrated the capabilities of the managerial organisation with teams of production workers on the shop floor (Lazonick, 1998).

(iii) The problem of eroding financial returns to portfolio investors
The inflation of the 1970s had an adverse impact on the returns of US households' financial portfolios, even as, in the corporate economy, many unionised workers retained the power to offset price increases by wage increases. The post-war boom, to which the growth, stability and benefits of corporate employment was central, had increased the number and proportion of US households with substantial financial assets. During the 1970s, as many of these households made the transition from employment to retirement, they saw their savings eroded by double-digit inflation. During the first half of the decade, government policy sought to restrain wage increases, most notably through wage–price controls. But in the last half of the 1970s, the policy focus was on the deregulation of financial markets to permit institutions and individuals to seek higher returns on their financial portfolios. It was within this context, that the institutional investor, abetted by permissive regulation such as the Employee Retirement Income Security Act (ERISA), became much more important in managing household savings.[1]

IV.f. 1980s: the corporate governance system in transition

(i) Contests for corporate control
During the early 1980s, there was a widespread attempt to undo the damage wrought by the centralisation of control in the conglomerate movement by separating constituent business units from conglomerate control by means of management buyouts (MBOs). Managers of conglomerate business units used the issue of debt, typically in the form of high risk, high yield 'junk' bonds, to finance the change in corporate control. Hence MBOs were also called leveraged buyouts, and the cost of the transformation of corporate control was that the new independent enterprises found themselves deeply in debt. The passage of the amendment to ERISA in 1978 had made it possible for pension funds to become holders of junk bonds, thus creating an important market for this financial instrument and hence facilitating the issue of junk bonds for MBOs.

During the first half of the 1980s the junk bond market grew rapidly, aided by the deregulation of the Savings & Loans companies (S&Ls) under the Garn–St Germain Act of 1982, which enabled these saving institutions, along with the pension funds, to absorb vast quantities of these high risk corporate securities. The willingness of

---

[1] Average annual real yields on NYSE common stock were 11·7% in the 1980s compared with –1·7% in the 1970s (Lazonick and O'Sullivan, 2000A, p. 27).

mutual funds, pension funds and S&Ls to hold junk bonds in their portfolios made it possible for corporate raiders to challenge for control of major corporations. Such contests for corporate control were highly organised efforts, with Michael Milken, the financier who during the 1970s had created the junk bond market, orchestrating the process. Milken would line up commitments from the financial institutions in his junk bond network to absorb the junk bonds issued by a target company when it was taken over. The cash thus raised could enable the corporate raider (typically hand-picked by Milken) to liquidate the controlling stock interest that had enabled him to take control of the company (Bruck, 1988). These corporate raids focused on 'stable-technology' industries with substantial natural resources (oil, timber) or mass market brand names (food, tobacco) (Hall, 1994). These were industries in which corporations could generate substantial revenues by depleting resources, selling off business units and downsizing the labour force.

In addition to the generation of such 'free cash flow' through a transformation of corporate control, throughout the 1980s and into the 1990s many large corporations, led by the example of Jack Welch at GE, maintained centralised control but drastically reduced the range of business activities in which they engaged. Such rationalisation helped reduce the overgrowth of the US industrial corporation that had developed in the post-World War II decades, and could potentially forge a greater integration between strategy and learning. But, as in the case of the hostile takeovers, this shift in corporate strategy was justified on the grounds that it could 'create value for shareholders'—an ideology that in the 1980s became a mantra of US corporate governance. CEOs of major corporations who ignored the shareholder value movement risked being ousted from their positions of corporate control, while those who embraced the movement had only unprecedented riches to gain. The shareholder value movement thus fostered a shift of old-line corporations from a regime of 'retain and reinvest' to one of 'downsize and distribute'. Managers who controlled the allocation of corporate resources and returns became focused on strategies to downsize the labour force and distribute corporate revenues to shareholders.

(ii) The persistence of 'retain and reinvest'

The market for corporate control did not directly affect old-line high technology companies in industries such as pharmaceuticals, automobiles or electronic equipment, where the integration of strategy and learning was of critical importance to product competition even in the shorter run. In these industries there was either a tendency to allocate resources to activities that could be done by 'narrow and concentrated' skill bases, or, in the case of automobiles, follow the Japanese example by broadening and deepening the skill base in which they invested. In automobiles, such skill base investments often entailed long-term relations with suppliers who undertook more of the development process in a shift toward 'modularity'. Alternatively, the decentralisation required for the integration of strategy and learning might be achieved by setting up a new independent division within the corporate structure, as was intended in the case of GM's Saturn.

Meanwhile, companies such as Intel, Microsoft and Sun Microsystems that were central to the rise of the 'new economy' emerged through their leadership in components and segments of the microcomputer industry. These original 'new economy' companies were highly specialised in complex technologies, the development and utilisation of which required a close integration of strategy and learning. Through a 'retain and reinvest' strategy that entailed investments in narrow and concentrated, but tightly integrated, skill bases of highly educated personnel, the companies were able to dominate their markets.

(iii) The redistribution of corporate returns
The market for corporate control created pressure on corporate executives of 'old economy' companies to favour resource allocation strategies that would make the corporation less vulnerable to hostile takeover. Employee lay-offs, even by highly profitable companies, was one response that tended to raise the market value of a company's shares and thus make it more expensive to acquire. Higher levels of cash distributions in the forms of both dividends and stock repurchases tended to raise a company's market value while reducing the amount of liquid assets to which a raider could lay claim after a takeover. Top executives who pursued such 'downsize and distribute' strategies reaped rewards in the forms of continued employment in positions of financial control and lucrative stock option packages, created ever-growing income inequality between those at the top and the rest of the organisation. The booming stock market and the explosion in CEO pay made shareholders and top managers of 'old economy' corporations the main beneficiaries of the new 'shareholder value' regime.[1]

Meanwhile, as had once been the case in the 'old economy', 'new economy' companies grew on the basis of retained earnings, paying little or no dividends to their shareholders. Many of these companies listed their stock on NASDAQ, with its less stringent listing requirements than the New York Stock Exchange. Compared with 'old economy' companies, the 'new economy' companies made more widespread and systematic use of stock options to recruit and retain employees. For example, in 1986, with just over 1000 employees and some $200 million in revenues, Microsoft's IPO raised $42 million for the corporate treasury and $16.7 million for owners such as Bill Gates and Paul Allen. But cash for enterprise expansion was not the prime purpose of the IPO. The microcomputer revolution of the early 1980s with its open architectures and opportunities for startups, had led firms to use stock options to recruit and retain highly mobile engineers and scientists. Microsoft did the IPO to make its stock options exercisable, and thus to keep its software engineers working for the company.[2] The exercise of stock options resulted in the dilution of the ownership stakes of existing shareholders. Yet the success of 'new economy' companies such as Intel and Microsoft in dominating their product markets meant that the stock market rewarded

---

[1] For an analysis that emphasises the impact of the inflationary 1970s on changing modes of corporate organisation and performance, see Dore, Lazonick and O'Sullivan (1999).

[2] Information from Microsoft's website: http://www.microsoft.com.

them with rising stock prices despite the fact that these companies paid little or no dividends and, initially at least, did not offset dilution through stock repurchases.

IV.g. 1990s: 'shareholder value' in the 'new economy'

A combination of the 'new economy' focus on investments in narrow and concentrated skill bases plus the 'old economy' focus on 'downsize and distribute' contributed to a persistent worsening of the US distribution of income, including declining real wages for the bottom half of the distribution, from the late 1970s to the late 1990s (Levy and Murnane, 1992; Moss, 2001). Having long since ceded unilateral control over corporate decision-making to management, the US union movement was not in a position to challenge the downsizing of the 'old economy' or to influence the breadth or depth of the skill base investments in the 'new economy'. With more affluent households shifting more and more of their retirement incomes into the stock market, there was little political opposition in the United States to the dominance of 'maximizing shareholder value' as an ideology of 'best-practice' corporate governance. During the 1990s corporate boards of directors insisted that top managers adhere to 'shareholder value' principles, and in some cases would replace corporate executives when a company's stock price performance was deemed inadequate.

In 'old economy' companies, the 1990s witnessed the continued explosion of CEO pay through the growing use of stock options, and the increasing separation of the rewards to CEOs from remuneration in the rest of the organisation. Between 1980 and 1995, CEO pay increased by 499%, factory wages by 70% and inflation by 85% (Lazonick and O'Sullivan, 2000A, p. 25). The downsizing of middle management as well as the blue-collar labour force marked the early part of the decade. Meanwhile stock repurchases were used on a larger and more systematic scale as a means, alongside dividends, of distributing returns to shareholders, with many corporate boards approving stock repurchase programmes that gave managers the authority to intervene strategically to support the stock prices of their companies. In most years in the 1990s, the extent of stock repurchases meant that, in aggregate, stock issues as a source of corporate funds were negative in the United States (O'Sullivan, 2000B, ch. 6).

Meanwhile, the 'new economy' practice of extending the use of stock options as a means of remuneration down into the corporate organisation meant that these companies, once they had grown large, had to worry about the dilution of the holdings of existing shareholders as and when large numbers of employees exercised their options. In order to diminish the dilution effect, by the last half of the 1990s, companies such as Intel and Microsoft were often spending billions of dollars—in many years even more than they spent on their substantial R&D efforts—to repurchase stock.[1] Precisely because these companies depend so heavily on stock options as a means of recruiting, motivating and retaining employees, they are under an imperative to keep

---

[1] Information from the Annual Reports of Intel and Microsoft, available on their websites (http://www.intel.com and http://www.microsoft.com). See also McGough *et al.* (2000), C1.

their stock market valuations not only high but also growing both absolutely and in relation to their labour competitors.

High stock prices also enable these companies to acquire other companies. In 'new economy' industries that are undergoing rapid technological and market change, the use of a company's stock price as an acquisition currency is now seen as a potent competitive weapon in the accumulation of innovative capabilities. The accumulation of capabilities by this means requires that companies maintain high and increasing stock prices in order to acquire companies for stock and to recruit and retain employees with stock options. Once some leading companies have embarked upon a path that relies on stock to accumulate capabilities, all companies wishing to compete for these capabilities must do so as well. In these types of industries, stock market valuation has become integral to technological competition, not because it enables a company to raise more cash to invest in innovative capabilities, but because it enables the company to use its stock as a currency to acquire other companies and gain access to key human resources.[1]

The Internet revolution of the last half of the 1990s shifted the balance of business activity in the United States from the 'old economy' model, with its emphasis on 'downsize and distribute', to the 'new economy' model, with its emphasis on 'retain and reinvest', but with the stock market now playing a central role in the accumulation of innovative capabilities through the use of stock as a currency for acquisitions and recruitment. The sustainability of the 'new economy' in the United States is now (in 2000–1) being tested. In particular, the use of the stock market as a mode of acquiring unproven start-up companies and of remunerating large numbers of employees may well impart financial instability and income inequality into the economic system in ways that the advanced economies, including the United States, have not thus far experienced. As already mentioned, from the late 1990s even older 'new economy' companies such as Intel and Microsoft have had to engage in massive stock repurchases to boost stock prices by offsetting the dilution of shareholding caused by the exercise of stock options and the use of stock as an acquisition currency. In the corporate allocation of resources and returns, such companies have to try to 'manage the stock market', a task that became increasingly difficult as, in late 2000 and early 2001, the prolonged speculative boom in 'new economy' shares faltered.

V. PENROSEAN THEORY IN HISTORICAL PERSPECTIVE

The Theory of the Growth of the Firm presents a theory of the central role of the industrial corporation in the allocation of resources and the accumulation of capabilities in a dynamic, changing economy. Penrose derived the theory from the evolution of a particular type of firm in a particular type of economy: the large industrial corporation that came into being in the late nineteenth century and which dominated the industrial organisation of the advanced capitalist economies at the time she wrote her book. Her thinking was, as we have seen, shaped primarily by the case of the US industrial

---

[1] For a case study of a 'new economy' industry in which some of the major players are 'old economy' corporations, see Carpenter and Lazonick (2001).

corporation, a business organisation in which managerial employees both controlled the allocation of resources and embodied the accumulation of capabilities.

As outlined in the last section, the organisation and performance of the US industrial corporation as well as the institutional environment in which it has operated have undergone substantial changes in the four decades since Penrose wrote her classic book. Specifically, the US industrial corporation has experienced (i) internal limits to its growth, (ii) challenges from new international competitors, (iii) pressures for higher returns on outstanding corporate securities, and (iv) the increasing role of the stock market in the accumulation of productive capabilities. In this section, I shall consider the implications of each of these changes for the continuing relevance of the Penrosean theory of the growth of the firm.

V.a. Penrosean theory and internal limits to corporate growth

Edith Penrose's theory of the growth of the firm captured the central characteristics of the US industrial corporation as it existed in the post-war decades. She took it as given that salaried managers embedded in the organisation exercised strategic control over corporate allocation decisions and that financial commitment was secured by the use of internal funds. Given these social conditions, she focused her analysis on the organisational processes that generated cumulative and collective learning both within and across lines of business that bore technological and/or market relations to one another. In effect, as the basis for understanding the growth of the corporate enterprise, Penrose fashioned the economic theory of what had in the 1950s come to be called the 'organization man' (Whyte, 1956).

Penrose was not the only important US scholar who focused on the organisation man (a species who, besides being male, was also typically White, Anglo-Saxon and Protestant) as the essential building block of the post-war US industrial corporation. Coming from different perspectives, this persona was also a central character in the work of C. Wright Mills (1951), Herbert Simon (1957), Alfred D. Chandler Jr (1962) and John Kenneth Galbraith (1967).[1] But only Penrose built a theory of corporate growth by identifying *organisational learning* as the essential contribution of the organisation man. She equated the 'firm' with its managerial organisation, and organisational learning with *managerial* learning. As Penrose (1995, p. xii) put it in the Foreword to the 1995 edition of her book: 'I elected to deal with what was called the "managerial firm"—a firm run by a management assumed to be committed to the long-run interest of the firm, the function of shareholders being simply to ensure the supply of equity capital.'[2]

---

[1] See also Penrose's (1995, pp. 233–4 n) references, contained in her footnote on the 'New Competition', to scholars who recognised the importance of the innovative capabilities of the mid-century US industrial corporation.

[2] Indeed, Penrose could have gone further by recognising that, in general, investment in a firm's shares did not raise capital for the firm but merely transferred ownership of the shares from one person to another. In the 'managerial firm', managers were not dependent on shareholders to raise capital. See Lazonick and O'Sullivan (2000A).

Penrose's emphasis on managerial organisation as defining the boundaries of organisational learning in the modern corporation reflected a fundamental reality of the US industrial enterprises that she studied in the mid-twentieth century. In the 1970s and 1980s, however, the capabilities of the 'organization man', and the viability of the US industrial corporation, came under severe pressure. One problem was the growth of the industrial corporation itself. Penrose argued that the availability of managerial resources limited the growth of the firm. But she did not envisage the situation in which the growth of the industrial corporation actually overreached these limits so that strategic control became segmented from organisational learning, and the performance of the firm, grown too large, deteriorated.

As Penrose (1995, p. 261) argued in drawing conclusions to *The Theory of the Growth of the Firm*:

There is no evidence to support the proposition often advanced that 'diseconomies of size' will arise at some point in a firm's growth and that the large firms will eventually become inefficient. They may reach a stage where their structure and behaviour become more akin to those of financial holding companies than of industrial firms, thus raising the question whether the two types of firm should be examined with the same type of analysis, but the efficiency of their productive activities need not suffer because of the change in organizational form.

For Penrose, the basis of this optimism was the notion that the growing importance of the multidivisional structure, with its decentralised organisation, made it possible to spin off existing businesses from the parent firm. As Penrose (1995, pp. 262–3) put it:

To the extent that the economies of size available to large firms are primarily economies of growth, no loss in efficiency would result if those activities that are already established and run more or less as separate 'businesses', were divorced from the parent firm and permitted to operate as independent businesses. There is every reason to presume that the 'new' firms thus created would continue growing, taking advantage of their own economies of growth, without any reduction in the efficiency of production or distribution of either the original firm or the new ones created. The decentralized type of organization that is becoming increasingly characteristic of the large firm facilitates the severance of a particular business from the rest; this is shown in the growing numbers of such 'businesses' that are bought and sold by large firms.

Penrose provided little, if any, evidence to support her claim that the buying and selling of business units would not adversely affect the innovative capabilities of these units. Yet there is considerable evidence that during the 1960s and into the 1970s, US industrial corporations grew too big, largely through conglomeration (Ravenscraft and Scherer, 1987; O'Sullivan, 2000B, ch. 5; see also Rumelt, 1986). The multidivisional structure was supposed to permit a growing enterprise to manage diversification into new technologies and markets that made use of the resources that it had accumulated in its prior growth. In practice, however, even this 'related diversification' often entailed a centralisation of authority and decentralisation of responsibility, which severed strategic decision-making from the processes of organisational learning. Such 'strategic segmentation' was generally even more pronounced in the unrelated diversifications that characterised the conglomerate movement of the 1960s. The result was the deconglomeration movement of the 1970s and early 1980s (including, as we have seen, the rise of the junk bond market for effecting management buyouts) to undo the damage of the previous overgrowth of the US industrial corporation.

The more general problem with the Penrosean theory in historical perspective is that Penrose assumed that the managerial structure based on the 'organization man'— that is, 'the firm'—was an integrated unit that would act as if the interests of one were the interests of all. In fact, she was justified in positing that a high degree of organisational integration was central to the success of the US industrial corporation in the post-war decades. In historical perspective, the 'organization man' of the 1950s was made, not born (Noble, 1977; Lazonick, 1985). Even insofar as, by the 1950s, the US industrial corporation integrated the interests of the thousands of administrative, technical and professional employees in the hierarchical and functional division of labour, there was no inherent reason to assume that this commonality of interests would persist. As I have outlined elsewhere (Lazonick, 2000), organisational integration entails a complex interaction of hierarchical, functional and strategic integration that, even when it is achieved, is liable to break down in the presence of industrial and institutional change. A theory of the growth of the industrial corporation must contemplate the possibility that, within a corporate enterprise, the organisational *integration* that supports organisational learning, and hence innovation, may give way to organisational *segmentation* that undermines this collective and cumulative process.

As a critical condition of innovative enterprise, moreover, organisational integration interacts with the conditions of financial commitment and strategic control. If in *The Theory of the Growth of the Firm* Penrose highlighted the importance of managerial learning, and hence the need for organisational integration to generate innovation, she did not explore the social conditions that determined financial commitment or strategic control in the US industrial corporation of the 1950s. Assuming both a stable structure of managerial control that favoured corporate growth and adequate finance for management's investment plans, she focused all of her analysis on organisational learning within an integrated managerial structure. Hence she did not ask how changes in the sources and extent of financial commitment or the loci and goals of strategic control might affect investments in organisational learning or the innovative performance of these investments.

V.b. Penrosean theory and the new international competition

The need for a theory of innovative enterprise that can link organisation, finance and strategy is of particular importance when a successful industrial enterprise is under pressure from new innovative competitors. As we have seen, Penrose saw 'big business' competition as a critical inducement to innovation and enabler of economies of growth. She saw possibilities for the emergence of new 'big business' competitors through the efforts of new firms that, having taken root in the 'interstices' of the corporate economy, successfully transformed themselves from new ventures into going concerns.

But the implicit assumption in *The Theory of the Growth of the Firm* was that these 'big business' competitors, whether old or new, functioned within and subject to the

same institutional, i.e., national, environment. Although the US industrial corporation that dominated the national and global economies of the time provided the
empirical foundation for her theoretical model, Penrose abstracted 'the growth of the
firm' from the specific national institutional environment in which it operated. Hence
she did not consider the role of changes in the innovative capabilities of international
competitors in certain industries in placing limits on the growth of firms that operate
subject to different organisational and institutional conditions.

In the case of the US industrial corporation, such an assumption reflected its
unprecedented dominance in global competition in the 1950s. But it was a dominance
that was soon to be challenged. The new competitors grew up in other national economies in which protected environments gave them the privileged access to finance and
markets that enabled them to accumulate the productive capabilities to enter into
global competition. For US industrial corporations in the decades after Penrose wrote
her book it was the Japanese who were the most formidable of the new competitors.
They chose to compete in those mass-production consumer durable and related
capital goods industries in which in the 1950s the US corporations had reigned
supreme. In these industries, Japanese corporations ultimately won out not only by
building more integrated managerial organisations than their once-dominant US
competitors but also by integrating production workers into the processes of organisational learning.

To recognise in historical retrospect the Japanese challenge to the US industrial
corporation by no means diminishes Penrose's insights, writing as she was at a time
when attention in the West was focused on the industrial recovery of Western Europe,
and the military power of the Soviets, with virtually no attention being paid to the
potential of Japan. On the contrary, Penrosean theory, with its focus on the growth of
the US industrial corporation, provides an indispensable foundation for a historical-
comparative analysis of the rise of a new international competitor such as Japan. The
fact is that, as an economist interested in the issues of economic development and
performance, Penrose dared to explore the internal workings of the industrial corporation—the central economic institution of the twentieth century. Yet such an inquiry is
an intellectual venture that economics as a professional discipline has been and
remains loath to undertake (Pitelis and Wahl, 1998). Her incisive theoretical perceptions into the process of organisational learning and enterprise growth provide foundations on which we can build. But for such analyses to be relevant to changing
historical circumstances, or even for understanding fully the growth of the corporate
enterprise and its limits in the post-World War II United States, requires that we go
beyond 'managerial learning' as the sole determinant of 'the growth of the firm'.

In comparative and historical perspective, the main internal weakness of the
Penrosean 'theory of the growth of the firm' for building a theory of innovative enterprise is its implicit assumption that organisational learning means managerial learning.
Such a perspective has difficulty explaining, for example, why most Japanese and
many European enterprises in the post-World War II decades extended organisational
learning to shop floor workers and independent suppliers, and how this development

and utilisation of broader and deeper skill bases affected international competitive advantage and national economic performance (Lazonick and O'Sullivan, 1996, 1997; Lazonick, 1998). It also has difficulty in explaining the resurgence of the innovative capabilities and competitiveness of US industrial corporations in the 1990s based on 'narrow and concentrated' skill bases of highly educated and specialised personnel, many of whom are highly mobile not only across firms but also across national boundaries.[1] Writing in the 1990s, Penrose (2001; also Penrose, 1995, pp. xviii–xx) correctly focused on the growing importance of strategic alliances and networks of firms in the innovation process. But an understanding of the internal organisation of the partners in these networks, including nodes of organisational integration, financial commitment and strategic control, may be critical for understanding the organisational learning by, and economic performance of, inter-firm alliances (see Lazonick and O'Sullivan, 2000B).

V.c. Penrosean theory and the financial revolution

In elaborating her theory, Penrose envisioned a 'retain and reinvest' regime of corporate governance: corporations retained corporate revenues and reinvested in corporate resources, thus driving, even if not ensuring, the growth of the firm. But especially when corporate growth has been rendered vulnerable by internal overextension and international competition, as was the case in the United States in the 1970s, a financial environment that seeks higher returns on corporate securities will create powerful pressures to shift to a regime of 'downsize and distribute'. Such a transformation in the US financial environment occurred during the 1970s, as a generation of the working population who, in large part through their participation in 'the growth of the firm', had accumulated savings in the post-war decades witnessed the erosion of these savings through inflation.

Penrose assumed throughout her book that the modern industrial corporation would always try to utilise the unused productive resources at its disposal. She understood that to make use of these available productive resources to enter new markets meant investing in new, complementary, productive resources, including reinvestment in the productive capabilities of current personnel. Starting in the 1970s and even more so in the 1980s and early 1990s, the changed financial environment led strategic decision-makers in the US industrial corporation to reassess the viability of reinvesting in the capabilities of the productive resources that the enterprises that they controlled had accumulated.

Penrose (1995, pp. 26–30) equated the profit motive and the growth motive in determining the investment strategy of the firm. But this equation only holds if those who control the allocation of corporate resources cannot or will not seek higher returns for the 'firm'—now defined as those who remain in the enterprise's employ-

[1] On the concept and potential significance of 'narrow and concentrated skill bases', see Lazonick and O'Sullivan (2001). For a recent study of the international flow of highly educated and specialised personnel, see Saxenian (1999).

ment, including themselves—by shedding 'unused productive resources'—that is, those 'human assets' whose services those who exercise strategic control deem to be no longer of value. The investment decision, moreover, can be based on new criteria for profitability—as indeed became the case in the last decades of the century. While holders of outstanding corporate securities demanded higher real returns, top managers who engaged in a strategy of 'downsize and distribute' to 'create value for share-holders' could themselves be remunerated at much higher levels, both absolutely and relative to others in their organisations, than had previously been the case.

Penrose's theory, elaborated in a very different financial environment, did not envision that a dramatic change in the distribution of rewards within the managerial structure could create the incentives of corporate strategists to shed rather than make use of the corporation's unused productive resources. Yet that is precisely what happened in the 'old' corporate economy of the 1980s and 1990s as manifested by the explosion in CEO pay and the concomitant downsizing of middle management, not to mention those further down the corporate hierarchy. Hence, the commonality of interests within the managerial organisation, that Penrosean theory took as given, ceased to exist. Specifically, lacking an explicit analysis of how organisational integration is affected by strategic control and financial commitment in the industrial corporation, the theory that Penrose constructed to explain corporate behaviour in the United States in the 1950s cannot comprehend the erosion of cohesive managerial organisation in major US industrial corporations that occurred in the 1980s and 1990s.

V.d. Penrosean theory and the rise of the 'new economy'

Yet the strategic orientation of US industrial corporations of the 1980s and 1990s was by no means characterised only by 'downsize and distribute'. In particular, as had been the case decades before with what ultimately became 'old economy' companies, 'retain and reinvest' strategies enabled many 'new economy' corporations of the 1970s, 1980s and 1990s to transform themselves from new ventures into dominant going concerns. As Penrose's theory would have predicted, many of these firms grew up in the 'interstices' of the old corporate economy, accumulating capabilities in markets and technologies that 'old economy' companies neglected (see Christensen, 1997). Even then, in historical perspective, there is a need to explain why, in many cases, existing corporations that developed new technologies that ultimately were successful failed to commercialise them—Xerox, with its many PC-related inventions, is a prominent example (see Smith and Alexander, 1988)—or why existing companies with an abundance of accumulated capabilities did not succeed in their attempts to enter new lines of business—GE, with its failures in computers and factory automation, in the 1960s and 1970s, is notable in this regard (see O'Sullivan, 2000B, ch. 5).

What distinguishes the 'old economy' from the 'new economy' is the mobility of people and money, or labour and capital, between firms, with the flows generally from the old corporations to the new corporations. In the case of labour, while Penrosean theory could contemplate the emergence of new ventures within the interstices of the old economy, the theory did not envision that human resources would be transferred from the old to the new firms. In effect, the theory of the growth of the firm did not foresee that an existing corporation might not be able to reinvest some of the 'unused' resources that it had accumulated in new lines of business because the people who were the bearers of these 'unused' resources might decide to exit the firm and use these resources elsewhere.

When such mobility of labour results in the creation of new firms, it results in a devolution of strategic control that, other things being equal, should result in a closer integration of strategy and learning. But the very mobility of labour from the 'old economy' to the 'new economy' that fosters the growth of new firms creates new problems for organisational integration for new ventures as well as going concerns. As we have seen in the case of Microsoft, innovative enterprises in the 'new economy' that depend on collective and cumulative learning for their success must seek to control the highly mobile labour market by extending stock-based remuneration further down the corporate hierarchy, often as a systematic compensation policy.[1] Meanwhile, older corporations, which had since the 1950s limited stock-based remuneration, have had to respond, typically in an *ad hoc* fashion on an individual basis, by using stock option packages to retain such highly mobile personnel (see Carpenter and Lazonick, 2001).

If the 'new economy' does not obviate the need for organisational integration to generate innovation, nor does it diminish the need for financial commitment. What was new in the US corporate economy of the 1990s was the increased role of the stock market, not for raising investment finance, for which (as we have seen) it has never been important in the twentieth century industrial corporation, but as a currency for the accumulation of capabilities through the acquisition of other companies. Indeed to manage such a currency has often required that companies spend huge sums of cash to buy back stock (Carpenter and Lazonick, 2001). 'Managing the stock market' has created new challenges for corporate growth.

Since Edith Penrose wrote her book over 40 years ago the social conditions of innovative enterprise have changed substantially. But there are underlying principles of corporate growth that one can draw from *The Theory of the Growth of the Firm* that remain relevant today. One such principle is that, in the US corporate economy of the 2000s as in that of the 1950s, the corporation grows by controlling market forces rather than being controlled by them (Lazonick, 1991).

VI. CONCLUSION: THEORY AND HISTORY

To recognise that Penrosean theory of the growth of the firm has its roots in a certain time and place is not to diminish its relevance to us today. Industrial innovation is

---

[1] For a description of the collective and cumulative character of the learning process at Microsoft, see Cusumano (2000).

a process of change—one in which the transformation of technology and markets interacts with evolving organisational and institutional conditions (Lazonick, 2000, 2001). To analyse the process of change, the economist needs a methodology that integrates theory and history. For theory to be relevant, it must be derived from the rigorous study of historical reality. Theoretical explanation must be based on an understanding of the historical process that is broad and deep enough to invoke confidence that the assumptions and relations that form the substance of the theory capture the essence of the reality to which they purport to be relevant. To develop relevant theory requires an iterative intellectual approach in which we derive theoretical postulates from the study of the historical record, and we use the resultant theory to analyse history as an ongoing and, in the present, unfolding process. Theory, therefore, serves as an abstract explanation of what we already know and as an analytical framework for researching what we need to know. In short, the integration of theory and history requires a 'historical-transformation' methodology, as distinct from the 'constrained-optimization' methodology to which economists are habituated (Lazonick, 2001B).

In the late 1980s, Edith Penrose (1989) wrote a highly perceptive essay entitled 'History, the Social Sciences and Economic "Theory", with Special Reference to Multinational Enterprise', in which she articulated the need for such an integrative methodology. After quoting Joseph Schumpeter's (1954, pp. 12–13) argument that 'the subject matter of economics is essentially a unique process in historical time', and that 'most of the fundamental errors currently committed in economic analysis are due to lack of historical experience more often than to any other shortcoming of the economist's equipment', Penrose (1989, p. 8) argued:

[S]ocial scientists ask questions of the present or the past, which are based on assumptions they think are likely to be applicable to those aspects of the societies that they want to analyse. With the aid of these assumptions they attempt to formulate hypotheses and to produce coherent explanations. In 'explaining' history, that is to say, in making the past intelligible, historians have to use the same type of assumptions, although perhaps not in an equally rigorous form. Thus historians borrow heavily from all the social sciences just as the latter borrow from historians; they live on each other to an extent greater than is realised by either. This symbiotic relationship is to a considerable extent responsible for the need both continually to re-write history and continually to reconsider the 'models' of the social sciences.

With particular reference to the discipline of economics, Penrose (1989, p. 10) observed that '[e]conomic historians, along with many economists, suspect that the present fashions in "pure" theory (whatever that is) are likely to lead to a dead end from their point of view', but warned that 'they should not allow this suspicion to encourage them to accept a distinction, far too commonly made, between a "real" world of history and a world of theory'. She continued:

It is impossible for economic historians to select and make sense of the 'facts' of history without the aid of the theories developed by students of economic affairs defined in the broader sense. Some of the 'theory' may be little more than dressed-up common-sense deductions from common observations and therefore not even recognised as such, but much of it has a deeper significance. Without theoretical analysis of cause and consequence one has no standard against which to appraise the significance of any given set of observations, for this significance is a question of what difference the observations make to what otherwise have been the historical interpretation.

Penrose (1989, p. 11) took particular exception to the arguments by the economic historian, Barry Supple (1989), in the introductory essay to the volume in which her essay was published, for his acceptance of the methodological position that 'history' and 'theory' were in opposition to each other, rather than, as she put it, 'a genuine complementarity'. As Penrose explained:

'Theory' is, by definition, a simplification of 'reality' but simplification is necessary in order to comprehend it at all, to make sense of 'history'. If each event, each institution, each fact, were really unique in all aspects, how could we understand, or claim to understand, anything at all about the past, or indeed the present for that matter? If, on the other hand, there are common characteristics, and if such characteristics are significant in the determination of the course of events, then it is necessary to analyse both the characteristics and their significance and 'theoretically' to isolate them for that purpose.

Penrose (1989, p. 11) concluded that 'universal truths without reference to time and space are unlikely to characterise economic affairs'. It is with this warning, and in this spirit, that we should continue to read and use *The Theory of the Growth of the Firm*. It is a great book that was written at a certain time and in a certain place. Understanding the evolution of the industrial corporation in the decades after she wrote the book is an enormous intellectual task. It cannot be done all at once or all alone.[1] Like the organisational learning that was at the core of Edith Penrose's theory of the growth of the firm, scholarly research that can comprehend the transformation and impact of the industrial corporation over the past 40 years and beyond is a cumulative and collective endeavour.

REFERENCES

BEST, M. (1990). *The New Competition: Institutions of Industrial Restructuring*, Cambridge, MA, Harvard University Press.

BEST, M. and GARNSEY, E. (1999). Edith Penrose, 1914–1996, *Economic Journal*, Vol. 109, F187–209.

BRUCK, C. (1988). *The Predators' Ball*, New York, Simon and Schuster.

CARPENTER, M. and LAZONICK, W. (2001). 'Innovation and Competition in the Era of "Shareholder Value": the Optical Networking Industry', INSEAD working paper.

CHANDLER, A. D. Jr (1962). *Strategy and Structure*, Cambridge, MA, MIT Press.

CHRISTENSEN, C. M. (1997). *The Innovator's Dilemma: When New Technologies Cause Great Firms to Fail*, Cambridge, MA, Harvard Business School Press.

CICCOLO, J. H. JR and BAUM, C. F. (1985). Changes in the balance sheet of the U.S. manufacturing sector, 1926–1977, in *Corporate Capital Structures in the United States*, edited by B. M. FRIEDMAN, Chicago, IL, University of Chicago Press.

---

[1] On Edith Penrose's enormously productive academic career, see Best and Garnsey (1999); Penrose and Pitelis, Chapter 2, this volume.

CUSUMANO, M. (2000). Making large teams work like small teams: product development at Microsoft, in *New Product Development and Production Networks* edited by U. JUERGENS, Berlin, Springer.

DORE, R., LAZONICK, W. and O'SULLIVAN, M. (1999). Varieties of capitalism in the twentieth century, *Oxford Review of Economic Policy*, Vol. 15, No. 4, 102–20.

FORTUNE, various years, 'Fortune 500'.

GALBRAITH, J. K. (1967). *The New Industrial State*, Boston, MA, Houghton Mifflin.

HALL, B. (1994). Corporate restructuring and investment horizons in the United States, 1976–1987, *Business History Review*, Vol. 68, No. 1, 110–43.

KAYSEN, C. (1996). *The American Corporation Today*, New York, Oxford University Press.

KOR, Y. Y. and MAHONEY, J. T. (2000). Penrose's resource-based approach: the process and product of research creativity, *Journal of Management Studies*, Vol. 37, No. 1, 109–139.

LAZONICK, W. (1985). Strategy, structure and management development in the United States and Britain, in *Development of Managerial Enterprise* edited by K. KOBAYASHI and H. MORIKAWA, Tokyo, University of Tokyo Press. [Reprinted in W. LAZONICK (1992). *Organization and Technology in Capitalist Development*, Cheltenham, Edward Elgar.]

LAZONICK, W. (1990). *Competitive Advantage on the Shop Floor*, Cambridge, MA, Harvard University Press.

LAZONICK, W. (1991). *Business Organization and the Myth of the Market Economy*, Cambridge, Cambridge University Press.

LAZONICK, W. (1992). Controlling the market for corporate control, *Industrial and Corporate Change*, Vol. 1, No. 3, 445–88.

LAZONICK, W. (1998). Organizational learning and international competition, in *Globalization, Growth, and Governance*, edited by J. MICHIE and J. GRIEVE SMITH, Oxford, Oxford University Press.

LAZONICK, W. (2000). 'From Innovative Enterprise to National Institutions: A Theoretical Perspective on the Governance of Economic Development', INSEAD working paper.

LAZONICK, W. (2001A). The theory of innovative enterprise, in *International Encyclopedia of Business and Management Handbook of Economics*, edited by W. LAZONICK, London, International Thomson.

LAZONICK, W. (2001B). Understanding innovative enterprise: toward the integration of economic theory and business history, in *Business History Around the World at the End of the Twentieth Century*, edited by F. AMATORI and G. JONES, Cambridge, Cambridge University Press.

LAZONICK, W. and O'SULLIVAN, M. (1996). Organization, finance, and international competition, *Industrial and Corporate Change*, Vol. 5, No. 1, 1–49.

LAZONICK, W. and O'SULLIVAN, M. (1997). Big business and skill formation in the wealthiest nations, in *Big Business and the Wealth of Nations*, edited by A. D. CHANDLER JR, F. AMATORI and T. HIKINO, Cambridge, Cambridge University Press.

LAZONICK, W. and O'SULLIVAN, M. (2000A). Maximizing shareholder value: a new ideology of corporate governance, *Economy and Society*, Vol. 29, No. 1, 13–35.

LAZONICK, W. and O'SULLIVAN, M. (2000B). 'Perspectives on Corporate Governance, Innovation, and Economic Performance', Report to the European Commission, DG-XII.

LAZONICK, W. and O'SULLIVAN, M., eds (2001). *Corporate Governance and Sustainable Prosperity*, London, Palgrave.

LEVY, F. and MURNANE, R. J. (1992). U.S. earnings levels and earnings inequality: a review of recent trends and proposed explanations, *Journal of Economic Literature*, Vol. 30, No. 3, 1333–81.

LEWELLEN, W. (1968). *Executive Compensation in Large Industrial Corporations*, New York, Columbia University Press.

LEWELLEN, W. (1971). *The Ownership Income of Management*, New York, Columbia University Press.

McGOUGH, R., McGEE, S. and BRYAN-LOW, C. (2000). Poof!, buyback binge now creates big hangover, *Wall Street Journal*, December 18.

MADDISON, A. (1994). Explaining the economic performance of nations, 1820–1989, in *Convergence of Productivity: Cross-National Studies and Historical Evidence*, edited by W. J. BAUMOL, R. R. NELSON and E. N. WOLFF, Oxford, Oxford University Press.

MILLS, C. W. (1951). *White Collar: The American Middle Classes*, Oxford, Oxford University Press.

MOSS, P. (2001). Earnings inequality and the quality of jobs: current research and a research agenda, in *Corporate Governance and Sustainable Prosperity*, edited by W. LAZONICK and M. O'SULLIVAN, London, Palgrave.

NOBLE, D. (1977). *America by Design: Science, Technology, and the Rise of Corporate Capitalism*, New York, Oxford University Press.

O'SULLIVAN, M. (2000A). The innovative enterprise and corporate governance, *Cambridge Journal of Economics*, Vol. 24, No. 4, 393–416.

O'SULLIVAN, M. (2000B). *Contests for Corporate Control: Corporate Governance and Economic Performance in the United States and Germany*, Oxford, Oxford University Press.

PENROSE, E. T. (1960). The growth of the firm—a case study: the Hercules Powder Company, *Business History Review*, Vol. XXXIV, Spring, 1–23.

PENROSE, E. (1989). History, the social sciences and economic 'theory', with special reference to multinational enterprise, in *Historical Studies in International Corporate Business*, edited by A. TEICHOVA, M. LÉVY-LEBOYER and H. NUSSBAUM, Cambridge, Cambridge University Press.

PENROSE, E. T. (1995). *The Theory of the Growth of the Firm*, 3rd edn, Oxford, Oxford University Press.

PENROSE, E. T. (2001). Growth of the firm and networking, in *International Encyclopedia of Business and Management Handbook of Economics*, edited by W. LAZONICK, London, International Thomson.

PITELIS, C. and WAHL, M. (1998). 'Edith Penrose and the Theory of (the Growth of) the Firm', Judge Institute of Management Studies working paper 5/98, University of Cambridge.

RAVENSCRAFT, D. and SCHERER, F. M. (1987). *Mergers, Sell-offs, and Economic Efficiency*, Washington DC, Brookings Institution.

RUMELT, R. (1986). *Strategy, Structure and Economic Performance*, revised edn, Boston, MA, Harvard Business School Press.

SAXENIAN, A. L. (1999). *Silicon Valley's New Immigrant Entrepreneurs*, San Francisco, Public Policy Institute of California.

SCHUMPETER, J. A. (1950). *Capitalism, Socialism, and Democracy*, 3rd edn, New York, Harper.

SCHUMPETER, J. A. (1954). *History of Economic Analysis*, New York, Oxford University Press.

SIMON, H. (1957). *Models of Man*, Chichester, Wiley.

SMITH, D. K. and ALEXANDER, R. C. (1988). *Fumbling the Future: How Xerox Invented then Ignored the First Personal Computer*, New York, William Morrow.

SOLO, C. S. (1951). Innovation in the capitalist process, *Quarterly Journal of Economics*, Vol. 65, No. 3, 417–28.

SUPPLE, B. (1989). Introduction: multinational enterprise, in *Historical Studies in International Corporate Business*, edited by A. TEICHOVA, M. LEVY-LOBOYER and H. NUSSBAUM, Cambridge, Cambridge University Press.

US BUREAU OF THE CENSUS (1976). *Historical Statistics of the United States from the Colonial Times to the Present*, US Government Printing Office.

WHYTE, W. H. JR (1956). *The Organization Man*, New York, Simon and Schuster.

WILLIAMSON, O. E. (1985). *The Economic Institutions of Capitalism*, New York, Free Press.

15

MANAGEMENT COMPETENCE, FIRM GROWTH AND ECONOMIC PROGRESS

SUMANTRA GHOSHAL[1], MARTIN HAHN[2] and PETER MORAN[1]

[1]*London Business School and* [2]*MIT Sloan School of Management*

Starting from the empirical observation of a positive correlation between the prosperity of an economy and the relative role of large firms operating in that economy, we propose that this correlation is an artifact of the positive influence of 'management competence' on both these variables. Drawing on Penrose's *The Theory of the Growth of the Firm*, we develop a theoretical framework that distinguishes between two aspects of management competence, i.e., entrepreneurial judgement and organisational capability. Both aspects relate to the process of value creation through the combination and exchange of economic resources. Whereas entrepreneurial judgement refers to the cognitive aspects of perceiving potential new resource combinations and exchanges, organisational capability is the ability to actually carry them out. As we show, the interplay of these two factors affects the speed at which firms expand their operations, the kind of expansion, and the process through which firms create value, not just for themselves, but for society as a whole.

This paper is motivated by a simple empirical observation: there appears to be a strong positive correlation between the prosperity of an economy and the relative role of large firms operating in that economy. The relationship seems to be reasonably robust to many different ways of operationalising both the prosperity of nations and the size of firms. At the extreme, Figure 1 represents this relationship when prosperity is measured simply by the GDP per capital of the country and when we consider only the very largest firms, namely those that find a place in *Fortune*'s list of the 500 largest firms in the world. Figure 2 shows the relationship between the GDP per capita of nations and the proportion of their working population employed by large companies, when firm size is measured at the level of firms earning in excess of US$10 million in revenues (which, though much smaller than *Fortune* 500 firms and often described as medium-sized rather than large, still figure at the right tail of the frequency distribution of firms by size in any society). Figure 3 shows the relationship between total expenditure on R&D (as a percentage of GDP) and the proportion of their working population employed by large companies (above US$10 million revenues). Collectively these different aspects of the relationship between the progress of nations and the extent of large firms operating in those nations render insufficient some obvious

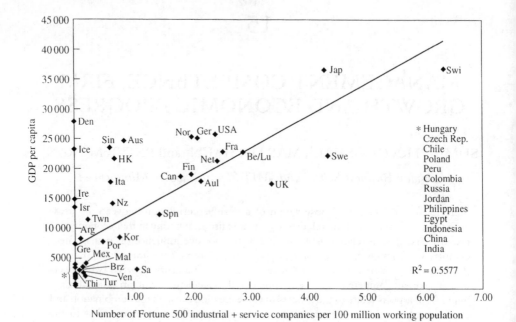

FIGURE 1. GDP per capita versus number of large companies for 48 countries (1994).
Source: World Economic Forum.

interpretations such as larger countries will have more large firms simply because there is more scope for such firms to exist. Clearly, something else is going on here.

If the persistence of this relationship across countries at very different stages of development can be considered to rule out sheer chance or other forms of spuriousness, three different causal explanations may exist. First, it may be that large nations provide a strength for local firms to grow, i.e., the relationship may represent a locational advantage for firm growth (e.g. Porter, 1990). Second, it may be that large firms help economic progress of societies relatively more than small firms constituting the same level of economic activity, i.e., the relationship may represent the societal dimensions of Galbraith's defence of large corporations. Third, both variables, i.e. economic prosperity of nations and the growth of firms to large size, may be influenced by a common factor.

Our objective in this paper is to present an argument for the third explanation. We suggest that 'management competence' is the factor that, on the one hand, leads to the growth of firms and, ultimately, to the emergence of a relatively higher proportion of large firms in any economy that is well endowed with this competence. The same factor, on the other hand, leads to the creation of more economic value out of a society's endowment of resources, thereby leading to that society's growth and prosperity. In other words, with management competence, a country simultaneously

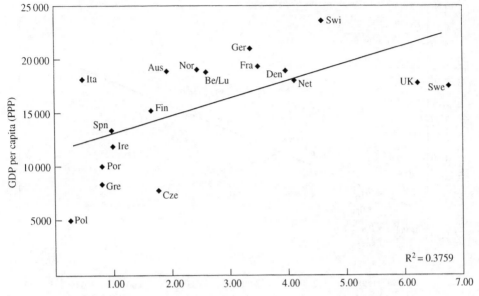

FIGURE 2. GDP per capita versus proportion of labour force employed by large countries for 18 European countries (1994). Sources: AMADEUS European Companies Database and World Economic Forum.

expands its income and the proportion of large firms operating in its economy, thereby establishing the correlation we have observed.

It needs to be noted that our use of the term 'large firm' may be somewhat at variance from the common conceptualisation of the term. We consider any firm with $10 million or more in annual revenues or with 100 or more employees as large—in the sense that the vast majority of business enterprises in any society tend to be considerably smaller in size. They are large also in the sense that effective allocation and usage of their resources require coordination and management to an extent that Adam Smith's butcher does not require. Therefore, firms that are often described as small and medium-sized enterprises (SMEs)—the German Mittelstadt, for example— are included in our definition of large firms. Our distinction is not between the SMEs and the corporate giants, in the traditional sense, but between complex organisations and the atomistic actors in the economists' ideal of a market economy: in essence the same distinction that Simon (1991) made between a market economy and an organ- isational economy. Management competence, we suggest, is what drives both the growth of firms and the growth of the overall economy in modern organisational economies.

We do not, as yet, offer this as anything more than a plausible theoretical specula- tion. But it is an interesting speculation. It is interesting, theoretically, because it

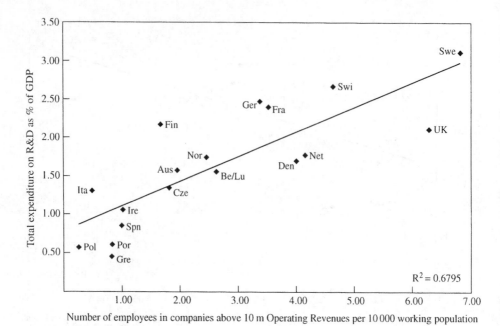

FIGURE 3. R&D expenditure as % of GDP versus proportion of labour force employed by large companies for 18 European countries (1994). Sources; AMADEUS European Companies Database and World Economic Forum.

suggests an important new dimension for 'growth theory' which has become an arena of almost frantic activity in the discipline of macroeconomics, and also because it opens up a new and potentially rewarding line of enquiry into the role of large firms within the field of management studies. It is also important, from a practical perspective, because it suggests some radically different approaches to modernising emerging economies—different from the controversial 'shock therapy' approach that has produced such a mixed record of outcomes—and also because of the implications it has for public policy with regard to the role and regulation of large firms and for the strategy and conduct of the firms themselves, and of their managers.

We begin by defining the term 'management competence' (section I) as the foundation of the arguments that follow. Subsequently, building on the arguments of Penrose (1959) and Schumpeter (1934), we show why and how management competence leads to the growth of firms and, ultimately, to an industrial structure in which large firms come to adopt a relatively greater role (section II). After that, we elaborate on the second half of our argument, namely, why management competence also leads to a better exploitation of an economy's stock of resources and, thereby, to its economic growth (section III). Finally, in section IV we put the two hypotheses into perspective and focus on the links between the size of firms and economic progress and discuss some of the theoretical and practical implications of our arguments.

I. MANAGEMENT COMPETENCE

The recognition that the performance of organisations depends to a large extent on the ability of those who administer their operations does not require much insight into the real world of the successes or failures of established enterprises. Yet, when it comes to the theoretical conceptualisation of organisational entities, such as firms, management competence does not usually occupy the centre stage. In this paper we wish to redress this deficiency.

Our analysis is grounded in three views on the role of the firm. First, following Knight (1921), Coase (1991) and Simon (1951), a firm can be viewed in terms of the employment relationship where the employer offers the worker a wage in return for a task that is to be specified by the employer. The contract between the two is necessarily incomplete as the employer may not be able to specify all of the tasks that will have to be carried out by the worker, partly because this is costly to do, and also because some of his/her perceptions on the nature of the project in question are not clearly developed and largely tacit. This view is significant not from the perspective of the employment relationship as a risk-sharing institution *per se*, but to underline the differences in the ability to perceive, evaluate and seize profitable opportunities. That is, the employer requires the worker to provide a complementary service that the worker may not be able to associate with the overall project of the employer (otherwise called the entrepreneur). As we shall explore later in this section, these differences in ability to perceive profitable opportunities together with certain costs of persuading and informing other individuals of the potential value of the undertaking suggests a way to define the role and tasks of firm management.

Second, and as an extension of the first view, a firm can be seen as a pool of more or less intangible and idiosyncratic resources. This position was first taken by Penrose (1959) and more recently developed in the evolutionary theory by Nelson and Winter (1982) and the increasingly influential 'resource-based view' that characterises the firm as a 'set of differentiated technological skills, complementary assets, and organisational routines and capacities' (Teece, 1988). This perspective on the firm has its origins in the work of Adam Smith and later Hayek (1945) who associated the division of labour with a corresponding division of knowledge. The knowledge that individuals gather is often tacit, i.e. it is difficult and costly (if not impossible) to be articulated (Polanyi, 1969). Moreover, individuals in organisations typically co-specialise in certain tasks and acquire a mutual understanding of this task that, over time, becomes idiosyncratic to the group. In other words, knowledge in an organisation is context-dependent and has reduced applicability, and thus value, outside the enterprise. Management competence, as we will argue later in this paper, is the factor that drives the acquisition, development and deployment of these context-dependent and idiosyncratic knowledge resources of the firm which in turn shape the firm's productive opportunities and, thereby, influence the direction of its growth.

Both the first and the second view of the firm are compatible and mutually supportive. The fact that the employer (entrepreneur) hires workers to perform complementary tasks that together contribute to the project envisioned by the employer implies some degree of specialisation on behalf of the employer to know what the task is and on the workers to perform the task. Moreover, several workers can be expected to specialise in different aspects of the project, which increases their specialist knowledge, but at the same time—given some degree of interdependencies between tasks—promotes the development of certain capabilities that are very much specific to the organisation in which they collectively operate.

The third view of the firm that we adopt is—compared to the first two—in a relatively underdeveloped stage: that is, the firm as an agent of discovery and economic progress, as first proposed by Schumpeter (1934). It is here that we wish to propose some ideas which, in combination with the first two perspectives, would suggest a more significant role of firms in general, and management competence in particular, with regards to the creation of economic wealth for society as a whole. Our use of the term 'management competence' is based on the first two views of the firm and bears fundamental importance for the third. In what follows, we will outline the dimensions of management competence in the context of our analysis.

I.a. The role of management: resource combination and exchange

The creation of economic value, be it by individuals or organisations (such as firms) in an economy, is a process which involves the use of economic resources. By the term resources we refer to any tangible or intangible object which can be transformed (by means of other resources) into an 'output' which, in itself, constitutes another resource. However, in order to produce something of value (i.e., a value that is greater than the value realisable from the resources themselves, if used independently), these economic resources must somehow be *combined*. In other words, combining resources provides the basis for the creation of new and better products, or new and better ways of making products (i.e., with higher productivity of given resources). In order to do so, (a) new resources have to be combined or (b) given resources have to be combined in new ways. Either way, as Schumpeter argued, 'to produce things, or the same things by a different method [or to] combine these materials and forces differently (1934, p. 66) constitutes 'economic development'. Such combinations represent 'simply the different employment of the economic system's existing supplies of productive means' (p. 68).

In order for economic actors to combine new resources or given resources in new ways they have to deploy the resources in question. That is, they must make them available for this purpose or, as Schumpeter might have put it, they must place the required materials and forces within reach. Each time resources are deployed in making new combinations, a new source of 'potential' value is created and added to the economic system. Yet this 'potential' value is of limited use to society and cannot contribute directly to its economic wealth. Economic value is realised when these resources are themselves used productively or reallocated to more productive uses (i.e., placed within the reach of more actors). This value realisation is accomplished largely by means of *exchange* (North and Thomas, 1975, p. 18). We will examine the role of resource combination and exchange in more detail in the third section of this paper.

The process of resource combination and exchange by any economic actor requires two distinct abilities: first, the ability to *perceive* the resource combination and exchange potentials, and second, the ability to actually *carry out* the resource combination and exchange in question. In a market environment where individual actors make independent choices as to the allocation of their available resources, the price mechanism is the common way of guiding people's behaviour—in a way that allocates resources to their most productive uses. In firms, on the other hand, besides a certain influence of the price system from 'outside' (e.g. in terms of input prices etc.), it is largely the function of management to influence the allocation of resources (Bower, 1970).

As outlined by the first view on the firm, the reason for this is that the employer/ entrepreneur possesses a unique insight into the firm's undertakings. Initially, when a firm is founded or relatively small, this insight may reside in a few or even only one individual. Obviously, as the organisation grows beyond an owner-manager stage and 'professional' managers are introduced and are given decision-making authority, many more individuals are likely to have their own 'unique insights'. Eventually, as each individual in the firm accumulates specific skills and a stock of knowledge about what he/she does, this 'unique insight' will evolve into a unique organisational view, comprising the insights of a multitude of different local experts (second view of the firm). Yet, it is still management that remains responsible for the ultimate allocation of the firm's resources, even though its influence may be either direct or indirect (Bower, 1970; Burgelman, 1983).

Thus, the two universal requirements necessary to create new resource combinations and exchanges (i.e., the ability to perceive the resource combinations and exchange potentials plus the ability to carry them out) define the role and primary responsibility of the management of a firm. The ability to combine and exchange economic resources effectively in firms can be associated with the competence of its management. We define the ability to perceive potential resource combinations and exchanges as *entrepreneurial judgement* and the ability to carry out any combination and exchange as *organisational capability*. In other words, management competence consists of entrepreneurial judgement and organisational capability. As we shall see, both aspects are crucial to the process of firm growth and to the value-creating role of firms in the economy.

I.b. Entrepreneurial judgement

The concept of entrepreneurial judgement is rooted in the literature on entrepreneurship. It is here that the relevance of the first view of the firm that we presented becomes apparent. In a world where novel possibilities for productively deploying resources are always emerging, the exercise of judgement is required to 'process complex and incomplete information usefully in an intuitive way' (Langlois, 1995, p. 83; see also Knight, 1921). We therefore define entrepreneurial judgement in a way that is consistent with the concept of entrepreneurship, as developed by Knight (1921), Schumpeter (1934) and Kirzner (1973), as the ability to discover new ways of dealing with known problems or 'new combinations' of given knowledge. There are two aspects of this concept that need highlighting. First, entrepreneurial judgement is contingent upon the initial stock of knowledge and insight within the firm. Second, entrepreneurial judgement embodies a strongly cognitive dimension such as is found in the literature on learning and innovation. The ability to see previously unrecognised opportunities, or to evaluate known opportunities differently must involve some degree of perceptive capability, which involves greater knowledge and insight about the relevant aspects of the task in question but must also include an ability to link this absorptive capacity (Cohen and Levinthal, 1990) to some overall vision or imagination of what might be possible (Hamel and Prahalad, 1994).

I.c. Organisational capability

As Schumpeter pointed out, entrepreneurship (or innovation) is significantly different from invention: 'The inventor produces ideas, the entrepreneur "gets things done" . . .' (1947, p. 152). To facilitate successful exploitation generally requires a firm to establish a context that accommodates the entrepreneur's judgement. This accommodation process implies a substitution of one institutional context for another (e.g. internalisation of some market or firm activity into another organisation—often a new firm). The ability to carry out new combinations typically entails some administrative reorganisation of economic activities that is needed to enhance one's ability to 'cope with the resistance and difficulties which action always meets with outside of the ruts of established practice' (Schumpeter, 1947, p. 152). We refer to the aspect of management competence that enables such administrative reorganisation, which in turn, facilitates the carrying out of new combinations, as 'organisational capability'.

Organisational capability is essentially a social phenomenon. As emphasised by Barnard (1938), it implies the existence of some insight into the behaviour and capabilities of those whose cooperation is needed to carry out the necessary resource deployments. To gain this insight takes time and close interaction. In addition, human actors cannot be assumed to have perfect foresight, nor do they always behave rationally (Simon, 1945). As a consequence, firms cannot fully anticipate the results of their actions and must, to some degree, rely on trial and error (Eliasson, 1990). By learning from past experience, firm members build up a stock of constantly evolving and changing knowledge about how to conduct certain activities. Through this ongoing process, organisations enjoy economies of learning arising from repeated and close social interaction (Nahapiet and Ghoshal, 1998). Therefore, beyond the aspects of formal structure, certain social dimensions such as trust, identification with the organisation, loyalty and sacrifice, are important elements of this component of management competence.

It is important to note that management competence is not restricted to 'managers'. The term which seems to be rooted in the first view of the firm that we presented (i.e., the ability of the employer to see opportunities, hire workers, and direct the production process), is in fact applicable to a much wider group. As we shall explore, the ultimate source of both entrepreneurial judgement and organisational capability potentially resides in any organisational member. Indeed, as we argue in the following section, whether or not a firm's managers are able to create an organisational context that accommodates changes in the management competence that resides throughout the firm determines, to a large degree, the extent to which the firm's collective entrepreneurial judgement is exercised within or outside the boundaries of the firm.

At the same time, both these components of management competence are shaped by the collective social organisation that constitutes the firm. As for entrepreneurial judgement, the process of organisational learning as a collective phenomenon enables the exchange of ideas and knowledge among individuals in a firm, widening the set of opportunities perceived as viable (see Moran and Ghoshal, 1999, for a more rigorous elaboration of this argument). Similarly, organisational capability (the ability to combine and exchange resources within organisations) is facilitated through collective interaction which creates capabilities that are embedded in the processes and 'routines' of organisational members (see Nelson and Winter, 1982). In effect, management competence is a social phenomenon that is linked much more to the second view of the firm. The following sections of the paper shall deal with the role of the two aspects of management competence in their influence on both the growth of firms and on the process of economic development (see Figure 4).

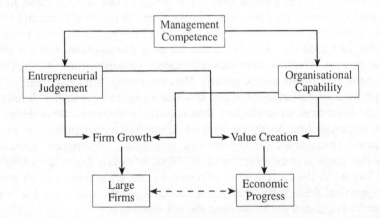

FIGURE 4. The dual role of management competence and the interaction between organisational capability and entrepreneurial judgement contributing to firm growth and economic progress.

II. FIRM GROWTH

Why do firms grow? What makes them expand their existing operations or develop activities into new fields? The argument put forward in this section is an attempt to answer these questions by looking at the way processes and procedures within the firm are organised, and not just how, but especially how well. In other words, we shall argue that management competence is a significant factor for the growth process of firms, determining not just the speed at which firms expand their activities, but also the success with which this growth can be achieved. Our analysis of firm growth consists of two parts. First, we look at the conditions under which entrepreneurial judgement produces incentives for firms to expand into similar or dissimilar activities and how organisational capability determines the locus of this growth (i.e., in new or existing firms). Second, we integrate the two processes into a conceptualisation of firm growth as an iterative process of expansion into related and different areas driven by the ongoing interaction of entrepreneurial judgement and organisational capability.

II.a. Penrosean growth

In her revolutionary book *The Theory of the Growth of the Firm*, Penrose (1959) opens the black box of the firm and takes a look at the internal processes that determine a firm's rate of growth. Arguably the most significant breakthrough in her work concerns the nature of the firm *per se*. The firm is viewed as a coherent administrative organisation which is characterised by resources that are in some way specific to it, even if only by their internal coherency (i.e., relative to all other things inside and outside the organisation). These resources 'provide both an inducement to expand and a limit to the rate of expansion' (Penrose, 1959, 3rd edn, p. xii). Because the value of any firm's productive opportunities depends on the resources available to that firm, these opportunities (even though they are external to the firm) are also necessarily firm-specific. Consequently, the value (to the firm) of the firm's internal resources, such as its managers, increases with the level of experience in their deployment with other resources inside the firm. Since this learning process takes time, it effectively limits the rate at which a firm can internalise resources into its organisational structure—the rate of the firm's growth. Yet, even though management as a factor poses a dynamic limitation to the speed at which a firm can expand, it simultaneously creates new incentives to apply this firm-specific management knowledge to new profitable opportunities, partly because of temporary under-utilisation of existing management. Thus, firms expand so they can utilise idiosyncratic organisational resources that come in indivisible 'bundles'. Note, however, that while a firm's *rate* of growth is limited by the current stock of resources that are within its reach, its *size, per se*, has no practical limit. What, if anything can limit a firm's current size over the long term? Penrose suggested the answer but did not develop it.

Responding to the ever widening inducement to expand—brought about by a growing firm's ever increasing (albeit, not continuous) availability of slack resources and its ever expanding opportunities to deploy these resources productively (particularly in a growing economy)—is, in itself, not enough to ensure the continued growth of a particular firm. Each successive wave of growth must be accommodated by a process that successfully assimilates the old with the new. That is, firms must 'undergo an administrative reorganisation to enable them to deal with the increasing growth' (Penrose, 1959, 3rd edn, p. xvii). Such reorganisation, however, is neither automatic nor assured. Indeed, as Penrose took care to emphasise, her theory explains the growth of *successful* firms (see Penrose, 1959, pp. 32–3). Later, she noted that, although she did not develop the supporting evidence, all the firms whose growth her theory was developed to explain 'seemed to undergo' such reorganisation in both their 'managerial function' and their 'basic administrative structure' (1959, 3rd edn, p. xvii). We submit that this accommodation process is the same change process that is commonly referred to in the literature as 'Schumpeterian integration' (e.g. see Langlois, 1995) and is used to describe the entrepreneurial process of 'carrying out new combinations' within new firms. As we argue below, Schumpeterian integration is necessary not only to enable an individual firm to grow over time but also for it to retain its size in any dynamic environment in which the forces of Schumpeterian integration also exist outside the firm.

II.b. *Schumpeterian integration*

The carrying out of new combinations requires an institutional context that motivates both the perception and execution of those combinations and any other resource deployments that are needed to support these combinations (Moran and Ghoshal, 1996). The context provided by existing markets and firms is unavoidably inappropriate for many such deployments. Schumpeter explained why: 'when confronted with innovations . . . [an individual] . . . needs guidance . . . While he swims with the stream in the circular flow which is familiar to him, he swims against the stream if he wishes to change its channel. What was formerly a help becomes a hindrance. What was a familiar datum becomes an unknown' (Schumpeter, 1934, pp. 79–80). In noting also that 'new combinations are, as a rule, embodied, as it were, in new firms which generally do not arise out of the old ones but start producing 'beside them', Schumpeter (1934, p. 66) also drives home the point that development is, in general, no easier nor any more likely to occur within existing firms than it is in existing markets.

No single institution, whether that institution is a state, a firm, a market or a system of markets can provide an institutional context that sufficiently enables the successful exploitation of more than a tiny fraction of the opportunities a society has for productively deploying its resources (Ghoshal, Moran and Almeida Costa, 1995). One reason, emphasised by Schumpeter (1934) is that the new combinations that must precede subsequent leaps in economic progress are *never* as productive as alternative deployments of the same resources that are also *available* at the time. Indeed, even though it may be shown retrospectively that a particular 'new combination', which subsequently triggered such progress may have been 'the "best" of the methods "possible" at the time', as Schumpeter (1934, p. 83) put it, this same combination, at the time it was carried out, was not likely to be 'the "best method" of producing in the theoretical sense', if 'best method' is defined as 'the most advantageous among the methods which [had] been empirically tested and [had] become familiar' at that time (Schumpeter, 1934, p. 83).

For a variety of reasons, it is difficult for others to evaluate the ability of the entrepreneur, or the profit opportunity itself. First, by definition, the entrepreneur is someone who can see things others cannot see. This has nothing to do with perfect foresight, and little to do with even a *relative* difference in overall capability or imagination. Rather, it stems from the coincidence of a unique resource set, which makes possible specific opportunities for deployment, and the necessary capability and imagination to see these particular opportunities when others with similar opportunities cannot. Second, in the absence of a consistent track record of discovering new projects, there is no way of estimating whether the entrepreneur is correct in his/her judgement. Third, and perhaps most importantly, the entrepreneur cannot trade his/her services through the market, because there are considerable costs involved in persuading and communicating with those who have complementary abilities or resources (Silver, 1984; Sah, 1991). Due to the idiosyncratic nature of the entrepreneur's knowledge, there exists a lack of 'receiver competence' (Eliasson, 1990) on the part of those who could gain from deploying their resources, but do not because they do not realise it or because they believe they might not be able to appropriate enough of the gain. By internalising these resources (and their associated opportunities for deployment) within the context of a new institution (i.e., as part of a new firm or newly organised old firm) the entrepreneur makes such deployments more salient (i.e., by enhancing their motivation and perception) and, therefore, more likely to occur.

It follows then that some change in organisational capability is generally required to accommodate the entrepreneur's vision and ability. This change can come in the form of a new firm, as Schumpeter emphasised, or as an 'administrative reorganisation' of an existing one, as is implicit in Penrose's theory. In either case, 'Schumpeterian integration', takes place as a substitution of the current organisational context (e.g. market, firm or sub-unit of a firm) by a new one that involves some administrative reorganisation. This internalisation of economic activity that Schumpeterian integration brings about arises not from any fact that the new organisation is *better* than alternatives. Rather, it is simply because of the lack of a supporting (i.e., enabling, motivating or perception stimulating) context in either the market or another firm that would divert resources *away* from theoretically 'best methods' towards the new combinations that may be more promising, but only appear so as perceived by the entrepreneur (see Moran and Ghoshal, 1999). This general unfamiliarity with means and the relative unavailability of supporting context implies that there exists an inherent 'non-contractible' nature of the coordination that is needed to carry out new combinations. It is this 'non-contractibility' that gives rise to the need for a non-market form of organisation and leads to the integration of at least some activities within a firm.

As a consequence, activities that need to be and can be complementary to the entrepreneur's competence are integrated inside the boundaries of a firm. The firm provides, via its administrative organisation, a coherent context for coordinating these activities in ways that ensure they become and remain complementary (and do not become substitutes). It does this, largely by influencing the perceptions and behaviour of its members in ways that are very different from that provided by any other context that may be 'within reach'. Note that this view drastically differs from the 'Comparative Institutional Analysis' of Williamson (1975, 1985A,B) which largely focuses on the role of asset specificity and its implications for opportunistic human behaviour. Instead, it is the idiosyncratic nature of knowledge and competence, i.e., the 'suitability to the innovation of the particular market institutions in place at the time in question' (Langlois, 1995, p. 86), that makes an entrepreneurial breakthrough very much context-dependent. It is also a manifestation of the differences in perceptive and cognitive capability that is at the centre of the first view of the firm presented in the last section of this paper. Consequently, it is the combination of differences in perceptive capabilities, on the one hand, and transaction- and information-costs of informing and persuading those with complementary abilities, on the other hand, that creates the forces which lead to the creation of firms and their expansion into previously non-complementary activities. In many cases, this will involve some form of vertical integration that internalises control over some of the activities of a supplier (upstream) or buyer (downstream) to enable the firm to pursue a certain kind of activity.

We are now ready to address the questions posed at the beginning of this section. While Penrose's theory explains why firms grow and the mechanisms through which growth is accomplished, continued growth over time requires Schumpeterian integration. Schumpeterian integration is needed to provide a context that is hospitable for carrying out new combinations. Without Schumpeterian integration firms will not be able to continue to grow indefinitely, particularly in prosperous economies where the opportunity cost of carrying large stocks of resources is likely to be greater for any one organisation. But if firms are, in general, no more likely than markets to provide the necessary context for new combinations, then what can account for the continuing growth of larger firms? Indeed, what enables these firms to forestall the migration of their resources to more productive uses by others? In other words, is there any reason why we should not expect to observe a long-term trend towards more and more Schumpeterian integration accommodated in new and relatively smaller (nor larger) firms? We suggest that 'management competence' contributes to a symbiotic process of persistent firm growth that counteracts the more natural process of the opportunities contributing to that growth being exploited outside of the firm.

II.c. The symbiotic process of growth

Because Penrosean growth and Schumpeterian integration are distinct processes, each driven by different drivers of growth, they tend to affect the direction of an individual firm's growth in different ways. For example, Penrosean growth, which stems from opportunities to enhance the productivity of a given set of resources, seeks, on the one hand, to extend the firm's existing organisational capabilities, often through horizontal expansion into fields that are in some way related to the existing firm-specific stock of knowledge. Schumpeterian integration, on the other hand, which is determined by the entrepreneur's demand for a 'redesign of [those] complementary activities' (Langlois, 1995, p. 90) that constitute the firm's organisational capability tends to induce growth by integrating and somehow changing one or more of the upstream or downstream activities in a production services value chain. While Penrosean growth is driven by *what the firm can do*, Schumpeterian integration is driven more by *what could be possible* with a different organisational capability. The interaction of these two forces can, perhaps, best be seen by tracing the emergence and growth of a hypothetical firm.

In the first instance, the inherent non-contractibility of the entrepreneurial judgement needed for 'carrying out new combinations' that generally cannot be accommodated by current organisational capacity leads to the creation or expansion of a firm via Schumpeterian integration. This creation or expansion implies immediate demand for improved organisational capability to encompass the newly integrated activities with the old. For example, by integrating vertically-related activities, firms provide the context for knowledge sharing and collective learning among previously competing entities. This process, therefore, contributes as much to an expansion into non-complementary activities (by redefining what is complementary) as it simultaneously sets the stage for improved internal coordination and adaptation, i.e., for the enhancement of organisational capability. As a result, firms will be able to (indeed, must to some extent) take on new related resources or tasks in order to utilise this enhanced organisational capability efficiently (i.e. Penrosean firm growth). Because each integration of activities leads in a similar fashion to more resources under the control of the firm and, consequently, an expanded set of productive opportunities within its reach, the firm faces a continuing cycle of inducements to initiate one form of growth or the other: first, to find more productive uses for its current resources (i.e. Penrosean growth), then to restructure its administrative organisation in order to coordinate these uses more efficiently (Schumpeterian integration) and finally, to bring in new resources to exploit these new coordinative capabilities, only to start the cycle once again.

As Cohen and Levinthal (1990) have documented, the ability of a firm to assimilate and use new knowledge is largely determined by its prior related knowledge. To the degree that the ability to perceive potential resource combinations is largely a function of knowledge (in addition to personality-related factors such as patience, and other virtues such as imagination, which may or may not be related to prior knowledge), the absorptive capacity of an organisation and that of its members arguably affects their entrepreneurial judgement. As Cohen and Levinthal argue, absorptive capacity is most likely the by-product of routine activities when '. . . the knowledge domain that the firm wishes to exploit is closely related to its current knowledge base' (p. 150). In other words, Penrosean firm growth promotes the development of cognitive capabilities, including those that are involved in the development of entrepreneurial judgement. At the same time, it is necessary to recognise that these consequences are not automatic and their actual realisation depends largely on the transfer of knowledge within the (grown) organisation. Thus, the limitations to the rate of growth as put forward by Edith Penrose apply equally to Schumpeterian integration, and thus to firm growth in the combined form that we have suggested. The ability of management to assimilate firm-specific knowledge is a crucial factor for the iterative and symbiotic cycle of Schumpeterian integration and Penrosean growth to be initiated and maintained.

Furthermore, as with management competence in general, this assimilative ability can reside in any organisational member and not just in the firm's managers. Cohen and Levinthal argued that the difference in the direction of the firm's expansion (i.e., into related versus unrelated activities), correlates with a difference in the ability to absorb new knowledge (i.e., firms find it harder to accumulate new knowledge when expanding into unrelated fields). We suggest that by redefining the basis for related-ness, Schumpeterian integration facilitates the absorption of new knowledge in new areas and, thereby, stimulates more unrelated expansion, as well. In other words, Schumpeterian integration can create the conditions for organisational learning and adaptation which in turn improves organisational capability and thus facilitates further lateral growth of the organisation to utilise existing, though diverse, capabilities of the organisation's members (for example into other geographical areas or through increased scale). This means, firm growth is very much path-dependent on the ability of the organisation to cope with successive stages of its growth process. It should be noted that, whereas Cohen and Levinthal focus on R&D activities, our discussion extends to all parts of the organisation.

There are various pieces of historical evidence to support this symbiotic characteri-sation of firm growth. For instance, consistent with Langlois and Robertson's account of the acquisition of Fisher Body by General Motors in 1926 (Langlois and Robert-son, 1989), had both Fisher and GM remained separate entities with divergent visions, the necessary acquisition of new related (but still non-complementary) knowledge that was needed to exploit GM's vision of the potential for closed-body vehicles, would probably not have occurred. Integration is what made the activities associated with acquiring this knowledge complementary. Moreover, the integration of Fisher Body also improved GM's internal production flow and thereby enabled workers to acquire skills about processes related to chassis assembly which were applicable in neighbouring activities. These skills, and the associated learning and adaptation of GM's production line workers, increased GM's overall productivity, which in turn created incentives to increase scale to accommodate these newly acquired skills. Thus, the ability to envisage a commercial solution (the closed-body vehicle), which made necessary the integration of vertically-related activities, sub-sequently increased the organisational capability to deal with this activity, and eventu-ally created incentives to broaden the scale (and scope) of this activity.

Similarly, Ford's acquisition of the Keim Mills initially sought to reduce uncertainty about product flows from its supplier. However, 'once the equipment was in the Ford plant, Ford engineers noticed ways to improve the technology and integrate it better into the rest of Ford's internal production system. This had the effect on the margin of biasing technological change in a systematic direction and further reinforcing Ford's trend towards vertical integration' (Langlois, 1995, p. 93). Initially, at least, this technological trajectory (Dosi, 1988) contributed to a coherent context for motivating and enhancing the salience of many resource deployment opportunities that were then within Ford's reach. This, in turn, exposed a need for and stimulated improvements in Ford's organisational capability, leading to a further raising of its production-cost advantage over its competitors. Thus, the firm's ability to acquire specific knowledge enabled it on the one hand to exploit economies of scope by moving into related markets, and on the other hand to expand geographically to exploit economies of scale (Chandler, 1990).

III. FIRMS AND ECONOMIC PROGRESS

There can be little doubt that the sources of economic progress for a society are rooted in the productive processes of its institutions—foremost are firms and their competitive interplay through markets. Given the self-evident importance of wealth creation to social and economic progress and its obvious dependence on the conduct of organisations in general, and business firms in particular, there is clearly a need for a theory that relates the roles of firms to the creation of wealth. Building on the work of Schumpeter (1934) and its subsequent development within the literature on the evolutionary theory of the firm (Nelson and Winter, 1982) and the resource-based view of strategy (Dierickx and Cool, 1989; Teece, Pisano and Shuen, 1997; Conner, 1991; Barney, 1991) Moran and Ghoshal (1996, 1999) have recently presented a theoretical justification for how firms may create economic value beyond the extent that can be achieved by markets alone and why many firms are needed to ensure the effective exploitation of the multitude of alternative productive opportunities inherent in any prosperous economy. In this section, we will recapitulate and build on their arguments, relating the process of value creation by firms and the economic progress it implies with the characteristic features of such progress with which all successful firms must cope.

III.a. *The conditions for resource combination and exchange*

In the second section of this paper we reiterated Schumpeter's argument that new resource combinations are the source of wealth and that their deployment determines whether, and the extent to which, additional value is added to the economic system. Here, we focus on two types of resource deployments, which account for the bulk of all wealth creation: (i) new resource combinations, as the source of new potential value; and (ii) exchange, which is necessary for the actual realisation of this potential value.

One can identify three conditions which must exist for any particular deployment of resources to occur intentionally and voluntarily. First, the opportunity to deploy the resources in question must obviously exist. Second, the parties with the opportunity to execute the particular deployment must be motivated to do so. That is, they must expect to appropriate some value from the deployment. Third, the resource deployment opportunity must also be perceived to exist by the relevant parties. Obviously, the satisfaction of these three conditions is interdependent. All purposive resource deployments imply the satisfaction of the last condition, which itself is greatly influenced by the existence of the other two.

Since resource combination and exchange are both means of deploying resources, the execution of either requires that all three necessary conditions for intentional and purposeful deployment be satisfied. That is, some potential opportunity for either must exist, and that opportunity must also be perceived and be motivated. In addition, since exchange, by definition, requires more than one party, an additional and potentially demanding condition is added, namely, that the three conditions must be satisfied by all parties whose resources are to be exchanged. This additional condition is commonly referred to as 'double coincidence'.

The alignment of these three conditions (i.e. the deployment opportunity and its motivation and perception) among actors capable of executing the resource deployment opportunities poses significant challenge for theory and practice. While exchange goes a long way towards overcoming the first constraint, exchange itself is largely inhibited by a failure to satisfy the second and third conditions. In a hypothetical world of no transaction costs, each opportunity or need for exchange would independently provide all the motivation that is necessary for any exchange to take place. However, in the real world, transaction costs probably manifest themselves by distorting an exchange opportunity's appropriability and/or one's perception of its appropriability or perception of the opportunity itself. Thus, the added constraint imposed by the 'double coincidence' is likely to accrue potentially large additional transaction costs in locating suitable exchange partners and in assessing the availability and usefulness of the resources in question. Under these circumstances, the institutional logic of markets as arenas of exchange imposes severe limits on the scope of value-adding resource deployments that are supported by exchange. There are two reasons for this.

III.b. The limitations of markets

First, markets may be 'missing', i.e., even though the opportunity or need for deployment is recognised, the market conditions or conventions necessary for exchange may be incomplete or entirely absent. Such situations can include exchanges for which pricing is difficult, money is inappropriate, rights are unclear, inadequately specified and protected/enforced, etc. The nature of these 'market failures' stems in part from the conservative standards that must be applied by most markets in establishing exchange viability and not from the distribution of resources and rights *per se*. In the markets of the most developed economies, conventions have evolved that support and reinforce an institutional logic which enables actors to enter into and exit from a variety of exchange relations at relatively little cost and, thereby, preserve their independence from all other actors. The very advantage of this independence of individual actors, which makes it easier (and therefore more efficient) for these market participants to adapt autonomously to changing conditions without the need to consult others, necessarily restrict the form of viability which must exist around each exchange transaction. Consequently, market exchanges must satisfy the condition of what Coleman (1990) has described as 'reciprocal viability', i.e. each exchange must be viable, by itself, for each of the participants to that exchange.

Second, many resource deployments are systematically discouraged by markets to the extent that, in the presence of transaction costs, *some* potentially value-creating resource deployments are always discouraged by any *single* organisation, whether that organisation comprises one market or one firm. In the presence of transaction costs, any institution (or set of institutions) that induces behaviour through a system of incentives (e.g. by allocating resources, assigning rights and restricting access) encourages the pursuit of some opportunities and necessarily discourages the pursuit of others. That is, each institution favours the conduct of a unique set of economic activities over all other activities—and the set of favoured activities comprises those that are more efficient, as defined by that institution (or set of institutions). Therefore, some resource deployments critical for future economic development may be unlikely to come about from market exchange, given the current distribution of resources, rights and individual perceptions (North, 1990). In other words, markets fail to adapt institutional incentives to new opportunities, which results in some degree of lock-in to the current set of opportunities that are motivated.

III.c. Value creation by firms

Consequently, for economic progress of societies (i.e. to enhance the fraction of potentially value-creating resource combinations and exchanges that are actually executed) some other institution, besides those associated with markets alone, is required. It is this role that organisations generally, and firms in particular, play and it is through playing such a role that new institutions are formed which help create value for society beyond what markets alone can create. Firms broaden the scope of exchange in ways that systematically address both of these limitations of markets by creating their own unique institutional logic for overcoming the market's stringent demands for viability and for circumventing (at least for a while) the severely constraining forces of static efficiency that exist in the market. Firms possess the ability to pursue resource-deployment strategies that are difficult (i.e. costly) or impossible to pursue in markets. Such strategies include those which require resources that are difficult to acquire or accumulate through market exchange (e.g. because prices or even markets are 'missing') or the use of which is difficult to coordinate among independent actors, subject to the stringent demands of 'reciprocal viability'.

Resource-deployment strategies also include those that cannot be created, accumulated or deployed in ways that viably satisfy the market's stringent demands for static efficiency yet appear promising to those with requisite local knowledge.

The organisation's advantage in overcoming the market's 'reciprocal viability' constraint dramatically broadens the scope of resources that are exchanged and considered for deployment within firms relative to markets. It is an organisation's internal institutional context that permits it to ensure the viability of exchange under less restrictive conditions. Two such less restrictive conditions of viability have been referred to by Coleman as 'independent viability' and 'global viability'. Whereas the former requires only that each actor has a positive account balance with the organisation as a whole and not with each of the other exchange parties, 'global viability' is even less restrictive, in that individual actors themselves do not all require a positive balance for it to be globally viable. Because the organisation itself is an 'implicit third party' to every exchange relation, members are able to enter into and maintain relations that may be beneficial to the organisation itself even if they are not directly beneficial to them (Coleman, 1990, pp. 428ff). Thus, by permitting individuals and groups to enter into voluntary exchanges that benefit the organisation but benefit themselves only indirectly, organisations open up and make accessible to members a much broader range of resource deployments (including exchanges) than would be possible were exchange required to satisfy the stringent condition of the 'double coincidence' and, hence, of 'reciprocal viability'.

The second advantage of organisations as additions to the market system lies in their ability to overcome the constraint that is present in any single institution (i.e., institution-specific efficiency). In structuring incentives, all institutions, whether market or organisational, and the conventions and norms that have evolved to support them, largely determine what economic activity is efficient and what is inefficient, given the institutional structure these institutions have helped to put in place. The conventions and norms of markets in most developed societies tend to be biased towards the achievement of 'allocative efficiency' (North, 1990). Instrumental in making a set of available options as efficient as possible by directing resources away from the less efficient and towards the more efficient uses, market adaptation tends to be guided by current relative efficiency and is independent of the efficiencies of future states. In other words, a highly efficient future state that must be preceded by the occurrence of relatively inefficient states is unlikely to be reached through market exchange, regardless of how efficient that future state may be (cf. Arthur, 1989).

Firms relax this constraint of market exchange by creating a unique subsidiary context within their boundaries—not an instrumental one that mirrors the market or responds to market failures (although, undeniably, some firms may do this)—but a coherent institutional context consisting of a combination of its own unique mix of incentives and muted market incentives that encourages the assimilation, sharing and combination of local knowledge in ways that are difficult to do under the alternative institutional context of the market. This unique context of each firm enables its members to actually defy (albeit, and importantly, only for a limited time) the relentless gale of market forces and thereby set the countervailing forces in motion necessary for a society to achieve 'adaptive efficiency' (North, 1990).

The role of the firm then is to transform the market context, which favours certain activities over others, into an alternative context that implicitly or explicitly favours economic activities that are otherwise unproductive, inefficient or simply not feasible in the market. The unique context of a firm affects, and is in turn affected by, both entrepreneurial judgement and organisational capability. Given that each individual in the firm is allowed to disregard the immediate and reciprocal viability requirements present in the market, the firm creates the incentives for each of its members to share their local knowledge (a form of resource deployment) to broaden the scope of their collective perception of opportunities. At the same time, by facilitating such sharing the firm also enhances its stock of knowledge and, thereby, its organisational capability to execute resource deployments and exchanges to pursue new opportunities. Arguably, it is this process, driven by the management competence of firms, that lies behind what several economists have highlighted as the on-the-job accumulation of human capital through 'learning-by-doing' (Arrow, 1961; Schultz, 1962; Minzer, 1962; Stokey, 1988; Young, 1991) and that Lucas (1993) has recently offered as a key explanation of the economic progress of societies.

Because of their greater size and diversity, large firms can do all these things, only with greater staying power. While the piper must ultimately be paid—i.e., the firm's defiance of the market straightjacket cannot continue indefinitely and must lead sooner or later to efficient (i.e., adaptable) behaviour either in the production of goods or services that conform to the market's evolving rules or that change the rules themselves—large firms can make more, bigger and longer lasting bets. Their advantage lies in their ability to integrate those resource deployment opportunities that are unfavoured (e.g. because the benefits are not easily appropriable) in markets. Hence they can reach much further than smaller firms beyond the 'prevailing wisdom' of current best practices while at the same time withstanding the strong selection forces imposed by markets (particularly prosperous ones) to adapt to the routines of current prices. While some, perhaps many, of these bets would undoubtedly go wrong and lead, ex-post, to complaints of waste and inefficiency (Williamson, 1991; Jensen, 1993), it is out of these bets that many of the value creating innovations (essentially new resource deployments) emerge that enhance the wealth of societies.

IV. DISCUSSION

In the last two sections we have made an attempt to explain the relationship between management competence and the emergence of large firms on the one hand, and economic progress of societies, on the other. So far, however, each half of our twin-hypothesis is largely independent of the other. In this section, we shall focus on the relationships between the two sides of the argument and their theoretical and practical implications.

In order to understand the possible linkages between firm growth and economic progress, we shall first examine the interrelationship between organisational capability and entrepreneurial judgement. As we have set out in the beginning, both are different (although related) aspects of the same concept. Both kinds of competence are non-contestable, path- and context-dependent, largely tacit, and difficult to imitate. Moreover, as we have seen in the discussions above, they are closely interlinked. As for the growth of firms, the non-contractability of entrepreneurial judgement in the presence of uncertainty and costs related to informing and persuading individuals with complementary capabilities/resources, leads to the emergence and subsequent expansion of the firm through Schumpeterian integration. Yet, once the firm is established, the shared knowledge out of collective learning enables the firm to take on new tasks in order to utilise this organisational capability (Penrosean firm growth). Similarly, the institutional context of the firm creates incentives to share local knowledge among its members which enhances its organisational capabilities, and simultaneously provides the context for collective 'learning-by-doing' which leads to improved entrepreneurial judgement. Together, the ability to perceive new resource combinations, carry them out, and subsequently realise their economic value through the means of exchange enhances the effectiveness with which the resources of a society are used, thus leading to its economic progress.

We should, however, be cautious as to the limits of these effects. Clearly, there are limitations to both firm growth and size. Moreover, there can be no doubt that these limitations also apply to economic progress. In other words, one giant firm that employs the whole population of a country (whether it makes profits or not) will almost certainly fail to enhance continuously the value that can be obtained from that country's endowment of resources. As we have seen, in the presence of transaction costs the advantage of those institutions, brought about by firms is their ability to relax the viability constraints of markets. Yet, this is achieved only at a cost. This cost manifests itself in the very nature of institutions as systems of conventions and norms.

IV.a. Institutional pluralism

Over centuries, institutions associated with both markets and organisations have evolved to help us cope with the ubiquity of transaction costs. By collectively establishing (largely through our institutions) routine ways of dealing with common activities, we have enabled ourselves to engage in a large number and variety of exchanges without the need to focus on every aspect of every single exchange. While the benefits of our habits and routines are many, they are accompanied by a significant cost associated with changing those routines.

The advantage of any single system of institutions, whether a decentralised system of prices, a centralised authority system or some other set of complementary conventions and norms, is its ability to focus on certain activities while ignoring others. The disadvantage is the cost in overcoming that focus to do other things. Thus, in what may sound paradoxical, firms, while reducing some of the constraints that markets place on individual transactions, actually increase constraints on individual perceptions. In other words, the disadvantage of institutions is the cost in overcoming those constraints in the presence of changing external conditions that challenge the very foundation of those institutions. In terms of our framework this means that the ability to support the scope of exchange that is necessary to promote all value-adding resource deployment opportunities in a systematic way gets lost in any *single* institutional system, whether it comprises a market or a firm. In the presence of transaction costs, no single institution or set of institutions by itself is capable of bringing about adaptive efficiency, because it seeks to adapt its constitution of resource rights to make them as efficient as possible, given the constitution of rights it must operate in (North, 1990).

Consequently, a variety of firms, by creating institutional contexts that specify different sets of opportunities that are motivated and perceived and which are unique to each, provide an institutional alternative that substantially broadens the scope for exchange relative to markets alone. Because the set of motivated and perceived opportunities differs from institution to institution, many more deployments than are likely to occur in a single institution can be expected in the presence of many institutions. In the process, the set of deployment opportunities motivated in the system as a whole continually adapt, in a relatively efficient manner, to opportunities as they are perceived.

In the face of these limitations of single organisations, that is, single firms or single markets, institutional pluralism remains the viable solution to either problem. Institutional pluralism constitutes itself in the multitude of firms, each representing a different context—i.e., a different set of convictions and bets—competing with one another with differing intensity. Institutional pluralism contributes to the process of achieving adaptive efficiency in several ways. First, the scope of exchange is broadened to include more opportunities that are not exploited elsewhere. Second, some resources that are currently deployed elsewhere are made available for deployment within the firm under a different set of motivating conditions. By replacing those motivating forces that encourage certain deployments with forces that motivate alternative patterns of deployment, firms make it easier for value-creating new combinations to be discovered. Thus, in the absence of institutional diversity, increasing firm size will create problems. First, larger size ultimately slows down the process of organisational learning (the basis for both aspects of management competence), because the ability to *share* a common stock of knowledge and capabilities is limited by the ability to *hold* it. Second, lacking institutional diversity reduces the diversity in entrepreneurial judgement, which reduces the scope of potential resources to be deployed in new ways and retards progress by '. . . restricting the number of independent sources of initiative' (Scherer and Ross, 1990, p. 660). This reduced scope, in turn, limits the quantity of resources actually exchanged, and thus the amount of value realised. The moral of the story is: individual rent-seeking firms must have the ability to create an internal context that relaxes certain market constraints and temporarily widens the scope of possible and actual exchanges, while at the same time there must exist sufficient competitive pressure by *other* firms which seek to appropriate some of that value created and thereby force the innovator to re-focus on ever-changing resource combinations.

While the firm thus benefits from innovative new resource-combinations that result either directly in better final products or in enhanced intermediate products, which translates into improved factor productivity (Grossman and Helpman, 1991) and, thus, into economic progress of the society, the inherent characteristics of innovation as information capital creates forces that will eventually lead outside firms to benefit from the innovation. Information capital, i.e., productive knowledge with value that derives from its utilisation in a given context, is very different from physical capital in that it is ultimately subject to imitation by competitors (e.g. Dosi, 1984). Thus, the process of internal resource combination and exchange inevitably leads to some of the newly-created value to spill over to other firms in a process that Schumpeter described as 'handing on'. However, this usually happens with a lag, so that, even if eventually all of the value is captured by imitators in the market, until that time, the innovating firm will have benefited from the rent stream that it had generated. Therefore, the more value a firm creates, the more likely it is to benefit from some of that value in the form of appropriable, if transient, rents. This competition can never reach a final outcome, as it is impossible to appropriate the full benefits of the innovation. As time passes, the information contents of the innovation will be revealed and diffused to competitors. Moreover, new innovations will undermine the value of the first in a never-ending process of 'creative destruction' (Schumpeter, 1950). In fact, this process constitutes the interaction between firm growth and economic progress in a dynamic framework. The rents that the firm keeps help it to acquire new resources and to grow; the part that is handed on contributes to societal growth. Thus, firms create new value and as markets force some of it to be handed on to society, both the firm and society grow, in a creative and constant tension, that drives economic progress.

In summary, the argument we have put forth in this paper is that the notion of 'management competence' provides a plausible explanation for the apparent disproportionate distribution of large firms in the most productive economies. We have suggested that 'management competence' is the factor that leads to the evolution of those institutional innovations that promote and sustain the continued growth of both prosperous nations and large firms in these nations. Institutional innovations are responsible, on the one hand, for the realisation of more economic value from a given society's endowment of resources and, thereby, its growth and prosperity. The emergence and growth of firms, where arguably the bulk of a society's management competence resides, are largely a manifestation of these institutional innovations and the prosperity they enable. The proliferation of institutional innovations and their accompanying prosperity, on the other hand, make it difficult for any individual firm, *and more difficult for larger firms*, to retain the level of management competence needed to continue their growth or even to sustain their size. To achieve continuing growth over time in such a challenging (i.e. high management competence) environment, firms must be able to attract, develop and retain high levels of management competence within their own institutional boundaries. Large firms in prosperous societies that are rich in management competence (i.e. those that have met the challenge), are more likely than similar sized firms in less richly endowed societies to discover and to implement institutional innovations that will enable them *and their societies* to explore better and to exploit the wealth-creating potential of their societies' resource base. In short, with management competence, a country simultaneously expands its income and the proportion of large firms operating in its economy, thereby establishing the correlation we have observed.

IV.b. *Implications for economic theory*

The issue of whether, and if so, how, the size of firms is related to the welfare of a country has been dealt with extensively by economists, particularly by those working in the fields of industrial organisation and institutional economics. The great majority of their arguments have tended to focus on factors such as market power and the link between innovative activity and firm size measured by profits or market share. It is only recently that studies have concentrated on the issue of 'creative destruction' as the source of economic development, and the findings seem to support the Schumpeterian hypothesis about the link between market structure and innovative activity (Caballero and Jaffe, 1993; Bughin and Jacques, 1994; Smulders and van de Klundert, 1995; Archibugi, Evangelista and Simonetti, 1995). Our arguments here are aligned to this line of enquiry and, taken together with Schumpeter's (1934, 1950) premise that 'perfect competition is incompatible with innovation', they suggest some potentially rewarding extensions in the areas of both industrial-organisation economics and growth theory.

For industrial organisation, one implication is the need of, and potential benefit from, looking more closely at the concept of management competence. Although the concept of management competence as a determinant of firm size has been formally introduced in the past (Lucas, 1978), main-stream economics has hardly been affected by it. Second, our arguments support the growing challenge to the perfect-competition formulations that have influenced past efforts at restructuring transitional economies. In the 'real world' of developing countries, there is now sufficient ground to argue that recent efforts of structural adjustment such as IMF/World Bank initiatives in the former Soviet Union, India, and other developing countries, have produced mixed results (Bock, 1994). In particular, one can now see that shock therapy attempts to create a *tabula rasa* of 'free' markets have in fact damaged the inherent management competence embedded in existing enterprises. The framework we have presented shows why, in the process of helping transitional economies, it may be necessary to focus on improving management competence within existing organisations while, at the same time, creating the markets that are necessary to enhance their innovative abilities. A further implication for industrial organisation, and in particular for transaction-cost economics, is the recognition that the existence of firms is not purely a manifestation of 'market failure' based on restricted assumptions about human behaviour (i.e. opportunism) and asset specificity as the determinants of the 'most appropriate' governance structure. Instead, as we have suggested here, it may be useful to develop a more 'positive' theory of the firm based on the recognition that organisations, such as firms, represent institutional contexts that have distinct advantages over markets in the process of identifying and exploiting new resource combinations.

For economic growth theory, a similar (and based on advances in industrial organisation) departure from perfect competition assumption has allowed the endogenisation of the process of innovation itself, which had so far been neglected. Some of the more recent contributions in the so-called neo-Schumpeterian growth models (Romer, 1990, 1994; Grossman and Helpman, 1991, 1994; Aghion and Howitt, 1992, 1994) have begun to focus on intentional and purposeful innovation. Most technological progress, which is undoubtedly the engine of growth (as first suggested by Schumpeter, 1934, and first empirically shown by Solow, 1970), requires an intentional investment by profit-seeking firms. This profit is initially earned by the innovator, but, at some stage, will be appropriated by another firm which discovers a better way of combining existing resources. There is empirical evidence that this resource-combining capability of firms has significant aggregate effects on macroeconomic growth. Carlsson (1991) and Eliasson (1991) have shown that the reallocation of resources within and between existing plants is a major contributor to total factor productivity growth. Our arguments also suggest that the concept of management competence has some important implications for the issue surrounding the measurement and comparison of productivity, in particular capital productivity. As has been found in a study by the McKinsey Global Institute (1996), the ability of American managers to allocate resources to their most efficient use has more than offset the considerably lower aggregate savings rate of the US economy compared to those of Germany and Japan.

Yet, despite recent advances in the adoption of more heterodox assumptions by economic theorists (such as imperfect competition, incomplete appropriability, and increasing returns to scale), there are still unexplored sides of the issue, some of which we have tried to identify. Most notably, theories of economic growth ought to incorporate the role of institutions and of transaction costs in the process of economic development (North, 1989; Aghion and Howitt, 1994). As we have argued, by looking at the institutional dimension of resource combination and exchange, economic development as the process of expanding the fraction of potential transactions that are realised, must necessarily rely on the institutional context provided by firms and markets. Moreover, by replacing physical capital with knowledge as the input for transaction creation, it should be possible to address the importance of specific idiosyncratic capability and competence—including management competence—embedded in those institutions as an essential factor in the growth of firms and the prosperity of nations.

IV.c. Implications for strategy theory

Two theoretical perspectives dominate the strategy literature and both focus on the appropriation of value by firms, rather than the creation of value by them, as the basis for explaining and predicting firm performance. In contrast, our arguments here will suggest that value creation rather than value appropriation must lie at the heart of strategy theory and, therefore, at the centre of how we conceptualise the role of management as the shapers of strategy in their firms.

Consider first those theories that are based on the perspective of traditional industrial organisation (IO) economics, which arguably represents the earliest and most rigorous efforts to date in formalising strategic concepts in theoretical terms. Standard economic theory holds that unless otherwise obstructed, the competitive forces that drive rivalry among firms in any given industry will also tend to force performance across industries and among firms within industries towards convergence at equilibrium. Firm differences, i.e. heterogeneities in their performance, that persist at equilibrium are attributed to the barriers to entry that characterise different industries (Bain, 1956) and mobility barriers that restrict rivalrous behaviour and promote strategic interaction among groups of firms within the same industry (Caves and Porter, 1977). The objective of strategic management, both as positive and as normative theory, according to this IO perspective, is one of gaining and maintaining market power to appropriate as much of the value that accrues from these economic rent-sustaining barriers as possible. Indeed, the prescriptions that flow from Porter's (1980, 1985) five-forces model of competitive strategy and value chain analysis are all centred around steps to gain competitive advantage by positioning a firm in its industry in ways that facilitate the appropriation of as much value as possible from the firm's suppliers, buyers, competitors, potential entrants (to its industry or strategic group) and producers and potential producers of substitute products and/or services.

The more recent emergence of the 'resource-based view' (RBV) of the firm has extended this IO perspective to explain behaviour at the level of the individual firm, particularly in the markets for a firm's factor inputs or resources (Rumelt, 1984; Wernerfelt, 1984). According to the general consensus of the resource-based view, isolating mechanisms (Rumelt, 1984, 1987) act as mobility barriers which restrict the extent to which essentially all firms are able to mimic any particular firm's behaviour and, thereby, replicate its performance and ultimately appropriate some or all of its rent streams. Note, that the strategic behaviour that is implied by this perspective (both prescriptive and normative) is very similar to the one that is implied by the IO perspective: the strategy of a firm focuses on appropriating the rents of other firms and preventing other firms from appropriating its rents and is not concerned with the sources of these rents.

The reason for this bias in favour of rent appropriation over rent creation stems not from any notion that appropriation is or should be preferred but reflects the prevailing view that purposive action is more usefully applied in the protection of rents than in their creation. Indeed, the RBV explicitly recognises heterogeneity among firms as the source of all rents but this view remains atheoretical in regard to explaining the value-creation process because it attributes the source of rents and, by association, the source of value to unexpected changes (Rumelt, 1984) or luck and foresight (Barney, 1986). The general consensus is that firms could not benefit from any recipe-like strategy for creating rents, even if one did exist, because once such a strategy is identified and implemented, its value would soon be eroded by its imitation by others. Therefore, no systematic theory of rent creation exists or can exist (see, for example, Barney, 1986; Schoemaker, 1990). It is not surprising, then, that the centre stage of strategy theorising has focused (indeed, is left with no alternative but to focus, if for no other reason but by default) on rent appropriation as the means of securing a position of sustained competitive advantage.

In the recent past, a number of scholars have begun to challenge this focus on value appropriation as the essence of strategy and, indeed, to contest the underlying theoretical framework that has been the source of this bias. Dierickx and Cool (1989), for example, have argued that to create sustainable competitive advantage, firms need to develop and accumulate strategic resources and capabilities; an argument that has been echoed and extended by Teece, Pisano and Shuen (1997) in their conceptualisation of 'dynamic capabilities' of firms. In a related but different stream of work, Conner and Prahalad (1996), Kogut and Zander (1992), Nonaka and Takeuchi (1995) and others have also begun to sketch out an argument on how firms can create value, with a particular emphasis on their ability to integrate, expand and exploit knowledge in ways that markets cannot.

While still incomplete in many ways, the theory we have put forward in this paper resonates with this relatively more positive view of firms as value-creating institutions. Such a view opens up the opportunities for developing some very different theories of strategy, based not on the task of locking-in the appropriablity of existing rent streams but on the continuous creation of new, if transient, rents through innovative new combinations of resources (Winter, 1995). Such a theory of strategy would also imply a very different and much more satisfying role of management: instead of being the agents that impede social welfare by preventing the Schumpeterian process of handing on, they would have to see themselves as the agents of social progress who drive the process of wealth creation.

REFERENCES

AGHION, P. and HOWITT, P. (1992). A model of growth through creative destruction, *Econometrica*, Vol. 60: 323–52.

AGHION, P. and HOWITT, P. (1994). Endogenous technical change: the Schumpeterian perspective, in *Economic Growth and the Structure of Long-term Development*, edited by L. L. PASINETTI and R. M. SOLOW, London, Macmillan.

ARCHIBUGI, D., EVANGELISTA, R. and SIMONETTI, R. (1995). Concentration, firm size and innovation: Evidence from innovation costs, *Technovation*, Vol. 15, No. 3, 153–63.

ARROW, K. (1961). The economic implications of learning by doing, *Review of Economic Studies*, Vol. 29, 155–73.

ARTHUR, W. B. (1989). Competing technologies, increasing returns, and lock-in by historical events, *Economic Journal*, Vol. 99, 116–31.

BAIN, J. S. (1956). *Barriers to New Competition*, Cambridge, MA, Harvard University Press.

BARNARD, C. I. (1938). *The Functions of the Executive*, Cambridge, MA, Harvard University Press.

BARNEY, J. B. (1986). Strategic factor markets: expectations, luck and business strategy, *Management Science*, Vol. 32, 1231–41.

BARNEY, J. B. (1991). Firm resources and sustained competitive advantage, *Journal of Management*, Vol. 17, 99–120.

BOCK, J. (1994). Innovation as creative destruction: the role of small businesses in the Commonwealth of Independent States (CIS), *International Journal of Technology Management*, Vol. 9, No. 8, 856–63.

BOWER, J. L. (1970). *Managing the Resource Allocation Process: A Study of Corporate Planning and Investment*, Division of Research, Graduate School of Business Administration, Harvard University, Boston.

BUGHIN, J. and JACQUES, J. M. (1994). Managerial efficiency and the Schumpeterian link between size, market structure and innovation revisited, *Research Policy*, Vol. 23, No. 6, 653–9.

BURGELMAN, R. A. (1983). A model of the interaction of strategic behaviour, corporate context, and the concept of strategy, *Academy of Management Review*, Vol. 8, 61–70.

CABALLERO, R. and JAFFE, A. (1993). How high are the giant's shoulders: an empirical assessment of knowledge spillovers and creative destruction in a model of economic growth, in *NBER Macroeconomics Annual*, Cambridge, MA, MIT Press, 15–74.

CARLSSON, B. (1991). Productivity analysis: a micro-to-macro perspective, in *Technology and Investment—Crucial Issues for the 1990s*, edited by E. DEIACO, E. HORNELL and G. VICKERY, London, Pinter.

CAVES, R. E. and PORTER, M. E. (1977). From entry barriers to mobility barriers to conjectural decisions and contrived deterrence to new competition. *Quarterly Journal of Economics*, Vol. 91, 241–61.

CHANDLER, A. D. JR. (1990). *Scale and Scope: The Dynamics of Industrial Capitalism*, Cambridge, MA, Belknap/Harvard University Press.

COASE, R. H. (1988). *The Firm, the Market, and the Law*, Chicago, University of Chicago Press.

COASE, R. H. (1991). The nature of the firm (1937), in *The Nature of the Firm: Origins, Evolution, and Development*, edited by O. E. WILLIAMSON and S. G. WINTER, New York, Oxford University Press, 18–33.

COHEN, W. and LEVINTHAL, D. (1990). Absorptive capacity: a new perspective on learning and innovation, *Administrative Science Quarterly*, Vol. 35, 128–52.

COLEMAN, J. S. (1990). *Foundations of Social Theory*, Cambridge, MA, Harvard University Press.

CONNER, K. R. (1991). A historical comparison of resource-based theory and five schools of thought within industrial organization economics: do we have a new theory of the firm? *Journal of Management*, Vol. 17, No. 1, 121–54.

CONNER, K. R. and PRAHALAD, C. K. (1996). A resource-based theory of the firm: knowledge versus opportunism, *Organisation Science*, Vol. 7, No. 5, 477–501.

DIERICKX, I. and COOL, K. (1989). Asset stock accumulation and sustainability of competitive advantage, *Management Science*, Vol. 35, No. 12, 1504–14.

DOSI, G. (1984). *Technical Change and Industrial Transformation*, London, Macmillan.

DOSI, G. (1988). Sources, procedures and microeconomic effects of innovation, *Journal of Economic Literature*, Vol. 26, 1120–71.

ELIASSON, G. (1990). The knowledge based information economy, in *The Knowledge Based Information Economy*, edited by G. ELIASSON *et al*, Stockholm, IUI.

ELIASSON, G. (1991). Deregulation, innovative entry and structural diversity as a source of stable and rapid economic growth, *Journal of Evolutionary Economics*, Vol. 1, 49–63.

GHOSHAL, S., MORAN, P. and ALMEIDA COSTA, L. (1995). The essence of the megacorporation: shared context, not structural hierarchy. *Journal of Institutional and Theoretical Economics*, Vol. 151, No. 4, 748–59.

GROSSMAN, G. M. and HELPMAN, E. (1991). *Innovation and Growth in the World Economy*, Cambridge, MA, MIT Press.

GROSSMAN, G. M. and HELPMAN, E. (1994). Endogenous innovation in the theory of growth, *Journal of Economic Perspectives*, Vol. 8, No. 1, 24–44.

HAMEL, G. and PRAHALAD, C. K. (1994). *Competing for the Future*, Boston, Harvard Business School Press.

HAYEK, F. A. VON (1945). The use of knowledge in society, *American Economic Review*, Vol. 35, No. 4, 519–30.

JENSEN, M. C. (1993). The modern industrial revolution: exit and failure of internal control systems. *The Journal of Finance*, Vol. 158, No. 3, 831–80.

KIRZNER, I. (1973). *Competition and Entrepreneurship*, Chicago, University of Chicago Press.

KNIGHT, F. H. (1921). *Risk, Uncertainty and Profit*, New York, Augustus M. Kelley [1965].

KOGUT, B. and ZANDER, U. (1992). Knowledge of the firm. Combinative capabilities and the replication of technology, *Organization Science*, August, 383–97.

LANGLOIS, R. N. (1995). Capabilities and coherence in firms and markets, in *Evolutionary and Resource-based Approaches to Strategy: A Synthesis*, edited by C. A. MONTGOMERY, Dordrecht, Kluwer Academic.

LANGLOIS, R. N. and ROBERTSON, P. L. (1989). Explaining vertical integration: lessons from the American automobile industry, *Journal of Economic History*, Vol. 49, No. 2, 361–75.

LUCAS, JR, R. E. (1978). On the size distribution of business firms, *The Bell Journal of Economics*, Vol. 9, No. 2, 508–23.

LUCAS, JR, R. E. (1993). Making a miracle, *Econometrica*, Vol. 61, No. 2, 251–72.

McKINSEY GLOBAL INSTITUTE (1996). *Capital Productivity*, Washington, D.C.

MINZER, J. (1962). On-the-job training: costs, returns, and some implications, *Journal of Political Economy*, Vol. 70, S50–S79.

MORAN, P. and GHOSHAL, S. (1996). Value creation by firms, in *Academy of Management Best Paper Proceedings*, edited by J. B. KEYS and L. N. DOSIER, Statesboro, GA, Georgia Southern University, 41–5.

MORAN, P. and GHOSHAL, S. (1999). Markets, firms and the process of economic development, *Academy of Management Review*, Vol. 24, No. 3, 390–412.

NAHAPIET, J. and GHOSHAL, S. (1998). Social capital, intellectual capital and the organizational advantage, *Academy of Management Review*, Vol. 23, No. 2, 242–66.

NELSON, R. and WINTER, S. (1982). *An Evolutionary Theory of Economic Change*, Cambridge, MA, Harvard University Press.

NONAKA, I. and TAKEUCHI, H. (1995). *The Knowledge Creating Company*, New York, Oxford University Press.

NORTH, D. C. (1989). Institutions and economic growth: an historical introduction, *World Development*, Vol. 17, 1319–32.

NORTH, D. C. (1990). *Institutions, Institutional Change and Economic Performance*, Cambridge, Cambridge University Press.

NORTH, D. C. and THOMAS, R. P. (1973). *The Rise of the Western World: A New Economic History*, Cambridge, Cambridge University Press.

PENROSE, E. (1959). *The Theory of the Growth of the Firm*, Oxford, Oxford University Press [1995].

POLANYI, M. (1969). *Knowing and Being*, Chicago, University of Chicago Press.

PORTER, M. E. (1980). *Competitive Strategy*, New York, The Free Press.

PORTER, M. E. (1985). *Competitive Advantage*, New York, The Free Press.

PORTER, M. E. (1990). *The Competitive Advantage of Nations*, New York, The Free Press.

ROMER, P. M. (1990). Endogenous technological change, *Journal of Political Economy*, Vol. 98, S71–S102.

ROMER, P. M. (1994). The origins of endogenous growth, *Journal of Economic Perspectives*, Vol. 8, No. 1, 2–22.

RUMELT, R. P. (1984). Towards a strategic theory of the firm, *Competitive Strategic Management*, edited by R. B. LAMB, Engelwood Cliffs, NJ, Prentice-Hall, 556–70.

RUMELT, R. P. (1987). Theory, strategy and entrepreneurship, *The Competitive Challenge*, edited by D. J. TEECE, Cambridge, MA, Ballinger, 137–58.

SAH, R. K. (1991). Fallibility in human organisations and political systems, *Journal of Economic Perspectives*, Vol. 5, No. 2, 67–88.

SCHERER, F. M. and ROSS, D. (1990). *Industrial Market Structure and Economic Performance*, 3rd edn, Boston, MA, Houghton Mifflin.

SCHOEMAKER, P. J. H. (1990). Strategy, complexity and economic rent, *Management Science*, Vol. 36, No. 10, 1178–92.

SCHULTZ, T. W. (1962). Reflections on investment in man, *Journal of Political Economy*, Vol. 70, S1–S8.

SCHUMPETER, J. A. (1934). *The Theory of Economic Development*, Cambridge, MA, Harvard University Press.

SCHUMPETER, J. A. (1947). The creative response in economic history, *The Journal of Economic History*, Vol. 7, No. 2, 149–59.

SCHUMPETER, J. A. (1950). *Capitalism, Socialism, and Democracy*, 3rd edn, New York, Harper.

SILVER, M. (1984). *Enterprise and the Scope of the Firm: the Role of Vertical Integration*, Oxford, Martin Robertson Press.

SIMON, H. A. (1945). *Administrative Behaviour*, New York, The Free Press.

SIMON, H. A. (1951). A formal theory of the employment relationship, *Econometrica*, Vol. 19, 293–305.

SIMON, H. A. (1991). Organizations and markets, *Journal of Economic Perspectives*, Vol. 5, No. 2, 25–44.

SMULDERS, S. and VAN DE KLUNDERT, T. (1995). Imperfect competition, concentration and growth with firm-specific R&D, *European Economic Review*, Vol. 39, No. 1, 139–60.

SOLOW, R. M. (1970). *Growth Theory: An Exposition*, Oxford, Oxford University Press.

STOKEY, N. L. (1988). Learning by doing and the introduction of new goods, *Journal of Political Economy*, Vol. 96, 701–17.

TEECE, D. J. (1988). Technological change and the nature of the firm, in *Technical Change and Economic Theory*, edited by G. DOSI, London, Macmillan.

TEECE, D., PISANO, G. and SHUEN, A. (1997). Dynamics capabilities and strategic management, *Strategic Management Journal*, Vol. 18, 509–34.

WERNERFELT, B. (1984). A resource-based view of the firm, *Strategic Management Journal*, Vol. 14, 4–12.

WILLIAMSON, O. E. (1975). *Markets and Hierarchies: Analysis and Antitrust Implications*, New York, Free Press.

WILLIAMSON, O. E. (1985A). Hierarchical control and optimum firm size, in *Markets and Hierarchies: Analysis and Antitrust Implications*, edited by O. E. WILLIAMSON, New York, The Free Press.

WILLIAMSON, O. E. (1985B). *The Economic Institutions of Capitalism*, New York, The Free Press.

WILLIAMSON, O. E. (1991). Strategizing, economizing, and economic organization, *Strategic Management Journal*, Vol. 12, 75–94.

WINTER, S. G. (1995). Four Rs of profitability: rents, resources, routines, and replication, in *Resources in an Evolutionary Perspective: A Synthesis of Evolutionary and Resource-based Approaches to Strategy*, edited by C. A. MONTGOMERY, Dordrecht, Kluwer Academic, 147–78.

YOUNG, A. (1991). Learning by doing and the dynamic effects of international trade, *Quarterly Journal of Economics*, Vol. 106, 369–406.

16

EDITH'S GARDEN AND A GLASS HALF-FULL: FURTHER ISSUES

CHRISTOS PITELIS

The Judge Institute of Management Studies and Queens' College,
University of Cambridge

I. INTRODUCTION

Great as it is, Penrose's contribution has its own 'missing links', and provides scope for potentially fruitful generalisations, extensions and syntheses. Perhaps uniquely interesting is Penrose's approach, in that the very lens that it provides helps to identify these very missing links and possibilities. The aim of this short concluding chapter is to deal with some such further issues. I hasten to add that this is not an exhaustive reference to points raised by others, for example in this volume. Rather I focus on issues that I consider crucial and less discussed.

I.a. Sell resources?

The first has to do with the issue of knowledge-related intra-firm resources. These are always taken by Penrose to lead to growth. This is not self-evident. As Penrose herself pointed out in the 1955 paper that preceded *TGF* and included its main points, there is always the possibility of selling the resources-advantages. Intriguingly, despite mentioning the possibility, Penrose did not pursue this further. The issue remains (and of course is well debated in the literature), why not sell your (intangible) assets in the open market? One has to return to some sort of 'market failure' argument to address this, either natural (transaction costs related) or structural (fear of creating a competitor) or both. Clearly, the argument can also be phrased in terms of differential capabilities; i.e., firms are better at making use of their own assets than other firms (i.e., the market!).[1] In any event, Penrose did not discuss transaction costs arguments.

---

[1] One can read arguments by Buckley and Casson (1976), and Kogut and Zander (1993), plus the importance of oligopolistic interaction (now recognised by most: see Pitelis, this volume) in this context. Whether one wishes to call this 'market failure' or 'differential capability', I believe is a semantic issue; the two are essentially the same. For example, Penrose herself explained vertical integration in terms of firms' differential ability to produce for their own needs. In part, she attributed this to difficulties in communicating to others a firm's exact needs. This can be interpreted as (dynamic) transaction costs due to tacit knowledge, or differential firm capabilities in transferring intra-firm tacit knowledge. We are talking about the same thing here.

Moreover, oligopolistic interaction is part and parcel of her story—yet is not explicitly discussed in the context of 'why not sell?'.

I.b. Shed resources?

A second, related, issue refers to the question 'why not shed excess resources?', e.g., labour. Penrose deals with the issue by claiming that she focuses on growing firms in a growing economy. This is all very well, but even successful firms in a growing economy may choose to shed excess resources, rather than use them for further expansion. While excess resources may have zero marginal costs, the costs of thinking what to do about them could be immense—as it is management time that is expended to do the thinking (see also Marris, this volume). In this sense, the issue is not as straightforward as it may appear at first sight. Shedding excess resources (labour, middle-level management) is not only an option; it is one that has been used extensively in the 1990s despite this being a period of unprecedented growth. The possibility of shedding resources clearly requires further elaboration.[1]

I.c. Intra-firm conflict?

A third issue has to do with a broad category of contributions related to what is now called 'agency', moral hazard, 'monitoring', residual claimants and the like. The literature on these issues is huge, notable contributions being Alchian and Demsetz (1972) and Jensen and Meckling (1976). In brief, the question here is that whenever there exists a 'principal–agent' relationship, as for example employer–employee, or shareholder–manager, and when the interests of the two parties are not *ex ante* fully aligned, agents may have discretion to pursue their own interests. When this is the case, it becomes important for principals to devise means for aligning the incentives of employees with their own.[2] The archetypal 'agency' problem is that of the Marxian notion of class struggle. In the context of the factory system, this takes the form of the capitalist employer being able to increase the rate of surplus value by intensifying

---

[1] Indeed one could go further and claim that there is an implicit theory of unemployment in Penrose; in brief, when resources exceed productive opportunity, unemployment, rather than expansion is more likely to be the outcome, at least domestically. This observation, of course, raises the most important issue of determining the conditions under which 'productive opportunity' falls short of the potential benefits of using excess resources *ceteris paribus* (i.e., even assuming zero marginal cost of using them, which need not be the case). This includes incompetent management to start with. (I am grateful to A. Pseiridis for pointing the latter out to me.)

[2] In the Alchian and Demsetz story, the question is applied to the very issue of obtaining the productivity benefits of the employment relationship and it is used for a conceptual justification of the need for private ownership and control of firms. Specifically, it is observed that in any team effort, shirking is likely to occur due to difficulties with measuring individual outputs. To ensure productivity is not thus prejudiced, a monitor of teamworkers is needed. However, the monitor needs to also be monitored. To avoid an infinite regress situation, it is best if the monitor is self-monitored, by becoming a residual claimant of any surplus left after expenses to other members of the team, etc., are paid. The contribution of Alchian and Demsetz is of extreme significance for numerous reasons, to which I cannot enter in this chapter. One of these, however, is that it addresses one of the most significant issues in economics, that of extracting labour from labour power, or more generally that of transforming work potential to work.

work and/or by introducing labour-saving technical progress. However, one does not have to go as far back as that. Besides Alchian and Demsetz, and Jensen and Meckling, managerial and behavioural theories of the firm are based on similar concerns. Managerial theories, as already suggested, rely on different utility functions by managers, plus control of their part, thus the ability to pursue non-profit (including growth!) objectives. The behavioural theory goes further, in suggesting the pervasiveness of conflicting objectives by different groups within firms, which, alongside bounded rationality, questions the very sacred cow, the profit motive. In view of conflicting intra-firm interests, and bounded rationality, 'satisficing' is more likely to be the firms' motive, see, e.g., Cyert and March (1963).

There is little of that in Penrose. Indeed there is no conflict of any kind within the production process. This is almost paradoxical concerning in particular the managerialist literature. As noted, Penrose's contribution has been linked with this literature. Her story has precious little to do with utility-maximising managers, favouring growth. Nevertheless, one might have expected that when reading this literature, the issue of potentially divergent objectives might have been given a fraction more prominence. This is a missing link with potential relevance for her analysis.[1]

I believe Penrose's analysis could be usefully enhanced by allowing some of these issues to enter it. In part this is because intra-firm divergence of interests can itself be a source of endogenous innovation and growth, as it motivates management (principals more generally) to devise creative means of addressing the problem.[2]

I.d. Intra-firm co-operation?

In addition to the above, and while in *TGF* intra-firm there is only cooperation, this is not to be found inter-firm. The main form of inter-firm cooperation allowed in *TGF* is the standard neoclassical one of price collusion. This is a missing link rectified by George Richardson. In his 1972 classic, Richardson showed how the Penrosean framework could be used to explain inter-firm cooperation. Richardson's focus was not on knowledge-creation through inter-firm cooperation (i.e., the Penrose story), as one might have expected; instead he focused on the division of labour between alternative means of organising production, i.e., market, firm and cooperation. He built explicitly on a Penrosean statement, to claim that the firm emerges when similarity (defined in terms of same underlying resources) and complementarity of activities

---

[1] Arguably, the lack of focus on intra-firm conflict can be justified in terms of different focus. Were there conflict within firms, moreover, management should be able to deal with it, for example by devising appropriate methods. The discussion parallels her assumption that management could/should devise methods for not allowing intra-firm unit costs to rise. The M-form she felt was such a method (see Penrose and Pitelis, this volume). While it is clear that no-one can deal with all issues, for her own purposes, Penrose might have needed to eschew intra-firm conflicts (which could well be a separate book).

[2] A perspective that has built on principals' efforts to have their own way, through devising appropriate institutions and organisations, has been fruitfully adopted by Douglass North (1981, 1991). The present author (Pitelis, 1991) has independently adopted the same heuristic to explain the genesis, growth, relative advantages and failures of capitalist institutions. Neither North nor I had anything reminiscent of endogenous innovation and growth.

co-exist. For dissimilar activities we have markets, for complementary but dissimilar activities we have cooperation.[1]

That Richardson built explicitly on Penrose suggests the extraordinary versatility of the Penrosean theme and contribution. This is despite the fact that, at least, as I believe, Richardson deals with only one aspect of the Penrosean story. The other, and equally fundamental is the extent to which inter-firm cooperation can replicate at least some of the advantages Penrose attributes to the firm (see Pseiridis, 2001). This takes us back to the very definition, the nature and the essence of the firm.

I.e. The employment contract and the 'essence' of firms

In his 1937 article Coase pleaded for clearer definitions and proceeded to define the firm as a multi-person hierarchy, its nature lying in the 'employment contract' between employers and employees. In this sense, Coase focused on the capitalist firm, as opposed to both non-capitalist firms and non-firm-like forms of early capitalist production, such as the 'putting-out' system. Coase's choice of focus might have been perfectly legitimate, in that he was at the time observing the rise and rise of the capitalist business firm. This, however, is not to say that non-capitalist production units are not firms, nor to say that single-person producers (i.e., no employment contract) are by definition not firms, nor even that an employment contract means a capitalist firm. Universities for one involve employment relations and employment contracts but are not capitalist firms.

Coase (1993) regretted his focus on the employment relationship. He said this limited focus on the issue of the 'nature', while one should also look at the issue of the 'essence' of firms, i.e., in 'running a business'. This involves not just an employment contract between capital and labour but also using other resources and one's own time and abilities to produce. Penrose had clearly dealt with the 'essence', having taken the 'nature' for granted.[2] The absence of an employment contract in Penrose is symptomatic of the more general absence of 'labour' in the original book. Indeed, the absence of any conflicting interests, to which we previously referred, is the outcome of the fact that to all intents and purposes the human resource *par excellence* is management. Labour is somewhere there in the background, but never a leading player.

As already pointed out, the Penrosean story can help explain the 'employment contract'. The latter is, I believe, crucial in understanding the capitalist firm, but not firms in general. Production units for own consumption existed since, by and large,

---

[1] One reason why Richardson's contribution is monumental, is that nobody else has dealt with the issue of the division of labour between alternative 'governance structures' in terms other than relative failure or relative success (which, as we said, is largely the same). Richardson's story goes beyond both market failures and differential capabilities. Indeed he provides a production-based explanation for both 'market-organisational failures' and differential success. This complements Penrose, but also the transaction costs, and all other stories, by addressing an important limitation.

[2] She also looked at the capitalist firm, which is once again a multiple-person hierarchy, under administrative coordination by human resources of both human and non-human resources, for production of a product for sale in the market. In this sense, there is, in her definition, all there is in Coase plus much more, minus the 'employment contract'.

the beginning of the human race. At a very generic level one could thus define the firm as any human activity that transforms inputs into outputs. In this sense, a person that collects freely available goods and consumes them directly does not qualify. One that expends labour time to transform inputs into outputs (including a hunter and fisher) does. I do not wish to attempt a historical account here (see Pitelis 1991, 1998 for more) but clearly one can claim that specialisation is likely to lead to division of labour, which will lead to exchange, to further specialisation and division of labour, etc., etc., i.e., to Adam Smith, Allyn Young and Edith Penrose. Production for exchange for own consumption is a more 'advanced' type of a firm (single producer). The next stage is production for exchange for sale in the market for a profit. This is the situation in the emerging merchant-capitalism—but it is not fully capitalist yet—it can also take place under feudalism.

For fully blown capitalist firms we need production of commodities by means of commodities (i.e., the commodity *par excellence*, labour power), for sale in the market for a profit by principals (capitalists, employees, entrepreneurs) who employ agents (labour, employees, workers, etc.) for this purpose. It is this multi-person hierarchy that Coase was looking at. In this sense, his focus on the employment contract was of utmost importance, and in need of explanation. In this same sense, the 'nature' of the firm Coase is looking at is the employment contract.

Firms, however, whether capitalist or non-capitalist, whether single-person or multi-person carry on doing what Coase called the 'essence' of their existence, i.e., 'running a business'. Focusing on the nature of the capitalist firm should not divert focus away from what firms in general do, and here it is that Penrose's contribution was vital.[1] That her insights help explain also the 'nature' adds weight to this argument.[2]

[1] The question of the single-producer firms was not alien to Penrose. Indeed, during her life in Waterbeach, she was particularly interested in two issues. The first is networking, an issue with which she dealt explicitly (see Penrose and Pitelis, this volume). The other, however, was arguably bothering her more. That was the extent to which the emergence of single-person–single-laptop-working-from-home 'firms' was questioning the very nature of her argument in the *TGF* and required a new theory of the firm. She raised this more than once in private conversations. I told her of my gut reaction to this question, which was flatly negative, but could provide no convincing reason. I still cannot, but here are some thoughts. First, Penrose was dealing with the growing capitalist enterprise, such enterprises are still around and going from strength to strength, so to the extent she failed to deal with previously existing and/or currently emerging single-person firms, so be it. It is a matter of focus. Secondly, however, the Penrosean insights are of relevance to such 'firms', too; in part because endogenous innovation and knowledge creation through the routes she described can be applied even to one person and his/her laptop (I understand from recent literature on technology that it can even be applied to the laptop it (her?) self: see *The Economist*, March 24–30, 2001); in part because some of the benefits Penrose identified can arguably be obtained also through cooperation between independent producers in the market. The contribution of Richardson is of relevance here, and relatedly, the more general issue of institutional specialisation, division of labour and teamwork (see below).

[2] I feel that the Coasean distinction between 'nature' and 'essence' is actually an afterthought, helping Coase to redress the balance between his focus on the *differentia specifica* of the capitalist firm (the capitalist employment contract) and what firms in general exist for and do towards this purpose, i.e., transform inputs into outputs, for own consumption (pre-capitalist firms), for exchange (feudal–early capitalists) or production of commodities by means of commodities for sale for a profit in the market (capitalist firms). While

footnote continued overleaf

Concerning the 'employment contract' (the 'nature') *per se*, I submit that this is one of the unresolved issues of the theory of the firm. As explanations abound, it is first worth looking at what the issue is. Among those who originally dealt with these issues, notably Marx, Knight (1921) and Coase, the issue was why independent producers agreed to work under dependent employment for an employer-capitalist. As is now well known, Marx stressed that capitalist coercion created the proletariat through the barrel of a gun, with the helpful hand of the Crown through, for example, the farm enclosures. Excellent accounts are in Stephen Hymer (1979) and in Heilbroner (1991). For Knight (1921) the reason is different attitudes towards risk. In effect, the risk-takers insure the timid and risk-averse by providing them with a relatively secure income-wage rate; it is, in effect, a division of tasks. For Coase (1937), it is efficiency gains due to reductions in market transaction costs.[1]

The Penrosean input to the 'nature' story can be knowledge-related advantages from intra-firm activities—clearly an efficiency/production-based complement to the Coasean insight.[2] In addition, Penrose looked at the broader issues—the objectives and the essence. However, going back to the 'employment contract', it is, and it is likely to remain, an unresolved issue why labour 'accepted' to work for capital. Efficiency benefits *per se* are not sufficient to explain why one should accept working for another voluntarily, whether the relationship is contractual or predatory, and also who will be the employee, who the employer and why. North (1981) for one has sided with Marx and Hymer in adopting the predatory (not contractual–cooperative perspective). I feel, with reality being what it is, that both elements were present, coercion and efficiency. Why and how, I have tried to explain elsewhere (Pitelis, 1991). In addition, I feel that Marx, Knight, Coase and Penrose are all right. In deciding whether and why to be an employee, coercion (thus restricted choice), potential benefits (from transaction costs and/or production efficiencies) and attitudes towards risk are relevant and useful; possibly others, too. One such is differential ability to exploit the benefits from specialisation, division of labour and teamwork. I discussed this Penrosean possibility in Pitelis (1991, 1993).[3]

footnote 2 continued from previous page

transformation defines the generic firm, the Coasean employment contract defines the capitalist—as it includes all that is distinct in capitalism, namely production for exchange (commodities) for a capitalist (residual claimant) profit. The above reinforces an earlier observation by this author (Pitelis, 1991) that one cannot fruitfully distinguish between nature and objectives, as 'nature' follows the objectives (at least of the principals) and indeed it supports the achievement of the objectives (to produce profitable commodities). Generalising this observation, 'objectives', 'nature' and 'essence' are inseparable, part and parcel of the same species. Part of the beauty of the Penrosean story is that objectives and essence are both there.

[1] It needs noting that while all the above deal with the capitalist employment contract, none of them deals with the overall issue of a combined objectives/nature/essence of the firm.

[2] In this context, and more generally it is extremely surprising that some authors, notably Rugman and Verbeke (2001) find Penrose's contribution to belong in the (market-)power-based tradition of Bain and others!

[3] In the 1993 article, in addition, I claimed that as compared with an original state of self-sufficiency, the emergence of the market, by definition, actually increases market transaction costs. In this sense, to explain markets (i.e., positive transaction costs) and firms one needed to look more deeply; differential abilities in

footnote continued overleaf

In a 1995 volume and building on an earlier 1988 contribution, Demsetz goes as far as suggesting that differential abilities in production brought about through knowledge-creation through the division of labour and teamwork explain the firm. Save for Penrose's very own endogenous innovation-knowledge, this is very much the Penrosean theme. It is not surprising then that Demsetz is also claimed to be one of the leading figures of the resource-based tradition (see for example Foss, 1997).

Are we reaching for a consensus? Perhaps so, but there are still missing pieces. First, the Penrose–Demsetz story is in effect the old Marshallian dictum that 'organisation aids knowledge' (see Loasby, this volume). However, markets too aid knowledge, as do families, states, international organisations, and more generally, institutions. It is institutions that aid knowledge. However, which institutions, when and for what type of knowledge? This is yet unanswered. The concept of specialisation, the division of labour and 'teamwork', as well as differential comparative advantage, can and need to be applied at the societal level at large, in explaining the economic institutions of capitalism. In this story, differential abilities in general will not do. One requires Richardson's contribution and more. As I have already noted, the Penrosean insights are of the utmost significance here, as they can provide a lens to inform this debate.[1]

II. SOME GENERALISATIONS

Besides missing links and related additions to the Penrosean themes, as discussed above, I believe there is scope for some generalisations. First, the Penrosean contribution does not require pre-existing excess resources for the arguments to follow. Secondly, it does not require the pre-eminence of management. Third, the recognition of intra-firm conflict can be of use in generalising the Penrosean insights of endogenous innovation-knowledge-growth, see Odagiri (1992), and help in part explain its direction. Lastly, I believe the Penrosean view can be generalised by incorporating in it insights and definitions from other contributions. I explain.

footnote 3 continued from previous page

production arising from specialisation, division of labour and teamwork, would be one answer, albeit certainly not the only, or a full, one. Edith attended the paper when presented at a Post-Keynesian Study Group meeting in London. She found it 'scholastic'. She also read my 1991 book and wondered why I needed to deal with everything everyone has ever said before getting to the point. She suggested starting from the point, and pursuing it relentlessly to its logical limits. That was, of course, her way. Yet I was gratified to see that few of the themes discussed in the 1993 article were independently discussed in much the same way in Demsetz (1995).

[1] Existing contributions in the mainstream are limited in their regard of institutions as 'constraints' (North 1981, 1991). Now, clearly institutions are constraints as much as they can be enabling. When a few years ago I invited Douglass North to visit Cambridge I asked him in a seminar this very question. His extremely candid response (which I only dare report because it was in a room full of others) was because he needed his theory to be neoclassical, i.e., optimisation under constraints, not because he did not recognise that institutions could aid knowledge! Suggesting that institutions enable clearly opens up a bag of worms; it re-raises all the issues discussed earlier in this book, about wealth creation versus resource allocation, about growth versus coordination, etc. I will not repeat my views, but state that viewing institutions as aiding knowledge leads to a far more exciting economics than the institutions-as-constraints view currently allows. This is my main criticism to what otherwise I consider to be the best application of transaction-costs thinking to the most important issue in economics, that of economic (under)development, by Douglass North.

II.a. The omnipresence of excess resources

First, the issue of excess resources. Building at the time on important insights by Babbage, Sargent Florence and others, Penrose (1955, 1959) considered it normal to expect excess resources to be present at all but very large scales of output, for reasons related to 'balance of processes', indivisibilities of resources, etc. All these are relevant but not required. Penrose can be generalised even by starting from a hypothetical situation of no excess resources and just observing knowledge generation within firms. This suffices. Indeed the very observation that human beings (and some machines) have ever-present unrealised (brain) potential will do. In effect, organisation aids potential, which (activates brain power and) releases resources. All other arguments are sufficient but not necessary conditions for the Penrosean story to follow.[1]

II.b. Pre-eminence of all human resources

The second theme is management. I believe the Penrosean focus that management is the human resource *par excellence* to be a product of its time, and not totally necessary or useful. To explain, let us first look at the other human resources, notably the labourer and the entrepreneur, even the (venture) capitalist. For Marx, but also for Adam Smith and the classicists, and for often-different reasons, the human resource *par excellence* is labour. In the classics, because of the labour theory of value—all value is created by labour so labour has to be the most important human resource, *n'est-ce pas*?

For Schumpeter the hero is the entrepreneur. Take the entrepreneur out and all you have is half-idiotic masses (a theme that Marx of all people liked) who would do precious little. So who creates wealth is the hero, and this is the entrepreneur. Even venture (for some vulture) capitalists are some people's heroes. Exit them, and so many good entrepreneurial ideas would be just that, ideas. Money makes the world go round, and this is provided by venture capitalists to whom we owe, at least in part, the recent resurgence of the 'new economy'.[2] The moral is clear. Truth is in the eye of the beholder. Claiming that managers are more important for the reasons Penrose did is neither convincing, nor essential, nor useful. Asserting the pre-eminence of all human resources is all these things together. Management's importance, and the 'Penrose effect', follow with no difficulties. Positing management's pre-eminence is not needed for the Penrosean story to follow.[3] Motivating, monitoring, rewarding, upgrading and

---

[1] Intriguingly, this observation provides support for the idea of (limited applicability of) the Penrosean insight to the single-person firm (who learns to work with him or herself, his/her laptop, telephone line, etc.), and in so doing s/he releases resources, etc.

[2] Venture capitalists are yet to be adopted as the hero of a big academic or management guru (at least to my knowledge), limited at the moment to the hearts of journalists and the press.

[3] It may be of some interest to note that the theme of the July 2001 biannual Global Business Colloquium of the Centre of International Business and Management (CIBAM) I direct, at Cambridge, was 'Human Resource Shortages' and was chosen by an advisory board of leading businesspeople, following a proposal by the board's Chair, Sir Martin Sorrell (CEO, WPP Group). For Sir Martin, competing for talent in general, not just managerial talent, is today's challenge for business in general and some businesses in particular.

maintaining talent is the name of the game. This is the old theme of extracting labour from labour power, work from work potential but with a twist. You may now need to bend backwards to exploit some people's potential, and sharing part of the 'residual' is one way.[1]

II.c. Enter intra-firm conflict

Intra-firm conflict is far too important and prevalent in all sorts of literature to be left out of Penrose. Importantly, it is of essence to the Penrosean story.

Penrose's approach was to look at the outside environment by first looking at what she called the 'nature' of the firm (not Coase's 'employment contract' but the very internal structure and constituent parts of the organisation she called a firm). She went on to posit a dynamic interaction between the internal and the external environments. The latter includes other firms, and Penrose went on to discuss the importance of oligopolistic interaction and inter-firm competition. In this sense, it is clearly in line with her own focus to look at intra-firm competition, too. As already discussed, this can take many forms and apply to many groups. Clearly, however, the conflict *par excellence* is that between employer and employee (Marx, Alchian and Demsetz, etc.). Recognising this can help Penrose in various ways. First, it provides an extra reason for thinking by entrepreneur-management of how to address this issue—this of course is innovation-knowledge generation. Second, it helps explain/predict, at least partly, the direction of this thinking. Last, but not least, conflict can lead to creative tension, and thus be a source of new information and knowledge.[2]

II.d. The nature of capitalism

The last issue I wish to consider is that of the very definition of the firm in general and the capitalist firm in particular. As I have pointed out elsewhere (Pitelis, 1991), even in books with the word in the title, the term capitalism is rarely defined. Defining capitalism as *production of commodities by means of commodities (labour power), under administrative co-ordination by one human resource (entrepreneurs) over the other human and non-human resources, for production for sale in the market for profit*, helps to generalise the Penrosean (and Coasean) definitions of the capitalist firm. As in Penrose, the capitalist firm is a bundle of resources, administered/controlled by one such resource, which produces commodities for a profit. The generalisation is afforded by the term commodity (products produced explicitly for exchange, already implicit in Penrose's view), and the concept of control of employers over employees, to include intra-firm

---

[1] As recently reported in the popular press, the concept of employee-dollar millionaires and their numbers are on the ascent. This may well be a limited phenomenon, not fundamentally questioning the nature of the system as we know it. However, our previous discussion reinforces the observation that putting equal weight to all human resources is safer.

[2] Considering a parallel, Marx is viewed in some circles as the best ally of capitalism; by pointing to its limitations he helped it address and in part solve them. Although he was exactly the opposite of Keynes (who was trying to protect capitalism from capitalists), Marx has helped to support and complement the Keynesian cause.

conflict. The reasons why these are both necessary and useful I have already discussed.

The focus on 'commodities' and the employer–employee relationship are clearly classical themes. Penrose was no Marxist, which in part explains the absence of employment contracts and intra-firm conflicts in her theory.[1]

While the 'commodity' has all but disappeared from conventional analysis, the employment contract and intra-firm conflict have not been, as we discussed already.

The absence of Marx (but also the behaviouralists, Alchian and Demsetz and Coase) from the Penrosean story, might in part explain her focus on success. Her glass was always a half-full one, rarely a half-empty one. Problems are there to be solved, and are (being) solved. Those who do not solve them, fail, but the focus is on those who succeed.[2]

III. EPILOGUE

The missing links and generalisations discussed here do not in any way detract from Penrose's fundamental contribution. Quite the opposite, for the reasons we have already discussed (i.e., firms and institutions aiding knowledge), her approach helps us to appreciate more clearly even the missing links, the missing themes, the needs for extensions. Generalising the Penrosean insights, accounting for the missing links, and complementing them, in particular by grafting dynamic transaction costs in the Penrosean story, is, I believe, no less than the future of economics. In this great future Penrose's contribution will, I believe, rightfully receive the place of the most influential economics and management book of the second half of the twentieth century. In part the aim of this volume was to support that view.[3]

---

[1] I feel she might, however, be amenable to accepting some of the arguments here. In a letter she wrote to Joan Robinson on February 1980 Penrose explains her critique of Robinson's views on development, as follows: 'Over and over again, I could not understand why you took the position you did. For example, at one point you agree that administration is useful; and *necessary*, but at the same time you insist that because it lives on the surplus, it reduces the growth of wealth. I find it impossible to follow. I have no particular love for the capitalist system, but I am not a Marxist. I accept that there is much that is valid and powerful in the Marxist approach. But I do not see why, in order to establish the positive and valid aspects one has to spend so much time setting what, to me in any case, are Aunt Sallys of orthodox theory in order to shoot them down. But do forgive the review. I am sure you will understand that one has no choice than to say what one thinks.' I believe that the commodity, employment relation and intra-firm conflict belongs to the 'valid and powerful'.

[2] I, myself, am more the half-empty type. Problems and constraints, I feel, can be useful in motivating change, innovation and the creation of knowledge. Half-full glasses may lead to complacency, half-empty ones can lead to demoralisation and inactivity. Perhaps what is needed is a creative tension between optimism and pessimism.

[3] I may have failed, but then again, who knows? If one person has been convinced, I would declare victory. My one concern is the length and style. Edith, among others, was not a fan of wasted space and cluttered exposition. In a letter to an eminent economist, she claimed no less than that his exposition might have helped him get things 'off his chest', but helped readers (including this one, i.e., her!) little! So much for public relations. But yet again that was Edith Penrose. After all, as Keynes hoped, it is ideas that count in the long run. That he went on to observe that when this comes about, we need not be around, it is of lesser concern to us here. Unlike us, ideas can stay. And Edith's most definitely will.

REFERENCES

ALCHIAN, A. and DEMSETZ, H. (1972). Production, information costs and economic organization, *American Economic Review*, Vol. LXII, No. 5, December, 777–95.

BUCKLEY, P. J. and CASSON, M. C. (1976). *The Future of Multinational Enterprise*, London, Macmillan.

COASE, R. H. (1937). The nature of the firm, *Economica*, Vol. IV, 386–405.

COASE, R. H. (1993). 1991 Nobel Lecture: The institutional structure of production, in *The Nature of the Firm*, edited by O. E. WILLIAMSON and S. G. WINTER, Oxford, Oxford University Press.

CYERT, R. M. and MARCH, J. G. (1963). *A Behavioural Theory of the Firm*, Englewood Cliffs, NJ, Prentice Hall.

DEMSETZ, H. (1995). *The Economics of the Business Firm: Seven Critical Commentaries*, Cambridge, Cambridge University Press.

FOSS, N. J., ed. (1997). *Resources, Firms and Strategies*, Oxford Management Readers, Oxford, Oxford University Press.

HEILBRONER, R. (1991). *The Worldly Philosophers*, 6th edn, London, Penguin Books.

HYMER, S. H. (1979). The multinational corporation and the international division of labour, in *The Multinational Corporation: A Radical Approach, Papers by Stephen Herbert Hymer*, edited by R. B. COHEN, N. FELTON, J. VAN LIERE and M. NKOSI, Cambridge, Cambridge University Press.

JENSEN, M. C. and MECKLING, W. (1976). Theory of the firm: managerial behaviour, agency costs and ownership structure, *Journal of Financial Economics*, Vol. 3, 304–60.

KNIGHT, F. H. (1921). *Risk, Uncertainty and Profit*, Boston, Houghton Mifflin.

KOGUT, B. and ZANDER, U. (1993). Knowledge of the firm and the evolutionary theory of the multinational corporation, *Journal of International Business Studies*, 4th quarter, 625–45.

NORTH, D. C. (1981). *Structure and Change in Economic History*, London and New York, Norton.

NORTH, D. C. (1991). Institutions, *Journal of Economic Perspectives*, Vol. 5, No. 1, 97–112.

ODAGIRI, H. (1992). *Growth through Competition, Competition through Growth*, Oxford, Clarendon Press.

PENROSE, E. T. (1955). Research on the business firms: limits to growth and size of firms, *American Economic Review*, Vol. XLV, No. 2.

PENROSE, E. T. (1959/1995). *The Theory of the Growth of the Firm*, Oxford, 3rd edn, Oxford University Press.

PITELIS, C. N. (1991). *Market and Non-Market Hierarchies*, Oxford, Basil Blackwell.

PITELIS, C. N., ed. (1993). *Transaction Costs, Markets and Hierarchies*, Oxford, Basil Blackwell.

PITELIS, C. N. (1998). Transaction cost economics and the historical evolution of the capitalist firm, *Journal of Economic Issues*, Vol. XXXII, December, 999–1017.

PSEIRIDIS, A. N. (2001). 'Competence Clusters as an Industrial Strategy', doctoral thesis, University of Cambridge, forthcoming.

RICHARDSON, G. (1972). The organisation of industry, *Economic Journal*, Vol. 82, 883–96.

RUGMAN, A. M. and VERBEKE, A. (2001). 'A Note on Edith Penrose's Contribution to the Resource-based View of Strategic Management', Mimeo.

INDEX